AF712733

Migmatites

QE
475
M5
M34
1985
DISCARDED

MIGMATITES

edited by

J.R. ASHWORTH

Lecturer in Geological Sciences
University of Aston

Blackie
Glasgow and London
Distributed in the USA by
Chapman and Hall
New York

Blackie & Son Limited,
Bishopbriggs, Glasgow G64 2NZ

Furnival House, 14–18 High Holborn, London WC1V 6BX

Distributed in the USA by
Chapman and Hall
in association with Methuen, Inc.
733 Third Avenue, New York, N.Y. 10017

© 1985 Blackie & Son Ltd
First published 1985

All rights reserved
No part of this publication may be reproduced, stored in a retrieval system, or transmitted, in any form or by any means, electronic, mechanical, recording or otherwise, without prior permission of the Publishers.

British Library Cataloguing in Publication Data

Migmatites.
1. Migmatite
I. Ashworth, J.R.
552'.4 QE475.M5

ISBN 0-216-91726-3

Library of Congress Cataloging in Publication Data
Main entry under title:

Migmatites.

Includes bibliographies and index.
1. Migmatite. I. Ashworth, J.R.
QE475.M5M54 1985 552'.4 84-27167
ISBN 0-412-00841-6

10 6/16/84

Photosetting by Thomson Press (India) Limited, New Delhi
Printed in Great Britain by Bell and Bain Ltd., Glasgow

Contributors

J.R. Ashworth, B.A., Ph.D., Lecturer in Geological Science,
Department of Geological Sciences, University of Aston, Gosta Green, Birmingham B4 7ET, UK.

David Barr, B.Sc., Ph.D., Higher Scientific Officer,
British Geological Survey, 19 Grange Terrace, Edinburgh EH9 2LF, UK.

James A. Grant, Ph.D., Professor of Geology,
Department of Geology, University of Minnesota, Duluth, MN 55812, USA.

W. Johannes, Dr. phil., Dr. rer. nat., Professor of Petrology,
Institut für Kristallographie und Petrographie, Universität Hannover, Welfengarten 1, 3000-Hannover-1, BRD.

E.L. McLellan, B.A., Ph.D., Assistant Professor of Geology,
Department of Geology, University of Maryland, College Park, MD 20742, USA.

Sakiko N. Olsen, B.Sc., Ph.D., Associate Research Scientist and Assistant Professor,
Department of Earth and Planetary Sciences, The Johns Hopkins University, Baltimore, MD 21218, USA.

J. Touret
Vrije Universiteit, Instituut voor Aardwetenschappen, de Boelelaan 1085, 1081 HV, Amsterdam, Netherlands.

Robert J. Tracy
Department of Geology and Geophysics, Yale University, New Haven, CT 06520, USA.

Contents

1 Introduction

J.R. ASHWORTH*

1.1 Scope of the book

This book comprises petrological review articles by a selection of current workers in the migmatite field. Such a collection of chapters cannot cover every aspect of the subject: the book is not intended to supersede, but rather to complement, the standard textbook of Mehnert (1968). The extended essays presented here were, to some extent, stimulated by the publication of an excellent symposium of research papers (Atherton and Gribble, 1983). Some of the problems encountered there are reviewed in more detail here. Both Atherton and Gribble (1983) and Mehnert (1968) treat granites as well as migmatites, whereas this book concentrates wholly on the latter. Chapters on specific methods of approaching migmatites are interleaved with two chapters on particular geographical regions in which there has been much recent research (New England and the Moines of Scotland). Each chapter represents the personal viewpoint(s) of the author(s), and this Introduction is a personal statement by one of them who happens also to be editor. Some difficulties and points of interest from the other chapters are brought together, and the editor takes the opportunity to air some of his own opinions (or, as some people would say, his own prejudices).

Recent progress in understanding migmatization has come largely from studies of simple cases. Some of these are spatially very localized occurrences of migmatites, such as migmatitic contact aureoles (e.g. Tyler and Ashworth, 1982; Grant, this volume). Simple geometries are selected where necessary for quantitative mass-balance models (Olsen, this volume). The relevance of such studies to complex terrains, with great varieties of migmatitic phenomena (e.g. Figure 1.1) may not immediately be obvious, but the aim of all the work is ultimately to make the complex cases interpretable. Barr (this volume) has made great progress in interpreting a polymigmatitic terrain in a framework of modern ideas about migmatite petrogenesis. Tracy (this volume) puts the

*Department of Geological Sciences, University of Aston, Gosta Green, Birmingham B4 7ET, UK.

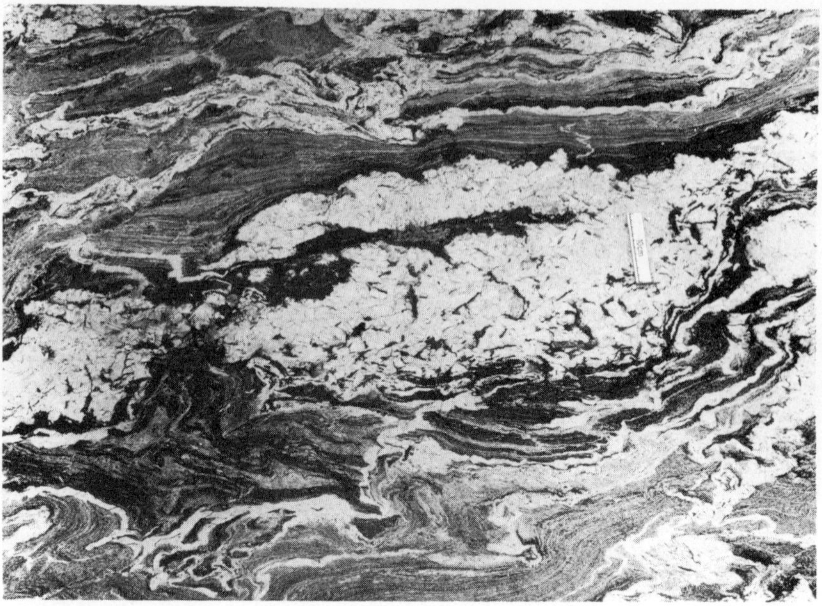

Figure 1.1 An example of the problem. Scale-bar 10 cm. Granholmen, SE of Runmarö, E of Stockholm, Sweden. Photograph kindly supplied by Prof. K.R. Mehnert.

migmatites of New England into the context of the numerous tectonic and metamorphic episodes recorded in that region.

The book is mostly about partial melting and its consequences. This emphasis reflects the current balance of opinion that *most* migmatites originate through partial melting. However, other origins are not to be neglected (e.g. Olsen, Ashworth and McLellan, Tracy, Barr, all in this volume). In this chapter, a section on the use of terms (both descriptive and genetic) is followed by a discussion of processes and the difficulty of erecting definitive criteria to prove or disprove partial melting.

1.2 Definitions

1.2.1 *Descriptive terms*

In the nineteenth century, there was already debate about the origin of 'lit-par-lit gneisses', now called stromatic migmatites. (Table 1.1 gives definitions of the terms used for the commonest structural types of migmatite. Mehnert (1968) defines, and beautifully illustrates, many further varieties.) The term *migmatite* was coined by the great Finnish geologist J.J. Sederholm, in a paper published in Swedish in 1907 (two other terms introduced in the same classic paper are 'anatexis' and 'ptygmatic'). His definition naturally uses terms which now seem

Table 1.1 Definitions of some important terms (largely after Mehnert, 1968)

Term	Definition
Migmatite	A rock, found in medium-grade to high-grade metamorphic areas, that is pervasively inhomogeneous on a macroscopic scale, one part being pale-coloured and consistently of quartzofeldspathic or feldspathic composition
Bodies within a migmatite	
Leucosome	A body of the pale-coloured, quartzofeldspathic or feldspathic lithology of a migmatite
Melanosome	A body of the dark-coloured lithology present in some migmatites, rich in mafic minerals, complementary to leucosome
Neosomes	Leucosome + melanosome
Mesosome (Henkes and Johannes, 1981) *Palaeosome* of some authors	A body of that part of a migmatite complex that is not neosome, generally having the appearance of an ordinary metamorphic rock, often intermediate in colour between leucosome and melanosome
Interpretative terms relating to bodies	
Palaeosome (of some authors) *Protolith* (of e.g. Johannes)	Hypothetical parent rock which developed into the neosomes
Restite	Residual body, from which more mobile material has been extracted
Structural types of migmatite	
Stromatic	Layered (usually rather irregularly)
Agmatic	Breccia-like
Schollen	Blocks or rafts of non-leucosome in leucosome
Schlieren	Streaks of non-leucosome in leucosome
Terms relating to melting	
Anatexis	Partial melting
Metatexis	Moderate degree of partial melting; leucosome subordinate to non-leucosome; pre-migmatization structures not disrupted
Diatexis	Extensive partial melting; leucosome volumetrically comparable with, or dominant over, non-leucosome; pre-migmatization structures disrupted: mafic minerals often prominent in leucosome

archaic, but essentially envisages migmatites as originating from a mixture of melt and solids. Mehnert (1968) rightly prefers a non-genetic definition, and my own effort (Table 1.1) is a modification of his. Brown (1973) would prefer to restrict the term 'migmatite' to stromatic types, excluding both agmatites and diatexites (see sections 1.2.2 and 1.6 below), but the broader definition is used in this book. Migmatites have parts which look superficially more or less like igneous rocks: 'of pegmatitic, aplitic, granitic or generally plutonitic appearance' (Mehnert, 1968). To avoid any implications as to whether it did form from a melt, a body of this material is descriptively called a *leucosome* (Table 1.1), after Mehnert. Generally there are other parts that look like ordinary metamorphic rocks, e.g. pelitic schist or dioritic gneiss: these have often been called *palaeosome* but the term has been objected to, on the grounds that it causes confusion with the parent rock from which a migmatite may form by a differentiation process. Logically, this hypothetical parent rock must no longer exist, and should not be assumed to be identical to the existing schist, gneiss etc. Therefore, Henkes and Johannes (1981) introduced the term *mesosome* for the latter. This change in terminology has not been universally accepted: for example, Olsen (this volume) uses 'palaeosome' in the sense just described. The word 'mesosome' is descriptive of a rock that is *intermediate*, between the pale-coloured leucosome and the dark *melanosome* (Table 1.1) that is sometimes present as a body spatially separating the leucosome from the mesosome. The boundaries of typical leucosomes are sharply defined (Mehnert, 1968, p. 60), with an obvious grain-size difference between coarser leucosome and finer mesosome or melanosome. (More diffuse varieties of leucosome are discussed by Mehnert, 1968, pp. 40–42, and Ashworth and McLellan, this volume.) The melanosome may be merely a mafic selvage, or a broader body with some felsic material, or both. Melanosomes pass gradationally into mesosomes (e.g. Ashworth, 1976; Dougan, 1979; McLellan, 1983). Grain-size relations between melanosomes and mesosomes are variable (Ashworth and McLellan, this volume). Melanosome and leucosome, both being products of the migmatization, can together be termed *neosomes* (Table 1.1; Mehnert, 1968). Palaeosome is etymologically the opposite of neosome, hence the tendency to apply it to the inferred, now non-existent rock parental to the neosomes (Mehnert, 1968, p. 356), but the phrase 'parent rock' is simple enough and perhaps preferable in the present state of terminological flux. Johannes (this volume) uses the term *protolith* for the parent rock.

The term melanosome becomes inappropriate in some migmatites with little mafic content (e.g. migmatized psammites). In cases where it is fairly certain that the migmatization involved segregation of the leucosome material from a residue, the residual rock can be called a *restite* (Table 1.1; Mehnert, 1968), whether or not it is mafic enough to be called a melanosome (e.g. Ashworth, 1976). (The corresponding term 'mobilizate' for the more mobile part of the migmatite is little used, since 'leucosome' covers the ground.)

1.2.2 *Processes*

Sederholm clearly viewed migmatites as intermediate between igneous and metamorphic rocks—the Greek 'migma' means mixture. Though at first he had contemplated a melt as the agency of migmatization (Sederholm, 1907), he later favoured a more tenuous migmatizing fluid ('ichor'). The 1907 paper and others documenting the subsequent evolution of his thought are accessible in English in a posthumous collected edition (Sederholm, 1967). Migmatizing fluids were supposed to emanate from nearby granites (e.g. Sederholm, 1967, p. 252). As noted by Olsen (this volume, section 4.1), the ensuing controversy with Holmquist (1921) contains all the essential elements of modern debate on migmatization processes. Holmquist thought that the fluids came out of the country rocks, not the granites; vapour segregation could produce migmatites

Table 1.2 Simplistic classification of migmatization processes

	Process requires open system	*Process does not require open system*
Process requires presence of melt	magmatic injection	anatexis
Process does not require presence of melt	metasomatism	metamorphic differentiation

at relatively low metamorphic grade, while partial melting could do so at higher grade. Table 1.2 puts modern names on the processes. As emphasized by Brown (1973), they are not mutually exclusive, i.e. more than one process can operate simultaneously. For example, one combination considered by Olsen (this volume) is anatexis with metasomatism. The scale of 'open' or 'closed' system under discussion is perhaps ~ 1 m, i.e. the scale encompassing one or a few leucosomes (e.g. Olsen, this volume, Figure 4.13). This level of interpretation does not yet address the cause of the process: metamorphic differentiation, for example, may be driven tectonically (through pressure gradients), by chemical potential differences among mineral assemblages, etc. (section 1.4).

Advanced stages of partial melting give rise to diatexites: the definition in Table 1.1 is modified from Mehnert (1968). Brown (1973) emphasizes the lack of continuous structure (stromatic banding) in diatexites. A complementary term for less advanced anatexis is *metatexis* (Table 1.1). Diatexites are often dominated by their leucosome, within which the non-leucosome lithologies occur disconnectedly, as blocks and rafts (*schollen*) or streaks (*schlieren*). The end-member of this metatexite–diatexite series is 'homogeneous diatexite' (e.g. Mehnert, 1968; Brown, 1979), which cannot logically be considered a migmatite (Mehnert, 1968, pp. 40–41) and is discussed in this context for the sole reason that it is closely associated with true migmatites.

1.3 Migmatites and granites

The field relations in the Finnish Precambrian indicated to Sederholm (1967) a connection between migmatization and the origin of granites (see also Hopgood and Bowes, 1978). Though the term *anatexis* has come to be used by migmatite specialists for partial melting *in situ* (with only small-scale segregation of the melt), Sederholm (1967, p. 49) originally linked the term with regional granite petrogenesis: 'a process proceeding from below upwards... a later consolidated magma derived from [an older granite] by rebirth'. Obviously, the term is still used in this sense in igneous petrology, as when a granite pluton is ascribed to 'crustal anatexis' (e.g. Wyllie, 1983).

Evidently there is a connection between some migmatites and some granites. Vein complexes fringing granitic intrusions are obvious examples of magmatic injection; not so obvious, perhaps, if the main intrusion is not exposed (e.g. Fettes and Macdonald, 1978). The reverse process of melts coalescing from source-rocks to form a mobile magma is of more fundamental interest, but more difficult to find. Approaching the problem from a migmatite petrologist's point of view, it is clear that the field relations of diatexites often indicate bulk flow. The classic work of Mehnert on the transition through metatexite to diatexite in the Black Forest is summarized in his book (Mehnert, 1968, pp. 246–276). The core of one dome-like structure is a homogeneous diatexite almost indistinguishable from a normal intrusive granite, and the dome structure itself is attributed to flow of melt. Brown (1979) describes a migmatitic terrain in NE Brittany, where the most advanced stages of diatexis produced obviously mobile rocks. A core of homogeneous diatexite is described as having 'punched through' an envelope of inhomogeneous diatexite and metatexite, showing clearly intrusive field relations against these and against unmigmatized metasediments at still shallower structural levels. In their geochemistry (Brown, 1979) as well as their field relations, the diatexites behave as igneous rocks, apparently having evolved by a degree of fractional crystallization.

To return to the classic terrain of Sederholm (1907), its most obvious property is extreme complexity. It lies in what is now called the Svecokarelian orogenic belt. The metamorphism is Precambrian (1800–1900 Ma: Hopgood *et al.*, 1983). To a mere petrologist, the structures are daunting (Figure 1.1) and it is not surprising that processes are difficult to unravel. In recent work on this terrain, the role of the leucosomes has been as time-markers in elucidating a sequence of structural events (Hopgood and Bowes, 1978). As to processes, both anatexis and metamorphic differentiation are invoked (Hopgood and Bowes, 1978). McLellan (1984) attributes some of the complexity to deformation in the partially molten state. Geothermometry indicates that parts of the terrain reached temperatures at which it is reasonable to suggest anatexis as a likely process (Campbell, 1980). It is not easy to deduce the origin of a migmatite after it has been plastically deformed several times: the evidence tends to have been wiped out.

Historically, prejudiced attitudes in the absence of good evidence led to the bitter and pointless 'granite controversy' (Read, 1957), in which migmatites became entangled through the granite-migmatite connection. In the writer's opinion, the granite controversy (as distinct from several migmatite controversies) is now over: granites are magmatic. There are, perhaps, abiding lessons from the controversy. For one thing, field relations can be ambiguous. It was possible to see a 'front of granitization' in the intrusive contact of a pluton or the unconformity between a basement gneiss and cover metasediments, if only field data were available, or if no account was taken of other evidence (such as the geochemistry of the pluton or the petrofabrics of the metamorphic rocks). Textures seen in thin-section were also misinterpreted, providing another lesson. The classic treatment of the subsolidus evolution of granite textures is by Tuttle (1952). Ashworth and McLellan (this volume) discuss the evolution of textures in melt-derived leucosomes.

More recently, as geologists have reacted against 'granitizationism', the granite-migmatite connection has unfortunately led to an extreme view that all regional migmatites must be anatectic. As a random example from the general geological literature, one reads in a recent textbook on fieldwork that 'migmatite means a mixture of rock-types, at least one of which is igneous' (Fry, 1984, p. 17). This statement is an oversimplification. For field use, the term migmatite must be defined descriptively, so as to avoid prejudicing interpretation by applying genetic terminology to the rocks at first sight. Any such descriptive definition will undoubtedly embrace some inhomogeneous rocks that have *no* melt-derived part. Anatectic and non-anatectic migmatites can be difficult to distinguish, and the distinction deserves some care (Ashworth and McLellan, this volume).

1.4 Melt-absent migmatization

Migmatites formed in the absence of a melt have been called *metamorphic migmatites* (e.g. McLellan, 1983). Processes to be considered are metasomatism and metamorphic differentiation (Table 1.2).

There does not appear to be much support for gross, regional metasomatism in the origin of migmatites. The metasomatism invoked by Olsen (this volume) is altogether subtler than 'granitization'. In the Moines, statistical geochemistry shows no bulk difference between migmatized and unmigmatized (lower-grade) parts of the same lithostratigraphic unit (Barr, this volume, section 7.2.1). Barr is critical of regional metasomatic hypotheses in two different settings in the Moines: a migmatite suite (Brown, 1971: see Barr, this volume, section 7.4.4) and a rather homogeneous granitic gneiss which has vaguely been attributed to metasomatic 'migmatization' (e.g. Dalziel and Johnson, 1963), but is now interpreted as a suite of deformed and metamorphosed intrusions (Barr, this volume, section 7.2.2).

Of course, prograde metamorphism has some effect on the composition of a pile of rocks. Apart from the obvious loss of volatile components (H_2O, CO_2), movements appear to be minor, though quite definite changes of a subtle nature have been established (Ferry, 1983; Haack *et al.*, 1984). When the escaping fluid is channelled along fractures, quartz veins are generally deposited if the metamorphic grade is low (e.g. Yardley, 1983). This represents local redistribution of the relatively soluble component SiO_2 in the aqueous fluid. With increasing grade, there is a tendency for feldspar to enter the vein assemblages (Vidale, 1974). The vein mineralogy is generally 'concordant' with the host-rock assemblage: thus, K-feldspar occurs in addition to plagioclase if it also occurs in the host rocks; in pelites, this situation becomes common only above the sillimanite + K-feldspar isograd. Below this onset of dehydration of muscovite + quartz, the commonest veins contain quartz + plagioclase.

Such veins are mentioned in this volume by Ashworth and McLellan (section 5.2.1) and Tracy (section 6.5.3). At temperatures approximating the onset of melting, experimental studies demonstrate that the aqueous fluid can transport quartz and feldspars (see Olsen, this volume, section 4.4.2). Like the quartz veins at lower grade, these metamorphic-hydrothermal veins are often coarse-grained (Vidale, 1974; Yardley, 1975), a feature which may help to distinguish them from anatectic leucosomes (Yardley, 1978). Some *pegmatitic* veins, then, are clearly due to a kind of metamorphic differentiation and, if abundant enough, may constitute leucosomes of a migmatite complex. Other pegmatite veins and dykes may, of course, be related to nearby granite intrusions. The question of a definition of 'pegmatite' in the migmatite context is addressed by Ashworth and McLellan (this volume, section 5.2.1). The term is unfortunately used rather loosely for any leucosome with grain size ~ 1 mm, so that many deep-seated anatectic leucosomes are called pegmatitic or pegmatoid (e.g. Mehnert, 1968).

Another kind of metamorphic differentiation operates under tectonic stress within solid, uncracked rock-masses. At low metamorphic grades, a very familiar phenomenon is the differentiation into quartz-rich and phyllosilicate-rich layers that accompanies the crenulation (microfolding) of a pre-existing foliation in pelites. Variants of a theory to account for this effect are available (e.g. Hobbs *et al.*, 1976, p. 251; Cosgrove, 1976; Robin, 1979). Essentially, in a rock with inhomogeneity of mechanical properties, the imposed stress sets up pressure gradients, and (as in the vein-forming process) the relatively mobile constituent quartz migrates by solution and re-deposition to the lower-pressure locations. Such a process operating at higher grades, and transporting feldspar in addition to quartz, can generate true leucosomes (McLellan, 1983). Although the pressure distribution may be more complicated than in the crenulation example, the absence of flow of a melt may lead to geometries differing from those of anatexites (McLellan, 1983) and the textures of the two kinds of migmatite (anatectic and non-anatectic) are found to develop differently in the same terrain (McLellan, 1983; Ashworth and McLellan, this volume). Tracy (this volume, section 6.5.4) discusses possible tectonically induced dehydration reactions producing leucosomes.

Metamorphic differentiation can operate without tectonic stress, as a result of reactions among assemblages of minerals. 'Flecky' gneisses (Loberg, 1963; Russell, 1969) have mafic clots surrounded by pale quartzose or quartzo-feldspathic haloes (leucosomes), set in a host mesosome. This geometry (Figure 1.2) is rather distinctive, though it is argued that stromatic geometries can be generated by the same processes (e.g. Loberg, 1963). Fisher (1973) develops a quantitative diffusion theory for the examples described by Loberg (1963). The cores consist of andalusite + biotite + quartz, the leucosomes are quartz + feldspar and the mesosomes sillimanite + biotite + quartz + feldspar. The overall reaction driving the differentiation is sillimanite $\rightarrow$ andalusite, with the product spatially separated from the reactant and Al behaving as

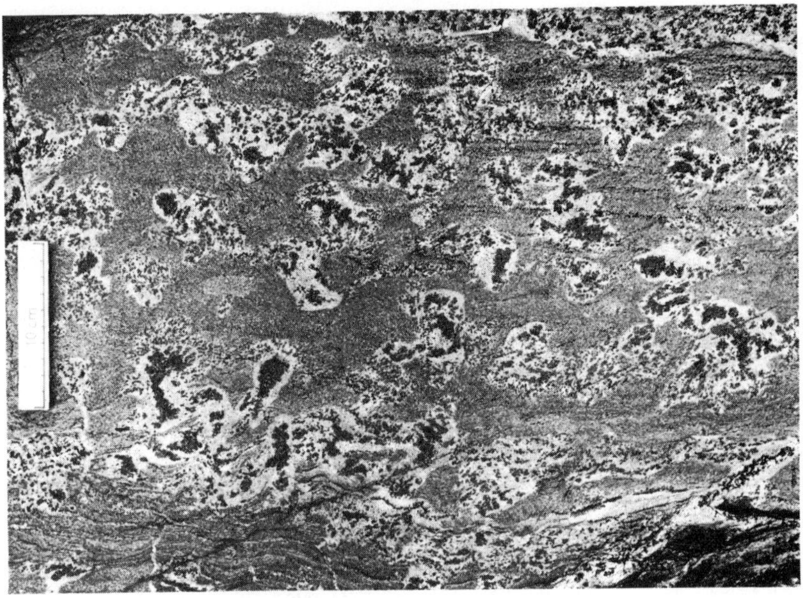

Figure 1.2 Flecky gneiss. Scale-bar 10 cm. Island of Krokö, SE of Västervik, Sweden. Photograph kindly supplied by Prof. K.R. Mehnert.

immobile. Under the resulting local gradients of chemical potential (Fisher, 1973), Fe and Mg diffuse into the core while K diffuses into the leucosome.

1.5 Compositions of anatectic leucosomes

Sections 1.5–1.9 are concerned primarily with the difficulty of devising objective criteria for or against anatexis. It is a paradox that most petrologists regard most migmatites as anatectic, while the very examples they cite are unable to pass obvious 'tests' of anatectic origin. The obvious 'tests' are regarded as oversimplified. Anatexis is deduced from a combination of various kinds of observations and, in many cases, the even greater implausibility of other interpretations. The first example of a 'test' which is often inapplicable is the proposition that anatectic leucosomes should have 'granite minimum' composition.

1.5.1 *The 'granite minimum'*

The classic work of Tuttle and Bowen (1958) established the 'minimum melt' composition in the system Qz–Ab–Or–H_2O, i.e. the composition of the lowest-temperature melt to form from an assemblage of quartz + alkali feldspar, water-saturated. Subsequent work extending to higher pressures, where the 'minimum' is a true eutectic because two alkali feldspars coexist with quartz and the initial melt in this system, shows that the 'minimum melt' composition varies slightly with pressure (Johannes, this volume, Figure 2.1). The effect of adding An to the system is still poorly known (Johannes, this volume, section 2.4), but it is the rocks with An-poorest plagioclases that have the lowest minimum-melting temperatures, and in these the effect of An on melt composition should be negligible. Tuttle and Bowen (1958) showed that the normative compositions of natural granites cluster around the experimentally determined minimum in the Qz–Ab–Or diagram. (This is usually taken as decisive evidence for the magmatic origin of granites.) In anatectic migmatites, incipient melting should obey the same phase diagram. The partly-melted rock is not likely to remain water-saturated (water-undersaturation is also important in granite petrogenesis—e.g. Wyllie, 1983). If there is a pure H_2O gas phase in the rock just before melting begins, then this phase will be consumed immediately by the melting reaction because it is much less abundant than the mineral phases participating in the reaction. More likely, there is a gas phase containing other components in addition to H_2O, and it will become H_2O-poorer as melting progresses with increasing temperature (Grant, this volume, section 3.5). The restricted supply of H_2O will constrain the amount of melt formed at given temperature, making it less likely (than under water-saturated conditions) that the melt composition will move away from the 'minimum' along a cotectic curve of the phase diagram. The proportions of Qz, Ab and Or in the water-undersaturated 'minimum' melt are not necessarily identical with those in the water-saturated experimental melt, but any difference should be small. Thus one might expect that leucosomes produced by incipient melting would approximate the composition of the 'granite minimum' melt.

Rarely, leucosomes with such simple compositions are found (e.g. Tobschall, 1971). In most anatectic migmatites, there are complications which can be discussed under these headings: advanced melting, incomplete segregation of melt, and scarcity or absence of K-feldspar.

1.5.2 *Advanced melting*

Diatexites are rocks of which a large proportion became molten. This requires departure of the melt composition from the 'granite minimum' unless the parent rocks fortuitously had granitic composition. As natural rock-masses progressively melt with increasing temperature, the common melt composition evolves from granitic towards granodioritic and eventually dioritic (Brown and Fyfe, 1972). The component An assumes more significance. Various attempts to measure the 'minimum melt' compositions in An-bearing systems are treated sceptically by Johannes (this volume), but it is agreed that the Qz:Ab:Or proportions vary: for example Or increases with increasing An. A further complication is the dissolution of unknown proportions of ferromagnesian minerals in the melt (Grant, this volume, section 3.3.1).

The evolution of melts studied by Brown and Fyfe (1972) involves the liberation of H_2O from hydrate minerals. Two important such reactions can be written, approximately (in an An-free system):

$$\text{muscovite} + \text{quartz} + \text{albite} \rightarrow \text{K-feldspar} + Al_2SiO_5 + \text{melt} \quad (1.1)$$

$$\begin{matrix} \text{biotite} + Al_2SiO_5 \\ + \text{quartz} + \text{albite} \end{matrix} \rightarrow \begin{matrix} \text{K-feldspar} + \text{cordierite and/or} \\ \text{garnet} + \text{melt} \end{matrix} \quad (1.2)$$

These are *dehydration-melting* reactions in the terminology used by Thompson (1982); Grant (this volume) prefers the term *vapour-absent* reactions, denoting that the gas phase does not appear in the reaction (at least in simple model systems, as in the above formulations). There may, of course, still be a gas phase present in the rock. Reactions of this type can generate much larger amounts of melt than the initial melting hitherto considered, which is restricted by the limited availability of free gaseous H_2O. It has been suggested that the consumption of heat by such reactions (notably 1.1) is sufficient to suppress the usual increase of metamorphic temperature with increasing depth in a terrain (Tracy, this volume, section 6.5.1), but in the classic ground of New England, Tracy (this volume, section 6.5.3) doubts the efficacy of such 'thermal buffering'. Since these reactions have positive slopes in a P–T diagram (Figure 1.3), 'abrupt' generation of melt by a dehydration-melting reaction can occur during decompression of a region (e.g. Thompson, 1982, Figure 8: Grant, this volume, section 3.8) as well as during prograde heating. Grant (this volume) thoroughly treats dehydration-melting reactions involving biotite in pelites. He works mostly in the simplified system K_2O–FeO–MgO–Al_2O_3–SiO_2–H_2O (KFMASH), where the absence of plagioclase means that reaction temperatures are higher than they would be with granitic melts; however, the *sequence of reactions* found in nature can be well modelled in such systems, and Grant (this volume, section 3.4.1) comments that petrogenetic grids in simplified systems (often with the additional simplification of degeneracies, i.e. collinearities among phases in the composition diagram) seem more naturalistic than the fully developed, complex grids of systems with many components. This is partly due to the restricted range of composition of common pelites, which limits the number of accessible equilibria and the extent of continuous reactions.

By themselves, the dehydration-melting reactions are unlikely to cause the melt composition to depart from 'granitic', but the generation of large proportions of melt makes it more probable that quartz or a feldspar may become exhausted from the residue so that the melt evolves away from the 'minimum' of the phase diagram. Quartz and plagioclase seem most likely to be eliminated, rather than K-feldspar, because the latter appears as a product of reactions (1.1) and (1.2) (but see section 1.5.4). Elimination of quartz is well-known in some contract-aureole migmatites (e.g. Grant, this volume).

Diatexite leucosome analyses scatter in the fields around the 'granite minimum' of the normative Qz–Or–Ab diagram (e.g. Brown, 1979; Torres-

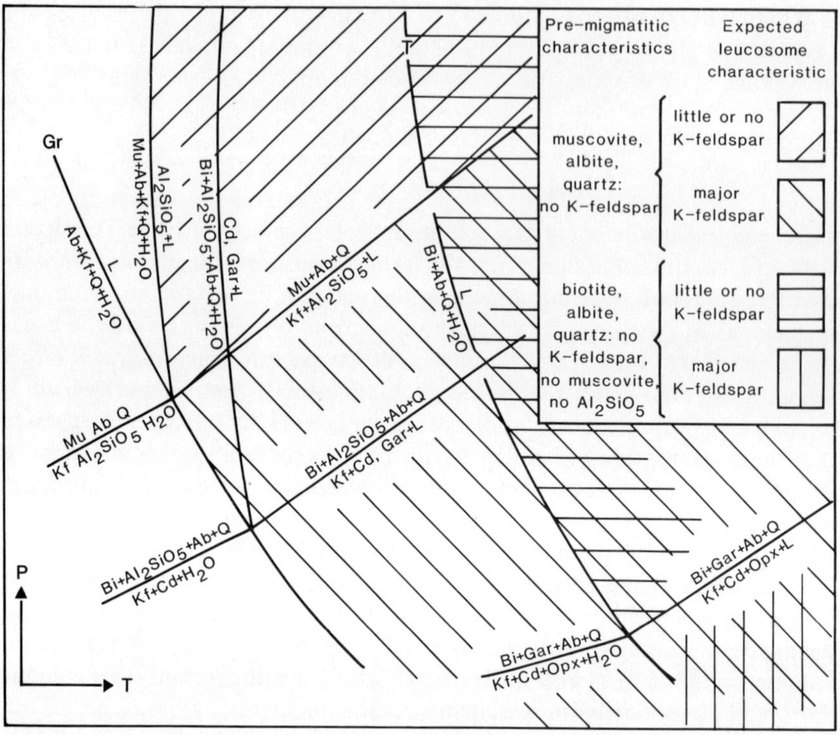

Figure 1.3 Crude cartoon to show the relative positions of equilibria modelling important boundaries in anatexis. Ab, albite. Bi, biotite. Cd, cordierite. Gar, garnet. Cd, Gar, cordierite and/or garnet. Kf, K-feldspar. Mu, muscovite. Opx, orthopyroxene. Q, quartz. Gr = granitic minimum melting curve. T = trondhjemitic minimum melting curve. For quantitative treatment of reactions involving muscovite see Thompson and Tracy (1979) and Thompson (1982). For higher-grade reactions see Thompson (1982) and Grant (this volume). For minimum melting curves see Johannes (this volume).

Roldan, 1983). Ashworth (1976) found the same effect, by modal analysis, at relatively advanced stages of melting. It is likely that not all the material being analysed was molten (see next section), i.e. the analyses are contaminated by residue. However, the departure of diatexites from the 'granite minimum' is perfectly possible as a simple consequence of progressive melting. Tracy (this volume, section 6.5.3) envisages anatectic melts reacting with, and partly assimilating (dissolving) pre-existing quartz-plagioclase leucosomes.

1.5.3 *Incomplete segregation of melt*

In studies of anatectic migmatites, it is becoming increasingly apparent that the simple equation of leucosome with melt and melanosome with residue is too naive.

Winkler and Breitbart (1978) analysed leucosomes and found a wide scatter about the 'granite minimum' composition. They concluded that not all the leucosome could have been molten at plausible migmatite-forming temperatures, and envisaged a mixture of melt and suspended crystals. One difficulty in studies of this type must be the practical problem of mechanically separating leucosome from non-leucosome for analysis. The most obvious rocks to select are leucosome-dominated migmatites such as diatexites: Winkler and Breitbart (1978) give no petrographic details, but the rocks selected from the area of Ashworth (1976) can be classed as diatexites, and presumably the others were also large, obvious leucosomes. Where leucosomes are small and intricately shaped, modal analysis is better than chemical analysis for separating the parts (Ashworth, 1976). It can be combined with microprobe data to measure major-element bulk compositions (Olsen, this volume). Even so, the modal data of Ashworth (1976) confirm the large departures from 'granite minimum' leucosome composition at advanced stages of melting, and the interpretation of Winkler and Breitbart (1978) is justified. The very difficulty of separating leucosome from schlieren, etc., for analysis is an aspect of the intimacy of admixture between melt and residue.

It follows that the compositions of these leucosomes cannot be used as evidence for anatectic origin (see criticism of Winkler and Breitbart, 1978, by Johannes, this volume, section 2.4.6). Johannes and co-workers (Henkes and Johannes, 1981; Gupta and Johannes 1982; Johannes and Gupta, 1982) argue for *in-situ* transformation to leucosome through partial melting without segregation of the melt. Minimal separation of biotite is invoked to produce the selvage-like melanosomes. McLellan (1983) infers a sequence from completely unsegregated to well-segregated anatexites. The unsegregated stage is represented by *ophthalmites*, with feldspar porphyroblasts but no true leucosomes (therefore these rocks are not migmatites within any usual definition of the term). There is a transition through incipiently differentiated rocks to stromatic migmatites with well-defined leucosomes (see also Ashworth and McLellan, this volume). Whether segregation occurs, and how efficiently, depends on deformation effects and extent of melting (Ashworth and McLellan, this volume; Barr, this volume). Thus a spectrum of 'contamination' of analytical data by unmelted rock occurs, from zero (clean segregation) to total obscuration of the melt composition where segregation failed completely. It is not the leucosome compositions of the poorly-segregated rocks that are being used to infer anatexis here.

1.5.4 *Scarcity or absence of K-feldspar*

The Or content of the 'granite minimum' melt in the Qz–Ab–Or diagram is approximately 25–30% (Johannes, this volume, Figure 2.1). There are leucosomes with far less Or than this, consisting essentially of quartz and plagioclase, often with no K-feldspar at all. It has been argued that such leucosomes cannot be anatectic (e.g. Amit and Eyal, 1976), but again there are counter-arguments. Several possibilities can be considered:

(1) K-feldspar was present but has been eliminated by later reactions;
(2) Melts were present which would have crystallized K-feldspar, but a fraction of the melt escaped from the rock so the K-feldspar is now somewhere else;
(3) K-feldspar was never present, because the rocks are K-poor. The temperatures were high enough to generate such K-poor melts;
(4) The K in the rock remained mostly locked in mica(s). Temperatures were high enough for the onset of water-saturated melting of quartz + plagioclase, but major breakdown reactions of the mica(s) would have required higher temperatures or lower activities of water. The mica components entered the melt to some extent, but the melt then merely crystallized mica(s) in addition to quartz and plagioclase, without crystallizing K-feldspar.

Suggestion (4) is the most controversial, and the others will be dealt with first.

(1) *Replacement of K-feldspar.* Some pseudomorphs of muscovite + quartz + myrmekite after K-feldspar are interpreted in terms of retrograde reaction in the presence of melt (Ashworth and McLellan, this volume, section 5.3.2). The inferred reaction is essentially reaction (1.1) backwards, and other such reversed dehydration-melting reactions ('rehydration-crystallization reactions', as they may be called) will have similar consequences. The 'blurred' boundaries of the pseudomorphs would make it difficult to reconstruct the original modal K-feldspar content for comparison with the 'granite minimum'. Moreover, the retrograde reaction probably involved transfer of material between leucosome and melanosome (section 1.9), modifying their bulk compositions. Retrograde muscovite formation may be aided by incoming aqueous fluids, e.g. from nearby granitic intrusions (Tracy, this volume, section 6.5.1).

If a migmatite is re-metamorphosed, K-feldspar may not be a member of the new stable assemblage and again may be replaced (Ashworth and McLellan, this volume).

(2) *Loss of melt or loss of K from melt.* In some anatectic complexes it appears that the melt compositions have been modified, becoming relatively K-poor, by interaction with the environment. These effects are strikingly invoked in some areas of tectonic instability that underwent rapid decompression (uplift) in the presence of melt, and the tectonics may perhaps have influenced the histories of the melts.

In the Central Gneiss Complex of British Columbia, Lappin and Hollister (1980) and Kenah and Hollister (1983) texturally infer biotite breakdown leaving amphibole in residues and producing a (presumably potassic) melt, but in one case the resulting K-feldspar is scarce and it is inferred that most of the melt left the rock (Kenah and Hollister, 1983, p. 149) and in another case all of the melt appears to have left, and there is no K-feldspar in the rock (Lappin and Hollister, 1980, p. 533). Both rocks seem to lack leucosomes. The postulated expulsion of melt is very plausible since the areas contain abundant parautochthonous intrusions. However, there is some doubt as to how much K-feldspar should be expected to crystallize from *in-situ* melts in these circumstances (see (4) below).

Decompressed terrains are studied by Brown (1983) in S Brittany and by Brown and Earle (1983) in Timor. Texturally-inferred biotite-consuming reactions, occurring during uplift, should produce K-feldspar as a mineral or as a component in a melt, but K-feldspar is not observed. The Timor example is non-migmatitic but modal compositions of some rocks look restitic, and it is suggested that a melt has been expelled, carrying the K-feldspar component with it (Brown and Earle, 1983, p. 191). In the migmatites of S Brittany, however, Brown (1983, p. 182) suggest migration of K^+, i.e. metasomatism in the presence of melt, and this idea is also entertained by Brown and Earle (1983, p. 191). It is not to be thought that reactions with melt during decompression always produce K-feldspar; Torres-Roldan (1983, p. 116) infers one that consumes the K-feldspar component.

(3) *'Tonalitic' and other melts in very K-poor systems.* Partial melting of K-poor rocks such as metabasites requires temperatures well above those of the water-saturated 'granite minimum' but, if it occurs at all, it will produce non-granitic melts. For example, trondhjemitic veins and small tonalite plutons in the Archaean of Zimbabwe are attributed by Condie and Allen (1980) to partial melting of amphibolites and ultramafic rocks. Oliver (1977) regards anorthosite veins in the Granulite Facies in New Zealand as partial melts from feldspathic gneiss. These lack quartz as well as K-feldspar.

(4) *'Trondhjemitoid' melting in pelites.* It has been doubted whether melts that will not crystallize K-feldspar can be generated from biotite-bearing rocks (e.g. Yardley, 1977). Ashworth (1976) invoked this effect in quartz–plagioclase–biotite migmatites in NE Scotland, calling them 'trondhjemitoid' to distinguish them from nearby, K-feldspar-bearing 'granitoid' migmatites.

The onset of water-saturated melting of quartz + plagioclase is comprehensively treated by Johannes (this volume). 'Trondhjemitoid' melting corresponds to the Ab-rich part of his 'tonalite system'. Melting is possible below 700 °C if $P > 4$ kbar and $x_{\mathrm{An}}^{\mathrm{Plag}} < 0.2$, or if $P > 8$ kbar and $x_{\mathrm{An}}^{\mathrm{Plag}} < 0.6$ (Johannes, this volume, Figure 2.4). Temperatures of beginning of melting for water-saturated quartz + sodic plagioclase are thus ~ 50 °C higher than the water-saturated 'granite minimum', but not impossibly high for discussion in the context of metamorphism in the uppermost Amphibolite Facies. The participation of additional minerals in the melting reaction can only lower (not raise) the melting temperatures. Such effects should be similar to those in granitic systems. Critically reviewing experimental data, Johannes (this volume, section 2.4.3) concludes that significant reduction of solidus temperatures by ferromagnesian minerals is unlikely. On the other hand, experimentally, incipient water-saturated melting of quartz + albite + biotite is reported at 650 °C with $P = 4$ kbar (Hoschek, 1976), and of quartz + plagioclase (An_{44}) + biotite + hornblende at approximately 690–660 °C, 3–7 kbar (Büsch *et al.*, 1974, Figure 20), suggesting depression of the solidus by 40–60 °C relative to the Fe, Mg-free system. Similarly, Grant (this volume, section 3.3.1) suggests that the addition of mafic phases may lower the Qz–Or–H_2O minimum-melting temperature by tens of degrees. The additional, fluxing effect of components such as B_2O_3 and F is likely to be negligible since they are very minor constituents in most of the rocks concerned (Johannes, this volume, section 2.4.2). A reverse effect is that water-undersaturation (due to presence of CO_2 for example) will raise the temperature of onset of melting.

The behaviour of biotite in melting has analogies with that of muscovite, and there are migmatites with muscovite in addition to biotite but no K-feldspar (e.g. Dougan, 1979; McLellan, 1983). The melting relations of muscovite have been more studied than those of biotite, and granitoid melting more than trondhjemitoid, so it is appropriate to attack the latter partly by analogy with the granitoid system in a simplified joint account of both (Figure 1.3). Plagioclase being treated as Ab for simplicity, the lowest melting curve represents the water-saturated minimum melting of 'haplogranite' (Johannes, this volume).

Water-saturated melting in the presence of muscovite but no K-feldspar sets in at a boundary with a steep slope; there is an aluminous phase (Al_2SiO_5) in the residue (Figure 1.3). Thompson and Algor (1977) place this boundary at approximately 650 °C. The melt should have a granitoid composition but K-feldspar as a mineral is not generated: on cooling, this melt will presumably crystallize muscovite + albite + quartz (Ashworth, 1979*a*), though it may be forced to evolve to K-feldspar-crystallizing compositions if the Al_2SiO_5 in the restite fails to react with it.

Melts that will crystallize major K-feldspar, and residues containing K-feldspar, are generated at slightly higher temperatures by the major dehydration-melting reaction (1.1).

Analogously, with Al_2SiO_5 + biotite, but no muscovite or K-feldspar, there are two steps to consider, one approximately water-saturated, the second a dehydration-melting reaction. The initial water-saturated melting produces a granitoid melt at slightly higher temperatures than in the muscovite-bearing system. Hoffer (1978) places this boundary at approximately 650–660 °C. The mica breaks down to leave a new phase in the residue: Hoffer (1978) found cordierite. In general, garnet also seems possible. Some biotite compositions may produce amphibole in the residue (Hoschek, 1976). Again, K-feldspar is not expected in residues but may crystallize eventually from evolving melts. The second step is reaction (1.2), giving K-feldspar in leucosome and restite.

In the melting reactions just considered, the behaviour of the hydrate mineral can be described as *incongruent* melting; the mineral breaks down to give a component in the melt that is different from the reactant minerals, plus a new mineral in residues. In the cases under consideration (incongruent melting of micas), the component produced in the melt is the K-feldspar component, and in pelites the product residual mineral is aluminous. The opposite type of behaviour is *congruent* melting, which would not produce a new (or abruptly increased) restite mineral (Johannes, this volume, section 2.4.7; Grant, this volume, section 3.3.3). Congruent behaviour, in this sense, does not preclude partition of elements with the melt so that the composition of the remaining mica may change. For example, Fe/Mg in biotite may be affected by the melting reaction (cf. Johannes, this volume, section 2.4.3), though it is rather unlikely that biotite subsequently crystallized in leucosomes would retain a composition distinct from that in intimately coexisting restite (section 1.9.3).

Micas behave congruently in the initial water-saturated melting of granitic assemblages (with two feldspars and quartz). Thus, if biotite and muscovite are present as additional reactants in the initial melting reaction labelled Gr in Figure 1.3, they will, at this stage, behave congruently (e.g. Thompson, 1982, Figure 7, curve 33′). Büsch *et al.* (1974) carefully describe congruent behaviour of biotite in initial granitic melting, and also note its failure to break down in *incipient* melting of a rock with no K-feldspar. This is a case where there is no Al_2SiO_5 to undergo the coupled reaction (1.2) with the biotite, and melting is delayed to a true trondhjemitic or tonalitic solidus, as shown in Figure 1.3. There is thus no reason to suspect that biotite behaves differently in *initial* melting with quartz + plagioclase from its (congruent) behaviour with quartz + two feldspars. Winkler (1979, p. 312) asserts that his experiments have produced tonalitic or trondhjemitic melts from quartz–plagioclase–biotite gneisses. This discussion is also consistent with the experimental results of Hoschek (1976). The composition of the initial melt, and its evolution during subsequent water-undersaturated melting, are not well known, and one may speculate that it could evolve gradually towards a granitoid composition. In the Granulite Facies (i.e. at relatively high temperatures), the observations of Lappin and Hollister (1980) and Kenah and Hollister (1983) suggest the production of mildly K-enriched melts in this manner, though the authors prefer to note the possibility of melt loss from the system (see (2) above). Major K-feldspar entry is likely when biotite + quartz dehydration-melting reactions not requiring Al_2SiO_5 are encountered, such as the one sketched in Figure 1.3 following Grant (this volume), which also produces orthopyroxene. According to Grant (this volume, Figure 3.16), such reaction is expected above 700 °C at $P \gtrsim 4$ kbar where melting had initially been granitic. In the initially trondhjemitic case, sufficient reduction of water activity, by progressive melting, to bring about the dehydrative reaction will be deferred to higher T

because the onset of melting was at higher T than in the granitoid case. Meanwhile, in some part of pressure–temperature–composition space, orthopyroxene can clearly appear without K-feldspar, giving the 'noritoid' leucosomes of Ashworth (1976). These are the classic 'cordierite-norites' of NE Scotland, forming agmatic and schollen structures with various refractory inclusions; they cannot be interpreted in terms of assimilation by gabbroic magmas, and are definitely products of anatexis of pelites (Gribble, 1970). In the K-feldspar-free rock experimentally melted by Büsch *et al.* (1974), at approximately 700 °C the biotite breaks down by a melting reaction involving the Ca-bearing plagioclase (approx. An_{44}) to produce Ca-amphibole (hornblende) and a granodiorite-like melt.

Ashworth (1979*a*) expected K-feldspar in all residues from the major dehydration-melting reactions, as is predicted by the crude model of Figure 1.3. However, the (unknown) composition of the melt will influence the composition of the residue, and absence of K-feldspar is conceivable (Grant, this volume, section 3.3.3). If K-feldspar is produced, it is more likely to survive (or at least the pseudomorphs after it may be more obvious) in melanosomes than in the leucosomes with their opportunities for retrograde evolution (see (1) above). Of course, all K-feldspar is liable to be removed if there is re-equilibration during very slow cooling or a second, subsolidus metamorphism.

Not the least crude aspect of Figure 1.3 is the neglect of plagioclase solid solution. Presence of Ca-bearing plagioclase blurs all boundaries into bivariant bands (Thompson and Tracy, 1979). Thus it is well known that muscovite + quartz coexist stably with K-feldspar + sillimanite + granitoid melt over an interval of temperature at given pressure. Such muscovite-granitoid melting is inferred by Ashworth (1976) alongside biotite-trondhjemitoid melting in interbedded rocks of different composition. This requires that muscovite can persist in the granitoids up to trondhjemitoid melting temperatures, around 700 °C. The very steep slope of the muscovite + quartz + plagioclase dehydration-melting reaction in the P–T diagram of Kerrick (1972, Figure 5) seemed to preclude this, but in the more thorough treatment of Thompson and Tracy (1979), the reaction occurs in a bivariant band whose slope is more shallow, so that the phenomenon inferred by Ashworth (1976) is possible at $P \gtrsim 5$ kbar. The activity of water, a_{H_2O}, must be high in the trondhjemitoids, indeed higher than in the granitoids where it has been reduced by the earlier onset of melting: a_{H_2O} must be controlled within small rock-systems. The initial approximately water-saturated trondhjemitoid melting reaction should produce less melt than the major dehydration-melting reaction (1.1) of the associated granitoids; this is consistent with the general field appearance of the area studied by Ashworth (1976).

Thus, trondhjemitoid melting in pelites seems possible under certain conditions. As evidence for it, Ashworth (1976) adduced good agreement between leucosome compositions (measured modally) and the cotectic in Ab–An–Qz–H_2O determined by Yoder (1968). This implies good segregation of melt, and could have been obscured by effects discussed above for granitoids. One major complication that is absent is the 'rehydration-crystallization' effect, (1) above.

Of course, there are trondhjemitoid migmatites of melt-absent origin (Ashworth and McLellan, this volume). They tend to have non-cotectic leucosome compositions (McLellan, 1983). Barr (this volume) makes out a case for regional, subsolidus trondhjemitoid migmatization in the Moines.

Even cotectic leucosome compositions are not, in isolation, proof of melt origin as distinct from a process of transport through aqueous fluid. The solute in the aqueous phase at migmatite-forming temperatures can approach the proportions of components that would be found in a melt (e.g. Burnham, 1967). Since non-cotectic compositions likewise do not disprove anatexis, it is not advisable to look at leucosome compositions alone, and other factors will now be discussed.

1.6 Textures and structures as possible indicators of melt presence

Igneous textures tend to differ from metamorphic ones. Anatectic leucosomes may have textures resembling those of plutonic igneous rocks, and some progress has been made towards putting this observation on an objective footing (Ashworth and McLellan, this volume). Obviously, primary textures tend to be obliterated during slow cooling, deformation or re-metamorphism, so textures are most informative in 'well-preserved' migmatites.

Segregation of melt is partly controlled by deformation (Ashworth and McLellan, this volume; Barr, this volume), and the partly-molten, partly-segregated rock-mass may in turn respond abnormally to the tectonic stress. This effect deserves further investigation. McLellan (1984) notes irregular folding in migmatites which are inferred to have been partly molten when folded, in contrast to the regular, harmonic folds of metamorphic migmatites in the same region. Ashworth (1979*b*) appeals to deformation in the presence of melt to explain the control of leucosome geometries by axial plane structures which are not otherwise represented by strong petrofabric elements. Differential behaviour of leucosome and non-leucosome is also inferred by Mehnert and Büsch (1982): the leucosome shows 'precrystalline deformation', the melanosome 'paracrystalline deformation', i.e. the deformation occurred while the leucosome was mostly liquid and the melanosome was recrystallizing. Many similar examples could be cited. Unsegregated anatexites may flow as a melt + crystal mush (McLellan, 1983, Figure 6). Brown (1973) regards typical agmatic structures as obviously melt-present: the magma has engulfed xenoliths (either cognate or foreign). The assumption that a melt will easily flow relative to coexisting solid bodies may not always be valid: Tracy (this volume, section 6.5.3) is working on the idea that (during fairly rapid deformation) layers of melt may behave as mechanically more competent than mica-schist.

The volume change on melting must be accommodated somehow, though it is doubtful whether the structural results can be disentangled from other complexities of migmatites. *Ptygmatic* structures have sometimes been interpreted in terms of volume changes, and have been regarded by some authors as diagnostic of melt presence. Ptygmatic folding is defined by Mehnert (1968, p. 27, summarized) as disharmonic and tortuous with no relation between thickness of the folded layer and position on crest or limb of the fold. Mehnert (1968, pp. 27–35) thoroughly reviews the problem in a historical manner. There is no justification for the idea that ptygmatic folding indicates a partly molten state, since the same style of folding is also found in completely different environments (Figure 1.4).

The study of post-migmatization deformation is a field in itself. Structural interpretation is more difficult in migmatites than in stratified metasediments: neosomes may have complicated shapes when first formed. While some new, anatectic leucosomes cut across old leucosomes (e.g. Barr, this volume, section 7.3.2), melt segregation may be controlled by pre-existing structure, including the geometry of pre-existing leucosomes (Tracy, this volume, section 6.5.3). The inference of timing of migmatization needs care (Ashworth and McLellan, this volume). That said, such relative age studies can bear on the origin of migmatites, if only by sorting them into different ages, which can be correlated with different metamorphic conditions (next section). Barr (this volume) distinguishes 'early' from 'late' migmatites in the polymetamorphic Moines,

Figure 1.4 Ptygmatically folded calcite vein in limestone. Width of field approximately 70 cm. Grindelwald, Switzerland. Photograph kindly supplied by Prof. K.R. Mehnert.

enabling him to infer different origins, in some cases for two migmatizations affecting the same exposure. On the other hand, a complicated structural sequence does not always mean polymetamorphism; the numerous fold-sets in the Svecokarelides represent one orogenic cycle, and the main tectonic and migmatite-forming events are radiometrically constrained to an interval of < 20 Ma (Hopgood *et al.*, 1983). McLellan (1984) doubts that all the fold-sets are genuinely distinct: re-folding can be mimicked by a single deformation of partly molten rocks. Geochronology is of fundamental importance in separating major tectonometamorphic episodes (Tracy, Barr, this volume). As radiometric techniques become ever more sophisticated, there are increasing possibilities for dating the crystallization of individual leucosomes, and resolution of small age differences between migmatites.

1.7 Estimation of P–T–a_{H_2O} conditions

In recent years, various methods of estimating metamorphic pressures and temperatures have become available (Essene, 1982). For melting, the activity of water is also important. Low a_{H_2O} suppresses melting.

Temperatures can be estimated, with various uncertainties (Essene, 1982) by a petrogenetic-grid approach, or by cation-exchange geothermometers most of which use Fe/Mg partition: popular mineral pairs are garnet–biotite and garnet–cordierite. Sources of uncertainty include errors in analysing the minerals, departures from equilibrium in the natural assemblages, errors in experimental calibrations (of which several rival ones are usually available), and activity models for the natural minerals which need to take into account the effects on the activities of Fe and Mg components of other components that are not present in simple experimental systems (e.g. the Ca component, grossular, in natural garnet: Hodges and Spear, 1982). In geobarometry, a petrogenetic-grid approach is usually employed, often with the complication that the relevant components are in very dilute solid solutions in the minerals, e.g. grossular in almandine-dominated garnet. Activity modelling is hazardous at high dilutions (Essene, 1982), and particularly problematic in garnet (e.g. Hodges and Spear, 1982) and plagioclase (Carpenter and Ferry, 1984). Fortunately, the problems are less severe at migmatite-forming temperatures than at lower metamorphic grades (where departures from ideal solid solution are generally greater). Geothermometers may be re-set by re-metamorphism, or during simple cooling: this appears to affect oxygen-isotope as well as Fe-Mg distribution geothermometers (Johannes, this volume, section 2.6.6). The inference of peak metamorphic conditions requires careful assessment of retrograde effects on mineral compositions (Tracy *et al.*, 1976).

The activity of water, a_{H_2O}, can be estimated by similar methods. Using the term gas to embrace supercritical fluids, a_{H_2O} in a gas that would be at equilibrium with a particular mineral assemblage can be estimated through the equilibrium constant for a dehydration reaction (Powell, 1978, p. 134), if the equilibrium P and T have been estimated. Uncertainties in P and T propagate through this calculation and make the uncertainty in a_{H_2O} the more severe. If the standard state is taken at the P and T of interest, then $a_{H_2O} = 1$ in a pure H_2O fluid and values < 1 represent water-undersaturation. Fluids are, in general, non-ideal solutions. Powell (1978) provides a simple approximate model of a_{H_2O} as a function of mole fraction, x_{H_2O} in H_2O–CO_2 fluids. Departures from ideality in H_2O–CO_2 at migmatite-forming temperatures and high x_{H_2O} are slight, so, to a fair approximation $a^{gas}_{H_2O} = x^{gas}_{H_2O}$ in this system. One advantage of working in terms of a_{H_2O} is that it can be defined in the absence of a gas phase, though Grant (this volume, section 3.5) finds that this may be unrealistic. Mixtures other than H_2O–CO_2 are reviewed by Ferry and Burt (1982). Grant (this volume) qualitatively treats the evolving composition of an H_2O–CO_2 gas phase during melting and crystallization reactions in simple model systems, in terms of $x^{gas}_{CO_2}$ (in his notation, $x^{v}_{CO_2}$). Other alternatives to activity in discussing the role of water are fugacity, f_{H_2O}, and partial pressure, P_{H_2O}. In the presence of a pure H_2O gas, $P_{H_2O} = P$ (as, for example, in the water-saturated experiments discussed by Johannes, this volume). If mixing is assumed to be ideal, the $P_{H_2O} = x_{H_2O} \cdot P$. Fugacity is measured in units of pressure, but takes into account the non-ideality of pure H_2O as well as its mixing behaviour. A thorough treatment of the meaning of activity, fugacity and partial pressure is given by Froese (1976). In this volume, a_{H_2O}, f_{H_2O} and P_{H_2O} are all used for the same purpose in qualitative discussions of cases where a_{H_2O} is < 1 and variable.

The dehydration reaction of muscovite + quartz measures a_{H_2O} at given P and T with or without melt (e.g. Tyler and Ashworth, 1982). At higher grades the dehydration reaction of biotite + sillimanite + quartz (cf. reaction 1.2) can be used: Holdaway and Lee (1977) treat the assemblage with both cordierite and garnet. There are unfortunate uncertainties about the behaviour of cordierite. The structural role of H_2O in the mineral is debated (Lonker, 1981; Martignole and Sisi, 1981). Another imponderable is the major Al, Si order-disorder transition; it is uncertain whether the natural cordierites are comparable with the experimental ones in their degree of ordering (Putnis, 1980). (Similar problems are better known in feldspars: e.g. Brown and Parsons, 1981; Carpenter and Ferry, 1984; cf. Johannes, this volume, section 2.4.4, on 'structural state'). Values of $a_{H_2O} \approx 0.5$–0.6 at $T \approx 700$–$720\,°C$ and $P \approx 4$–6 kbar, obtained by Ashworth and Chinner (1978) for aureole migmatites, seem reasonable for melting in terms of the diagram showing 'granite minimum' P–T curves for various $x^{gas}_{H_2O}$ given by Kerrick (1972, Figure 5). However, Johannes (this volume, Figure 2.3) presents new calculations by J.R. Holloway indicating that $x^{gas}_{H_2O}$ over a haplogranitic melt decreases relatively slowly with increasing T. This suggests that high a_{H_2O} (corresponding to $x^{gas}_{H_2O} \gtrsim 0.7$) is required for granitic melting at $T \approx 700\,°C$, and casts doubt on the results of the Holdaway–Lee method of estimating a_{H_2O}. Other equilibria can be used, if sufficient thermochemical or experimental data are available and activities in the minerals can be modelled. In the Amphibolite–Granulite Facies transition at Broken Hill, Australia, Phillips (1980) uses biotite + sillimanite + quartz $\rightleftharpoons$ garnet + K-feldspar + H_2O (thus avoiding cordierite) to estimate $a_{H_2O} \approx 0.6$–0.4 at $T \approx 740$–$800\,°C$ and $P \approx 4$–6 kbar, a result which seems to permit melting. In the Granulite Facies, amphibole–pyroxene equilibria have been tried, but encounter problems with activity models for the solids (e.g. Phillips, 1980; Barnicoat, 1983). In summary, although T estimates for many migmatite terrains lie within the melting range, it is important also to consider a_{H_2O}, for which the methods of estimation remain relatively crude.

1.8 The Granulite Facies

This is an area of current controversy in which the writer lacks personal experience. The Granulite Facies is characterized by anhydrous minerals though there are usually some hydrated ones still present. The controversy centres on whether dehydration was accompanied by wholesale melting. Temperature estimates are generally $> 700\,°C$, so melting is expected *if* a_{H_2O} *is high enough,* and the discussion follows on from the last section.

Most Granulite Facies terrains are of deep-seated origin and old (Precambrian), with attendant complications of protracted excavation history and polymetamorphism. As mentioned in the last section, Phillips (1980) infers melt-producing conditions in the classic region around Broken Hill (which is described as migmatitic). Release of the water from the crystallizing melts is held partly responsible for widespread retrograde metamorphism (Corbett and Phillips, 1981). In some terrains, Granulite Facies relics (interpreted as restites) are contained in granitoid masses (interpreted as anatexites) that have evolved back into the Amphibolite Facies (Turner, 1981, p. 466). The above examples are Proterozoic. Much interest has been focused on the earliest part of Earth history, the Archaean, aspects of whose tectonics and petrology seem to indicate different processes from those operating in later times. It has been suggested that Granulite Facies terrains are regional residues from which partial melts have been extracted, and that they are commoner in the Archaean than in younger regions because geothermal gradients were higher in the Archaean, so that melting temperatures were attained at relatively shallow depths in the Archaean continental crust (Brown and Fyfe, 1972). There are geochemical depletions in the Granulite Facies, relative to lower-grade rocks, which have been interpreted in terms of removal of partial melts; however, the geochemical data in detail do not seem to favour this process, suggesting metasomatism instead (Weaver and Tarney, 1983). The statistical sampling difficulties in such studies are huge, a point which has been noted, along with the complexity of natural melting reactions, by authors who wish to set aside the conclusions of the geochemists (e.g. Powell, 1983*a*; Barnicoat, 1983).

Possible causes of high or low a_{H_2O} in the Granulite Facies have been proposed. In some meta-igneous rocks, it is possible to argue that low a_{H_2O} is inherited from the magmatic stage (e.g. Wells, 1979). In terms of processes understood in younger terrains, it is difficult to envisage a *metasedimentary* pile entering the Granulite Facies *without* melting: it starts wet; prograde metamorphism will drive out much water, but successive dehydration reactions will force a_{H_2O} upwards towards 1 so that melting seems inevitable once the 'granite minimum' or other appropriate beginning-of-melting curve is reached in P–T space, and further melt generation will occur largely by dehydration-melting reactions (Brown and Fyfe, 1972; Powell, 1983*a*). The way to suppress melting is to reduce a_{H_2O}, bringing about thorough dehydration below the solidus while raising the beginning-of-melting temperatures to unattainably high values. This drying-out could be attributed to a previous metamorphism, but only if that metamorphism had itself reached the Granulite Facies so as to eliminate the major reservoir of H_2O, the hydrated minerals. Thus a polymetamorphic hypothesis to account for low a_{H_2O} merely moves the problem back one step in time. A currently popular idea is that CO_2-rich fluids 'flushed' the Archaean crust, causing the geochemical depletions by gas-phase transport, and producing the Granulite-facies assemblages without melting. The evidence for this hypothesis is largely field evidence, tending to arouse scepticism in the present writer: however, the field relations in question are very striking. Patchy and ramifying areas of diffuse dehydration to charnockite are figured by Janardhan *et al.* (1982), Holt and Wightman (1983) and Friend (1983), from the classic localities in S India. There is migmatization nearby: the field relations appear to indicate that an H_2O-rich fluid moved, ahead of the CO_2-rich one, through parts of the rocks, promoting anatexis which was then arrested by arrival of the CO_2-rich phase (Janardhan *et al.*, 1982; Holt and Wightman, 1983; Friend, 1983). The ensuing 'carbonic metamorphism' (Newton *et al.*, 1980) is further discussed by Grant (this volume, sections 3.7 and 3.8) from a theoretical standpoint. The anatectic stage produces diatexites in places, and may be responsible for generation of large-scale intrusive magma (Closepet granites: Friend, 1983). The CO_2 is implicitly of deep-seated origin; it is debated whether the source is in the Crust or Mantle (Newton *et al.*, 1980). This interpretation of the Archaean Granulite Facies, with its fluid-controlled 'fronts' sweeping through the rocks, is faintly reminiscent of the regional metasomatism hypotheses of yesteryear, but should not be condemned for that. Gross metasomatism of major components other than H_2O and CO_2, and perhaps alkalis, is not implied. The new 'ichor' is a well-defined carbonic fluid; the granites in the story are igneous, and the migmatites anatectic.

CO_2-rich fluid inclusions in Granulite Facies rocks, first reported by Touret (1971), are often cited as evidence for the 'flushing' hypothesis, but can be explained anatectically as the H_2O-depleted fluids coexisting with melts. Touret and Olsen (this volume) further consider fluid inclusions. Powell

(1983*a*) points out the abundance, in the Granulite Facies, of low-variance assemblages which appear to buffer a_{H_2O} (see Grant, this volume, for an extended treatment of the buffering effect). Buffering is inconsistent with the simplest form of 'flushing', in which a low value of a_{H_2O} would be uniformly imposed by the externally derived carbonic fluid, rather than being locally controlled by buffering assemblages. Powell (1983*a*) argues that buffering supports an anatectic interpretation, the buffering equilibria being water-deficient melting reactions. The analogous effect with the muscovite + quartz dehydration-melting reaction in the Amphibolite Facies (Powell, 1983*b*) has been demonstrated in detail in natural assemblages by Tracy (1978). However, subsolidus dehydration reactions can also buffer a_{H_2O} (Grant, this volume). Thus, in the Proterozoic of E Antarctica, at estimated P–T conditions of $700 \pm 30\,^{\circ}C$ and 5.5 ± 1 kbar, Granulite Facies assemblages are interpreted by Grew (1981) as buffering a_{H_2O} but at low values, too low for melting. There is migmatitic or migmatite-like layering but it predates the observed mineral assemblages. Probably, some agency 'dried out' the terrain prior to the observed metamorphism. Grew (1981) adduces the buffering effect as evidence against a 'flushing' interpretation for the low a_{H_2O}. On the other hand, buffering is by no means incompatible with large-scale fluid movements, since it seems to occur frequently during regional dehydration in ordinary, subsolidus metamorphism (Grant, this volume, section 3.7). In subsolidus dehydration, buffering causes increasing a_{H_2O} with increasing grade; the reverse variation in a_{H_2O} inferred by Phillips (1980) at Broken Hill clearly favours melting (Powell, 1983*a*). Thus the interpretation of the Granulite Facies will probably hinge on the development of reasonably accurate methods of estimating a_{H_2O} and its variation with grade.

1.9 Mineral compositions in leucosome and melanosome: the plagioclase problem

On very well-known experimental grounds (Bowen, 1913) the normative plagioclase in a partial melt is expected to be more sodic than the residual plagioclase in the restite. If the melt segregates and crystallizes in its own closed system, leucosome plagioclase should be much more sodic than that in the melanosome. The effect is actually observed in some migmatites (e.g. Tobschall, 1971). More usually, leusocome plagioclase may, on average, be slightly more sodic than in melanosomes, but the effect is much slighter than the phase diagrams predict (Mehnert, 1968, p. 259). Until recently, it was possible to speculate that the effect might be much reduced in complicated systems compared with the simple Ab–An system studied by Bowen (1913), but now Johannes (this volume) has experimentally confirmed a large effect in quartz–feldspar–H_2O systems.

Some authors have held that migmatites lacking the plagioclase-differentiation effect cannot be anatectic (e.g. Misch, 1968; Yardley, 1977,

1978). However, others have studied migmatites which, on combined evidence of leucosome compositions, textures and field relations are interpreted as anatectic, but lack the plagioclase-differentiation effect (e.g. Mehnert, 1968; Ashworth, 1976). The latter school have some explaining to do. Several ideas have emerged.

1.9.1 *Subsolidus re-equilibration*

Plagioclase in Amphibolite Facies rocks is often homogeneous. Given that the prograde metamorphism has involved a sequence of reactions tending to change the plagioclase composition (e.g. Crawford, 1974), the absence of zoning suggests that diffusive homogenization of plagioclase is possible under Amphibolite Facies conditions, despite the very slow kinetics of diffusion in this mineral (discussed by Johannes, this volume). The leucosome and restite of a stromatic migmatite are in intimate contact and share some common metamorphic history (retrograde, at least). It seems possible that the plagioclase is homogenized between the two bodies, after the crystallization of the anatectic leucosome, thus reducing the total free energy of the system. This effect is most likely where the neosomes share a prograde history in addition to a retrograde one, i.e. in re-metamorphosed migmatites. It cannot have been fully effective where growth zoning, dating from the melt-crystallization stage, is preserved in some leucosomes (Ashworth and McLellan, this volume, section 5.3.1).

1.9.2 *'Unstable' melting*

Johannes (1978, 1980, this volume) reports great difficulties in studying melting at realistic temperatures in systems with Ca-bearing plagioclase, due to very slow kinetics. The cause is now regarded as slow diffusion within the plagioclase (Johannes, this volume, section 2.3.3). The expected plagioclase differentiation effect develops only in runs of long duration. The kinetically-controlled melting reaction without differentiation has been called 'metastable' melting but is now called 'unstable' by Johannes (this volume), as seems reasonable since a continuum of disequilibrium, incompletely differentiated states is involved rather than a metastable free-energy minimum. Extrapolations by Johannes (1978) suggest that $\sim 10^5$ years may be required for equilibrium to be attained at 730 °C. At likely temperatures of initial melting in migmatites (< 700 °C), Johannes' (1983) calculations suggest that equilibrium might never be reached on any geological timescale. These results have been used by some authors (e.g. Mehnert and Büsch, 1982) in interpreting the natural migmatites, the idea being that plagioclase differentiation is suppressed by disequilibrium melting in nature. This interpretation is attractive, particularly for migmatites with relatively short petrogenetic histories (contact-aureole and shallow-level regional settings), but Johannes (this volume) himself does not favour it. The estimation of diffusion parameters from the low-temperature experiments is difficult, and the extrapolations to geological timescales are subject to large possible errors. According to Johannes (this volume), the main importance of unstable melting is as a hindrance to experimental determination of equilibrium phase relations, the latter being regarded as the true controls in nature.

1.9.3 *Reaction with residue during crystallization of melt*

If plagioclase can be homogenized metamorphically (section 1.9.1), it can also be homogenized during retrograde reaction between melt and crystals if cooling is slow enough. Slight differentiation and slight zoning might still prevail, and marked zoning in the outer parts of some leucosome grains (Ashworth and McLellan, this volume, section 5.3.1) may represent failure of reaction at a late stage as pockets of melt become effectively isolated, and diffusion is hindered by declining temperature. This is the interpretation favoured by Johannes (this volume). In the words of Mehnert (1968, p. 259): 'If... the melt crystallizes anew, the former plagioclase is re-established in the solid phase'. Ashworth (1976) invoked this effect. It is especially likely where there is evidence that the retrograde mineralogical and textural evolution of granitoid leucosomes (Ashworth and McLellan, this volume) involved reaction across the leucosome-melanosome boundary, as often seems to be the case. Plagioclase is not the only mineral to have its composition 're-established'. Muscovite and biotite compositions are usually indistinguishable between leucosome and melanosome, even where these minerals, though present at the climax of melting, have also clearly grown in part by retrograde rehydration-crystallization reactions.

As well as major elements, trace elements are often observed to be homogeneously distributed, rather than following igneous-like differentiation patterns that would be expected if the melt could not react with the restite (e.g. Mehnert and Büsch, 1982). Dougan (1981) finds homogeneous trace element contents in the minerals of some of his specimens, perhaps surprisingly since the melts seem to have segregated efficiently (Ashworth and McLellan, this volume). In others, however, the same mineral has different trace-element contents in leucosome and melanosome (Dougan, 1981); the melt presumably became 'detached' from the restite. Interestingly, these migmatites are the 'discordant' ones (with different mineral assemblages in leucosome and melanosome: Olsen, this volume, section 4.5.2). The partitioning of trace elements in migmatization is certainly not fully understood, and Johannes (this volume, section 2.6.7) pleads for more experimentation on the kinetics of their exchange. Trace-element studies should be most informative where there is major-element differentiation (e.g. in plagioclase). Geochemistry, indeed, comes into its own when the melt becomes thoroughly separated from its source-rock (e.g. Condie and Allen, 1980).

Reaction between melt and residue, then, is facilitated if segregation is incomplete. The virtual absence of segregation in the cases studied by Henkes and Johannes (1981) and Gupta and Johannes (1982) allowed complete 're-establishment' of plagioclase compositions. Maintenance of contact between melt and residue is the key to Johannes' solution (this volume) of the plagioclase problem, and can be seen from the preceding sections to be an explanation for some other differences between anatectic leucosomes and normal igneous rocks. Sections 1.5–1.9 should have given an appropriate impression of the advocates of anatexis wriggling round awkward problems. In many cases, however, the evidence for anatexis can be shown to outweigh the negative results of 'tests' for melting, provided that the history of the rocks after partial melting is borne in mind.

1.10 Open and closed systems

The non-dilational character of some agmatites and vein-complexes has been much discussed. The criterion for this effect, which is defined in the field and requires careful examination of good exposures in three dimensions, is that structures in the country-rocks (the blocks in the agmatite or wall-rocks of the veins) show no displacement across the leucosome. The contrary (dilational) case, with displacement due to forcing apart of the boundaries of the leucosomes, is simply interpreted in terms of injection, usually of a magma (Mehnert, 1968, p. 9), and so is clearly open-system. The non-dilational case seems to require replacement of the country-rock by the neosome and so has also been regarded as open-system (metasomatic). However, an alternative possibility here is that leucosomes grew by abstraction of material from a contracting country-rock mass (Mehnert, *loc. cit.*). Barr (this volume) distinguishes dilational and non-dilational pegmatites, interpreting some of the latter as subsolidus segregations (section 7.3.4).

The question of whether neosomes are locally derived is often addressed by measuring compositions of leucosomes, melanosomes and mesosomes, to assess whether the mesosome can be modelled by mixing the two neosomes. Stromatic migmatites are relatively easy to sample, serially perpendicular to leucosome contacts (e.g. Dougan, 1979, Figure 4). Systematic collinearity in sets of two-component plots (e.g. Dougan, 1979, Figure 7) or three-component triangular diagrams (e.g. Mehnert and Büsch, 1982, Figure 2) is unlikely to arise by chance and is interpreted as meaning that the mesosome is representative of the material that differentiated to produce the neosomes. Ashworth (1976) found an absence of collinearity and was not surprised: the observed mesosome is not the same piece of rock as the parent rock of the neosomes, so it is allowed to have different composition by random chance. Mehnert (1968, Figure 93), in his very thorough study of the Black Forest migmatites, finds *statistically* approximate collinearity between *average* leucosome, mesosome and melanosome in a three-component plot (quartz–

feldspar–mafics). This suggests that the proportions of these three components do not control migmatization, i.e. are not *systematically* different in differentiated and undifferentiated material, though the predictable random variation is present.

An improved approach is to measure also the volume proportions of melanosome and leucosome, and statistically assess whether mixing them in the observed proportions would approximately reproduce the local mesosome, considering all components simultaneously in a least-squares model (Olsen, this volume). *Some* migmatites studied by Olsen can be interpreted as closed-system differentiates produced from a rock approximating the observed mesosome. In view of the likely random effects just mentioned, the types of study pioneered by Olsen and Dougan (reviewed by Olsen, this volume) will be possible only where the mesosomes are fairly homogeneous.

If the mesosome is the same as the parent rock that differentiated into neosomes, then a legitimate question is: why did one rock become differentiated while an adjacent, indistinguishable one did not? Searches for subtle differences in mineral compositions have largely proved negative. The plagioclase An contents may range to higher values in mesosomes but there is overlap with neosomes (e.g. Ashworth, 1976; see Johannes, this volume, section 2.5). The sites of segregation of anatectic leucosomes may be deformationally controlled (e.g. Barr, this volume, section 7.3). Alternatively, Ashworth (1976) suggested that a_{H_2O} varied subtly (unmeasurably) from place to place, and melting set in where a_{H_2O} was highest. This idea seems to be supported by Johannes (this volume, section 2.4.2). Further incremental melts produced with increasing temperature might then migrate into the existing leucosome pools. The initial melting will tend to desiccate adjacent rock, suppressing its melting. On the other hand a_{H_2O} in and near the leucosome may fall below that in unmelted mesosome, and diffusive exchange of material may be driven by the resulting gradient in water activity (or fugacity: Olsen, this volume), again causing the existing leucosome to grow.

Johannes and Gupta (1982) and Gupta and Johannes (1982) deal with poorly segregated migmatites which have more K-feldspar in neosome than in mesosome. The transformation to neosome by partial melting affected only the K-feldspar-rich layers of the parental metasediments. In the area studied by Johannes and Gupta (1982), this alternation of K-feldspar-rich and K-feldspar-poor layers can actually be observed directly in the unmigmatized equivalents of the (higher-grade) migmatites. Now, the temperature of beginning of melting is not affected by the proportions of the minerals (Johannes, this volume), a point ably demonstrated by direct experiments on the rocks in question (Henkes and Johannes, 1981; Gupta and Johannes, 1982), so the transformation to neosome must be controlled by the *amount* of melt generated (Henkes and Johannes, 1981, p. 127; Johannes, this volume, section 2.4.2), even though simple water-saturated experiments may fail to reproduce this difference in behaviour between neosome and mesosome

(Johannes, this volume, Figure 2.15). Obviously, this modal control on migmatization means that mesosomes are not intermediate in composition between leucosome and melanosome (if present). Such 'isochemical' (closed-system) migmatites would fail the test for a closed system used by Olsen (this volume). Olsen's criterion requires initially rather homogeneous rocks; Johannes and Gupta's mechanism of migmatization requires initially inhomogeneous ones.

Olsen (this volume) invokes introduction of material into migmatites that fail her closed-system test (and have homogeneous or weakly layered mesosomes). The volume proportions of leucosome exceed those predicted by a closed-system calculation. This effect, occurring alone, could be due to the observed mesosome having lost a small melt fraction to the leucosome, i.e. being slightly restitic rather than pristine parent rock. However, the non-collinearity effect is also observed, and there is a systematic pattern of composition of 'excess' material required to model each type of open-system leucosome. In the case described in Olsen's 1984 paper, mesosomes are layered but vary only slightly in composition. Leucosomes cut across the layering, ruling out an interpretation along the lines of Johannes and Gupta (1982). The leucosomes cannot be adequately modelled in a closed system with melanosome plus any combination of the layers (Olsen, this volume and 1984, Figure 3). The inferred introduced material does not have the composition of a melt and is envisaged as entering the system in aqueous solution, which *caused* anatexis. In another case of inferred open systems (Olsen, 1982 and this volume), the introduced material can be modelled as an injected melt, which proceeded to react with the wall rocks.

The open-system interpretation always depends on the assumption that observed mesosome is representative of the parent rock for the locally-derived portion of the neosomes. In view of the above discussion, there are clearly pitfalls for the unwary here. However, Olsen's (this volume) interpretations of systematic phenomena stand up to commonsense. It is not likely that migmatization always proceeds simply by segregation in a direction perpendicular to the boundary of the leucosome, on a scale < 1 m. Interactions on larger scales, and particularly transport parallel to stromatic layering, are easily envisaged. To assess the net direction and distance of transport would be a formidable 'tactical problem' (Olsen, this volume). Johannes (this volume, section 2.5) discussing the migmatites regarded by Johannes and Gupta (1982) as closed-system, now admits the possibility of some alkali metasomatism.

In studying mass balance, then, Olsen (this volume) is concerned with distinguishing closed from open systems and, separately, distinguishing melt-independent transport from movement of melt. Dougan (reviewed by Olsen, this volume, section 4.5), having established the case for closed systems, is concerned with how much melt has been extracted from a melanosome and the consequent modal changes, which are to be used in inferring melting reactions and interpreting other data such as textural observations (Ashworth and McLellan, this volume). Barr (this volume, Table 7.2) deals with incompletely segregated anatexites. The amount of melt expelled from one rock is constrained using an observed differentiation effect in plagioclase. In another case the melt:residue proportions and compositions are estimated in both leucosome and melanosome using modal constraints.

1.11 Conclusions

Processes of migmatization provide a challenging field of petrological interpretation. No single criterion for anatexis is reliable. Interpretations are best made by using a combination of modal and/or chemical compositions of bodies, textures (unless original textures have been obliterated by later events), field relations, and mineral compositions (with a proviso as for textures). Mineral compositions should particularly be used, where possible, for geothermometry and estimation of a_{H_2O}. Early migmatites in polymetamorphic terrains may not retain clear evidence of the processes of their origin, and may have to be interpreted by analogy with less modified migmatites. Criteria for a closed system are based on assumptions which cannot be tested, but a good case has been made out for open-system behaviour in some stromatic migmatites, including metasomatic processes accompanying anatexis. The segregation of melt in anatexites, even in closed systems, is subject to subtle controls which deserve further scrutiny. The need for further research is a recurring theme in this volume. It is hoped that the book provides some pointers along roads that are not dead ends.

Acknowledgements

The editing of this book as a whole was greatly aided by the secretarial help of Mrs J.E. Davis and Mrs C.A. Bick, and the advice and encouragement of the publishers. My sketches were re-drawn properly by B.A. Parker. Dr D.J. Vaughan improved the English in places. All remaining blunders are to be blamed on the editor, who meanwhile thanks all the authors for their hard and thoughtful work, especially those who met the deadline.

This book was conceived by Dr M. Brown. In editing it, I merely tried to execute his plan.

References

Amit, O. and Eyal, Y. (1976) The genesis of Wadi Magrish migmatites (N-E Sinai). *Contrib. Miner. Petrol.* **59**, 95–110.

Ashworth, J.R. (1976) Petrogenesis of migmatites in the Huntly-Portsoy area, north-east Scotland. *Mineral. Mag.* **40**, 661–682.

Ashworth, J.R. (1979*a*) Genesis of the Skagit Gneiss migmatites, Washington, and the distinction between possible mechanisms of migmatization: Discussion. *Geol. Soc. Am. Bull.* **90**, 887–888.

Ashworth, J.R. (1979*b*) Comparative petrography of deformed and undeformed migmatites from the Grampian Highlands of Scotland. *Geol. Mag.* **116**, 445–456.

Ashworth, J.R. and Chinner, G.A. (1978) Coexisting garnet and cordierite in migmatites from the Scottish Caledonides. *Contrib. Miner. Petrol.* **65**, 379–394.

Atherton, M.P. and Gribble, C.D. (eds.) (1983) *Migmatites, Melting and Metamorphism.* Shiva, Nantwich.

Barnicoat, A.C. (1983) Metamorphism of the Scourian Complex, NW Scotland. *J. Metamorphic Geol.* **1**, 163–182.

Bowen, N.L. (1913) The melting phenomena of the plagioclase feldspars. *Am. J. Sci.* **35**, 577–599.

Brown, P.E. (1971) The origin of the granitic sheets and veins in the Loch Coire migmatites, Scotland. *Mineral. Mag.* **38**, 446–450.

Brown, M. (1973) The definition of metatexis, diatexis and migmatite. *Proc. Geol. Ass.* **84**, 371–382.

Brown, M. (1979) The petrogenesis of the St. Malo migmatite belt, Armorican Massif, France, with particular reference to the diatexites. *Neues Jb. Miner. Abh.* **135**, 48–74.

Brown, M. (1983) The petrogenesis of some migmatites from the Presqu'ile de Rhuys, southern Brittany, France. In *Migmatites, Melting and Metamorphism*. eds. Atherton, M.P., and Gribble, C.D., Shiva, Nantwich, 174–200.

Brown, M. and Earle, M.M. (1983) Cordierite-bearing schists and gneisses from Timor, eastern Indonesia: *P-T* conditions of metamorphism and tectonic implications. *J. Metamorphic Geol.* **1**, 183–203.

Brown, G.C. and Fyfe, W.S. (1972) The transition from metamorphism to melting: status of the granulite and eclogite facies. *24th Internat. Geol. Congr.* Section 2, 27–34.

Brown, W.L. and Parsons, I. (1981) Towards a more practical two-feldspar geothermometer. *Contrib. Miner. Petrol.* **76**, 369–377.

Burnham, C.W. (1967) Hydrothermal fluids at the magmatic stage. In *Geochemistry of Hydrothermal Ore Deposits*, ed. Barnes, H.L., Holt, Rinehart and Winston, New York, 34–76.

Büsch, W., Schneider, G. and Mehnert, K.R. (1974) Initial melting at grain boundaries. Part II: Melting in rocks of granodioritic, quartzdioritic and tonalitic composition. *Neues Jb. Miner. Mh.* 345–370.

Campbell, D.S. (1980) Structural and metamorphic development of migmatites in the Svecokarelides, near Tampere, Finland. *Trans. R. Soc. Edinb. Earth Sci.* **71**, 185–200.

Carpenter, M.A. and Ferry, J.M. (1984) Constraints on the thermodynamic mixing properties of plagioclase feldspars. *Contrib. Miner. Petrol.* **87**, 138–148.

Condie, K.C. and Allen, P. (1980) Origin of Archaean migmatites from the Gwenoro Dam area, Zimbabwe-Rhodesia. *Contrib. Miner. Petrol.* **74**, 35–43.

Corbett, G.J. and Phillips, G.N. (1981) Regional retrograde metamorphism of a high grade terrain: the Willyama Complex, Broken Hill, Australia. *Lithos* **14**, 59–73.

Cosgrove, J.W. (1976) The formation of crenulation cleavage. *J. Geol. Soc. Lond.* **132**, 155–178.

Crawford, M.L. (1974) Calcium zoning in almandine: a model based on plagioclase equilibria. In *The Feldspars*, ed. MacKenzie, W.S. and Zussman, J., Manchester University Press, Manchester, 629–644.

Dalziel, I.W.D. and Johnson, M.R.W. (1963) Evidence for the geological dating of the granitic gneiss of western Ardgour. *Geol. Mag.* **100**, 244–254.

Dougan, T.W. (1979) Compositional and modal relationships and melting reactions in some migmatitic metapelites from New Hampshire and Maine. *Am. J. Sci.* **279**, 897–935.

Dougan, T.W. (1981) Melting reactions and trace element relationships in selected specimens of migmatitic pelites from New Hampshire and Maine. *Contrib. Miner. Petrol.* **78**, 337–344.

Essene, E.J. (1982) Geologic thermometry and barometry. In *Characterization of Metamorphism through Mineral Equilibria*, ed. Ferry, J.M. *Reviews in Mineralogy*, **10**, Mineralogical Society of America, 153–206.

Ferry, J.M. (1983) Regional metamorphism of the Vassalboro Formation, south-central Maine, USA: a case study of the role of fluid in metamorphic petrogenesis. *J. Geol. Soc. Lond.* **140**, 551–576.

Ferry, J.M. and Burt, D.M. (1982) Characterization of metamorphic fluid composition through mineral equilibria. In *Characterization of Metamorphism through Mineral Equilibria*, ed. Ferry, J.M., *Reviews in Mineralogy* **10**, Mineralogical Society of America 207–262.

Fettes, D. and Macdonald, R. (1978) Glen Garry vein complex. *Scott. J. Geol.* **14**, 335–358.

Fisher, G.W. (1973) Nonequilibrium thermodynamics as a model for diffusion-controlled metamorphic processes. *Am. J. Sci.* **273**, 897–924.

Friend, C.R.L. (1983) The link between charnockite formation and granite production: evidence from Kabbaldurga, Karnataka, southern India. In *Migmatites, Melting and Metamorphism*. eds. Atherton, M.P. and Gribble, C.D. Shiva, Nantwich, 264–276.

Froese, E. (1976) Applications of thermodynamics in metamorphic petrology. *Geol. Surv. Canada Paper* **75–43**.

Fry, N. (1984) *The Field Description of Metamorphic Rocks.* Open University Press, Milton Keynes.

Grew, E.S. (1981) Granulite-facies metamorphism at Molodezhnaya Station, East Antarctica. *J. Petrol.* **22**, 297–336.

Gribble, C.D. (1970) The role of partial fusion in the genesis of certain cordierite-bearing rocks. *Scott. J. Geol.* **6**, 75–82.

Gupta, L.N. and Johannes, W. (1982) Petrogenesis of a stromatic migmatite (Nelaug, southern Norway). *J. Petrol.* **23**, 548–567.

Haack, U., Heinrichs, H., Boneß, M. and Schneider, A. (1984) Loss of metals from pelites during regional metamorphism. *Contrib. Miner. Petrol.* **85**, 116–132.

Henkes, L. and Johannes, W. (1981) The petrology of a migmatite (Arvika, Värmland, western Sweden). *Neues Jb. Miner. Abh.* **141**, 113–133.

Hobbs, B.E., Means, W.D. and Williams, P.F. (1976) *An Outline of Structural Geology.* Wiley, New York.

Hodges, K.V. and Spear, F.S. (1982) Geothermometry, geobarometry and the Al_2SiO_5 triple point at Mt. Moosilauke, New Hampshire. *Am. Miner.* **67**, 1118–1134.

Hoffer, E. (1978) Melting reactions in aluminous metapelites: Stability limits of biotite + sillimanite + quartz in the presence of albite. *Neues Jb. Miner. Mh.* 396–407.

Holdaway, M.J. and Lee, S.M. (1977) Fe-Mg cordierite stability in high-grade pelitic rocks based on experimental, theoretical and natural observations. *Contrib. Miner. Petrol.* **63**, 175–198.

Holmquist, P.J. (1921) Typen und Nomenklatur der Adergesteine. *Geol. För. Stockh. Förh.* **43**, 612–631.

Holt, R.W. and Wightman, R.T. (1983) The role of fluids in the development of a granulite facies transition zone in S India. *J. Geol. Soc. Lond.* **140**, 651–656.

Hopgood, A.M. and Bowes, D.R. (1978) Neosomes of polyphase agmatites as time-markers in complexly deformed migmatites. *Geol. Rdsch.* **67**, 313–330.

Hopgood, A.M., Bowes, D.R., Kouvo, O. and Halliday, A.N. (1983) U–Pb and Rb–Sr isotopic study of polyphase deformed migmatites in the Svecokarelides, southern Finland. In *Migmatites, Melting and Metamorphism*, eds. Atherton, M.P. and Gribble, C.D., Shiva, Nantwich, 80–92.

Hoschek, G. (1976) Melting relations of biotite + plagioclase + quartz. *Neues Jb. Miner. Mh.* 79–83.

Janardhan, A.S., Newton, R.C. and Hansen, E.C. (1982) The transformation of amphibolite facies gneiss to charnockite in southern Karnataka and northern Tamil Nadu, India. *Contrib. Miner. Petrol.* **79**, 130–149.

Johannes, W. (1978) Melting of plagioclase in the systems Ab–An–H_2O and Qz–Ab–An–H_2O at $P_{H_2O} = 5$ kbars, an equilibrium problem. *Contrib. Miner. Petrol.* **66**, 295–303.

Johannes, W. (1980) Metastable melting in the granite system Qz–Or–Ab–An–H_2O. *Contrib. Miner. Petrol.* **72**, 73–80.

Johannes, W. (1983) Metastable melting in granite and related systems. In *Migmatites, Melting and Metamorphism*, eds. Atherton, M.P. and Gribble, C.D., Shiva, Nantwich, 27–36.

Johannes, W. and Gupta, L.N. (1982) Origin and evolution of a migmatite. *Contrib. Miner. Petrol.* **79**, 114–123.

Kenah, C. and Hollister, L.S. (1983) Anatexis in the Central Gneiss Complex, British Columbia. In *Migmatites, Melting and Metamorphism*, eds. Atherton, M.P. and Gribble, C.D., Shiva, Nantwich, 142–162.

Kerrick, D.M. (1972) Experimental determination of muscovite + quartz stability with $P_{H_2O} < P_{total}$. *Am. J. Sci.* **272**, 946–958.

Lappin, A.R. and Hollister, L.S. (1980) Partial melting in the Central Gneiss Complex near Prince Rupert, British Columbia. *Am. J. Sci.* **280**, 518–545.

Loberg, B. (1963) The formation of a flecky gneiss and similar phenomena in relation to the migmatite and vein gneiss problem. *Geol. För. Stockh. Förh.* **85**, 3–109.

Lonker, S.W. (1981) The P–T–X relations of the cordierite–garnet–sillimanite–quartz equilibrium. *Am. J. Sci.* **281**, 1056–1090.

Martignole, J. and Sisi, J.-C (1981) Cordierite–garnet–H_2O equilibrium: a geological thermometer, barometer and water fugacity indicator. *Contrib. Miner. Petrol.* **77**, 38–46.

McLellan, E.L. (1983) Contrasting textures in metamorphic and anatectic migmatites: an example from the Scottish Caledonides. *J. Metamorphic Geol.* **1**, 241–262.

McLellan, E. (1984) Deformational behaviour of migmatites and problems of structural analysis in migmatite terrains *Geol. Mag.* **121**, 339–345.

Mehnert, K.R. (1968) *Migmatites and the Origin of Granitic Rocks.* Elsevier, Amsterdam.

Mehnert, K.R. and Büsch, W. (1982) The initial stage of migmatite formation. *Neues Jb. Miner. Abh.* **145**, 211–238.

Misch, P. (1968) Plagioclase compositions and non-anatectic origin of migmatitic gneisses in Northern Cascade Mountains of Washington State. *Contrib. Miner. Petrol.* **17**, 1–70.

Newton, R.C., Smith, J.V. and Windley, B.F. (1980) Carbonic metamorphism, granulites and crustal growth. *Nature* **288**, 45–50.

Oliver, G.J.H. (1977) Feldspathic hornblende and garnet granulites and associated anorthosite pegmatites from Doubtful Sound, Fiordland, New Zealand. *Contrib. Miner. Petrol.* **65**, 111–121.

Olsen, S.N. (1982) Open- and closed-system migmatites in the Front Range, Colorado. *Am. J. Sci.* **282**, 1596–1622.

Olsen, S.N. (1984) Mass-balance and mass-transfer in migmatites from the Colorado Front Range. *Contrib. Miner. Petrol.* **85**, 30–44.

Phillips, G.N. (1980) Water activity changes across an amphibolite-granulite facies transition, Broken Hill, Australia. *Contrib. Miner. Petrol.* **75**, 377–386.

Powell, R. (1978) *Equilibrium Thermodynamics in Petrology.* Harper and Row, London.

Powell, R. (1983*a*) Processes in granulite-facies metamorphism. In *Migmatites, Melting and Metamorphism*, eds. Atherton, M.P., and Gribble, C.D. Shiva, Nantwich, 127–139.

Powell, R. (1983*b*) Fluids and melting under upper amphibolite facies conditions. *J. Geol. Soc. Lond.* **140**, 629–633.

Putnis, A. (1980) Order-modulated structures and the thermodynamics of cordierite reactions. *Nature* **287**, 128–131.

Read, H.H. (1957) *The Granite Controversy.* Murby, London.

Robin, P.-Y.F. (1979) Theory of metamorphic segregation and related processes. *Geochim. Cosmochim. Acta* **43**, 1587–1600.

Russell, R.V. (1969) Porphyroblastic differentiation in fleck gneiss from Västervik, Sweden. *Geol. För. Stockh. Förh.* **91**, 217–282.

Sederholm, J.J. (1907) Om granit och gneis. *Bull. Comm. Géol. Finl.* 23.

Sederholm. J.J. (1967) *Selected Works. Granites and Migmatites.* Oliver and Boyd, Edinburgh.

Thompson, A.B. (1982) Dehydration melting of pelitic rocks and the generation of H_2O-undersaturated granitic liquids. *Am. J. Sci.* **282**, 1567–1595.

Thompson, A.B. and Algor, J.R. (1977) Model systems for anatexis of pelitic rocks I. Theory of melting reactions in the system $KAlO_2$–$NaAlO_2$–Al_2O_3–SiO_2–H_2O. *Contrib. Miner. Petrol.* **63**, 247–269.

Thompson, A.B. and Tracy, R.J. (1979) Model systems for anatexis of pelitic rocks II. Facies series melting and reactions in the system CaO–$KAlO_2$–$NaAlO_2$–Al_2O_3–SiO_2–H_2O. *Contrib. Miner. Petrol.* **70**, 429–438.

Tobschall, H.J. (1971) Zur Genese der Migmatite des Beaume-Tales (Mittlere Cévennen, Dep. Ardèche). *Contrib. Miner. Petrol.* **32**, 93–111.

Torres-Roldan, R.L. (1983) Fractionated melting of metapelite and further crystal-melt equilibria—the example of the Blanca Unit migmatite complex, north of Estepona (southern Spain). *Tectonophysics* **96**, 95–123.

Touret, J. (1971) Le facies granulite en Norvège meridionale, 2. Les inclusions fluides. *Lithos* **4**, 423–436.

Tracy, R.J. (1978) High grade metamorphic reactions and partial melting in pelitic schist, west-central Massachusetts. *Am. J. Sci.* **278**, 150–178.

Tracy, R.J., Robinson, P. and Thompson, A.B. (1976) Garnet composition and zoning in the determination of temperature and pressure of metamorphism, central Massachusetts. *Am. Miner.* **61**, 762–775.

Turner, F.J. (1981) *Metamorphic Petrology.* 2nd edn., McGraw-Hill, New York.

Tuttle, O.F. (1952) Origin of the contrasting mineralogy of extrusive and plutonic salic rocks. *J. Geol.* **60**, 107–124.

Tuttle, O.F. and Bowen, N.L. (1958) Origin of granite in the light of experimental studies in the system $NaAlSi_3O_8$–$KAlSi_3O_8$–SiO_2–H_2O. *Geol. Soc. Am. Mem.* **74**.

Tyler, I.M. and Ashworth, J.R. (1982) Sillimanite-potash feldspar assemblages in graphitic pelites, Strontian area, Scotland. *Contrib. Miner. Petrol.* **81**, 18–29.

Vidale, R.J. (1974) Vein assemblages and metamorphism in Dutchess County, New York. *Geol. Soc. Am. Bull.* **85**, 303–306.

Weaver, B.L. and Tarney, J. (1983) Elemental depletion in Archaean granulite-facies rocks. In *Migmatites, Melting and Metamorphism*, eds. Atherton, M.P. and Gribble, C.D., Shiva, Nantwich, 250–263.

Wells, P.R.A. (1979) Chemical and thermal evolution of Archaean sialic crust, south west Greenland. *J. Petrol.* **20**, 187–226.

Winkler, H.G.F. (1979) *Petrogenesis of Metamorphic Rocks*, 5th edn. Springer-Verlag, New York, etc.

Winkler, H.G.F. and Breitbart, R. (1978) New aspects of granitic magmas. *Neues Jb. Miner. Mh.* 463–480.

Wyllie, P.J. (1983) Experimental studies on biotite- and muscovite-granites and some crustal magmatic sources. In *Migmatites, Melting and Metamorphism*, eds. Atherton, M.P. and Gribble, C.D., Shiva, Nantwich, 12–26.

Yardley, B.W.D. (1975) On some quartz-plagioclase veins in the Connemara Schists, Ireland. *Geol. Mag.* **112**, 183–190.

Yardley, B.W.D. (1977) Petrogenesis of migmatites in the Huntly-Portsoy area, north-east Scotland—a discussion. *Mineral. Mag.* **41**, 292–294.

Yardley, B.W.D. (1978) Genesis of the Skagit Gneiss migmatites, Washington, and the distinction between possible mechanisms of migmatization. *Geol. Soc. Am. Bull.* **89**, 941–951.

Yardley, B.W.D. (1983) Quartz veins and devolatilization during metamorphism. *J. Geol. Soc. Lond.* **140**, 657–663.

Yoder, H.S. (1968) Albite–anorthite–quartz–water at 5 kb. *Yb. Carnegie Instn Wash.* **66**, 477–478.

2 The significance of experimental studies for the formation of migmatites

W. JOHANNES*

2.1 Introduction

The aim of this chapter is:

(a) to summarize the results of experimental studies which have been used to explain the formation of migmatites;
(b) to evaluate the significance of these results;
(c) to discuss cotectic melting of granitic and tonalitic assemblages and to explain the detected unstable melting in experiments with plagioclases of intermediate composition;
(d) to give a reasonable explanation for the observed very similar composition of plagioclases in the different layers of a migmatite;
(e) to make suggestions for further experimental and petrographic investigations.

What type of experiments have been performed so far to elucidate the formation of migmatites? Different processes have been discussed in order to explain the formation of migmatites, e.g. magmatic intrusion, metasomatism, metamorphic differentiation (segregation) and partial melting (or different combinations of these). Of these processes only partial melting can easily be simulated in high pressure–high temperature experiments. Consequently there have been a large number of investigations dealing with partial melting in granitic and related systems, and with crystallization of synthetic melts.

The similarities between the results of melting experiments and the compositions of leucosomes (= leucocratic layers of a migmatite) seem to prove the anatectic formation of migmatites, and have often been considered as proof for partial melting as the migmatite-forming process (Winkler and Breitbart, 1978; Winkler, 1979). However, it has become obvious that experimental

*Institut für Kristallographie and Petrographie, Universität Hannover, Welfengarten 1, 3000 Hannover 1, BRD.

results deserve a more critical evaluation. This will be done in the present chapter.

Most migmatites are of granitic, granodioritic, or tonalitic composition. Almost all of them are the result of high-grade metamorphism and have formed in deeper parts of the continental crust. This contribution will therefore be confined to rocks of granitic to tonalitic compositions and their response to crustal conditions.

Unstable melting may happen in hydrothermal melting experiments with mineral assemblages containing plagioclase of intermediate composition. It is less probable in melting experiments with (An-free) haplogranitic mineralogy. So it is reasonable to begin the summary and evaluation of the experimental studies with:

(1) the haplogranite system Qz*–Or–Ab–H_2O, followed by
(2) the tonalite system Qz–Ab–An–H_2O—melting behaviour of plagioclases can be investigated at relatively low temperatures in this system; furthermore, results obtained from this system can be incorporated into the more complicated
(3) granite system Qz–Or–Ab–An–H_2O, which can easily be extended to normal granites and granodiorites.

2.2 Experimental studies in the haplogranite system Qz–Or–Ab–H_2O

Basic data which have been obtained to elucidate the genesis of granites are also the fundamental data of migmatite research. The most important results for interpretation of both granite and migmatite genesis are the findings of Tuttle and Bowen (1958), who deal with the experimental determination of phase-equilibrium relations in the haplogranite system Qz–Or–Ab–H_2O at P_{H_2O} up to 4 kbar, and with the application of the results to petrological problems. The findings of this classic publication are well known and will not be discussed in the present paper. However, it is intended to discuss whether the results of Tuttle and Bowen (*op. cit.*) represent equilibrium (or near-equilibrium) conditions.

The H_2O-saturated liquidus field boundaries for 1, 2 and 3 kbar of Tuttle and Bowen (1958), for 5 and 10 kbar of Luth *et al.* (1964) and for 20 kbar of Huang and Wyllie (1975) are shown in Figure 2.1. The solidus temperatures determined in the haplogranite system by Tuttle and Bowen are given in Figure 2.2 along with the data of Luth *et al.* (1964) and Johannes (1984). Additional phase equilibrium data for the pure haplogranite system at 4 kbar (Steiner *et al.*, 1975) and at high pressures of 10–35 kbar (Merrill *et al.*, 1970) are also available in the literature.

The phase equilibria determined in the system Qz–Or–Ab–H_2O form the

*Qz, SiO_2; Or, $KAlSi_3O_8$; Ab, $NaAlSi_3O_8$; An, $CaAl_2Si_2O_8$.

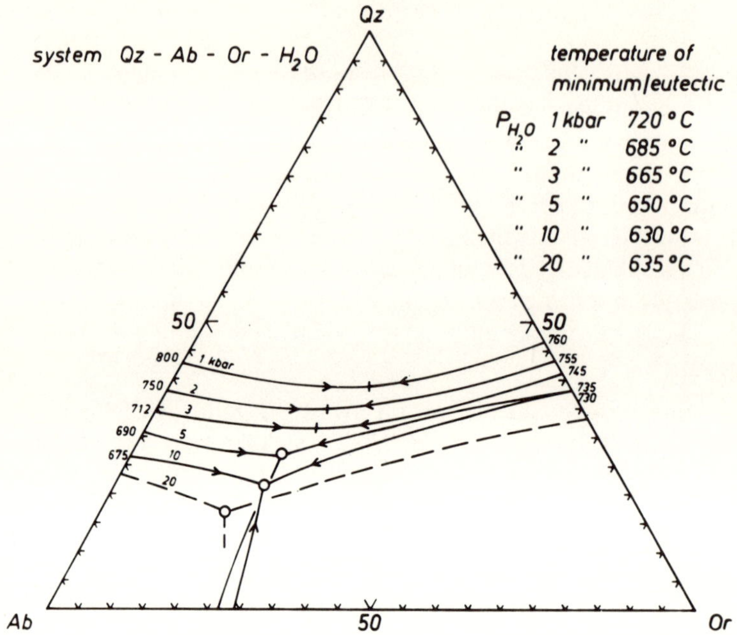

Figure 2.1 H_2O-saturated liquidus field boundaries in the system Qz–Ab–Or–H_2O for various water pressures. 1, 2, 3, kbar data: Tuttle and Bowen (1958); 5, 10 kbar: Luth *et al.* (1964); 20 kbar: Huang and Wyllie (1975). At 20 kbar jadeite is stable instead of an Na-rich alkali feldspar. Qz:SiO_2; Ab:$NaAlSi_3O_8$; Or:$KAlSi_3O_8$.

basis for later experimental studies in more complicated granite–granodiorite systems. They are the key data for interpretation of the genesis of magmatic rocks having haplogranitic, granitic, or granodioritic compositions (like the leucosomes of many migmatites). It is therefore important to know if the data published by Tuttle and Bowen represent equilibrium (or near-equilibrium) results or how far the data may deviate from equilibrium compositions. What controls the reaction rate of melting reactions? Or more precisely: what controls the reaction time necessary to reach equilibrium compositions? We have to look for the slowest process occurring in the reaction between quartz, alkali feldspars, a melt, and a water-rich fluid phase. The most sluggish process, controlling the kinetics of melting, is most probably the diffusion of components within the feldspars. In the case of alkali feldspars, only Na and K have to be exchanged to change the feldspar composition. The reaction time necessary to reach equilibrium composition can be estimated if the interdiffusion coefficients for alkali feldspar, for the appropriate temperature and grain size of the feldspars, are known. The diffusion coefficients determined in ideal or almost ideal crystals under dry conditions allow us to calculate the maximum time necessary to reach equilibrium composition. The interdiffusion of K and Na in alkali feldspars was investigated by Brady and Yund (1983) in

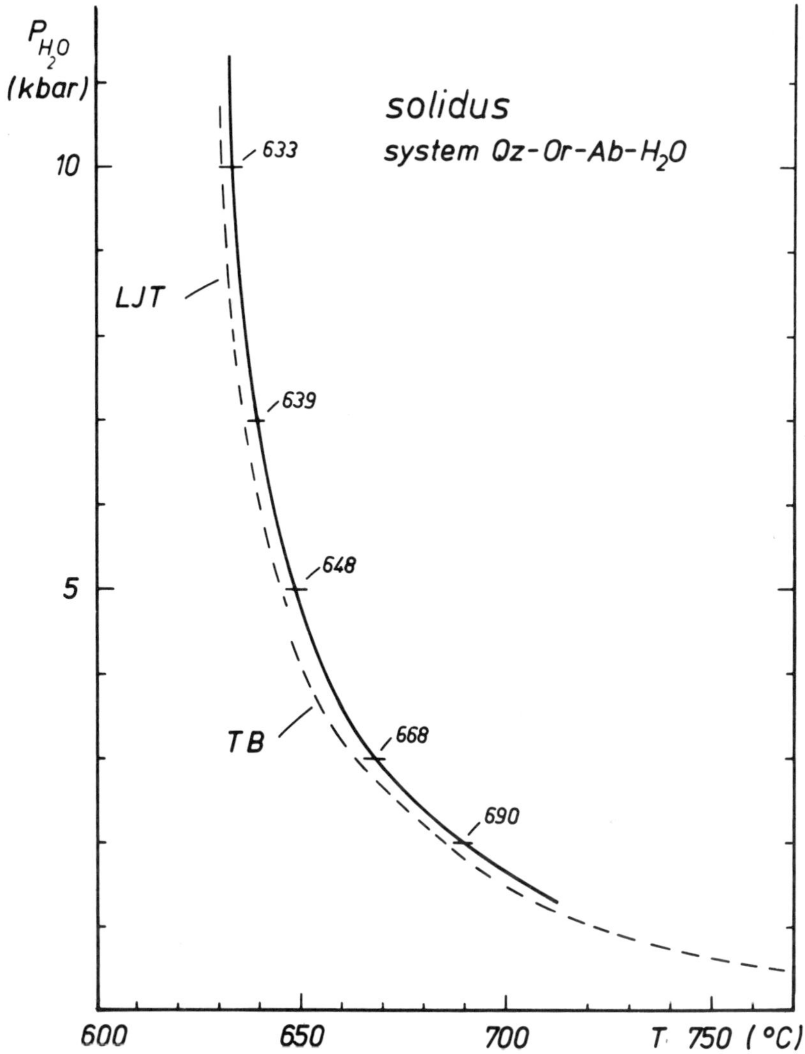

Figure 2.2 Beginning of melting temperatures for minimum melt compositions at low pressures and eutectic compositions at high pressures. Numbers: temperatures in °C from Johannes (1984); TB: Tuttle and Bowen (1958); LJT: Luth *et al.* (1964).

homogenization experiments at 1 atmosphere. The rates of homogenization of coherent exsolution lamellae in cryptoperthites were used by the authors to evaluate K–Na interdiffusion. An 'average' interdiffusion coefficient ($\tilde{D}$) was found to be between 2 and $4 \times 10^{-17}\,cm^2\,sec^{-1}$ for a synthetic cryptoperthite at 600 °C (in the composition range Or_{23}–Or_{53}) and between 6 and $12 \times 10^{-16}\,cm^2\,sec^{-1}$ for a natural cryptoperthite at 650 °C in the composition range Or_0–Or_{85}. At 5 and 15 kbar confining pressure the interdiffusion of K and Na was studied by Christofferson *et al.* (1983) by diffusion-couple experiments performed at 900 and 1000 °C. The alkali interdiffusion was found to be fastest on surfaces normal to (001). For 700 °C, the interdiffusion coefficient ($\tilde{D}$) was calculated to be between 10^{-14} and 7×10^{-14} cm^2/sec for the composition range Or_{20}–Or_{70} (using the thermodynamic mixing data for 1 kbar).

The grain sizes of the starting materials used by Tuttle and Bowen are not known. Intending to use finely crystalline products and not large crystals as starting materials, these workers crystallized the feldspars from finely ground materials at 700 °C. According to the present author's experience, the grains of alkali feldspars crystallized in such a way have a size around 1 μm. 'Well'-crystallized alkali feldspars show a grain size around 10μm, but are badly shaped and have holes (see Johannes, 1979). Assuming a grain size of 1 μm, and accepting (for homogenization of spheres) $\tilde{D} = \rho_o^2/(4t)$ where ρ_o = radius of spheres in cm and t = time in seconds, and taking an interdiffusion coefficient of 2.5×10^{-14} cm^2/sec (for 700°C, see above), we obtain the equilibration time as follows:

$$4t = (10^{-4}\,\text{cm})^2/2.5 \times 10^{-14}\,\text{cm}^2/\text{sec}$$
$$t = 10^5\,\text{sec} = 27.8\,\text{h}.$$

In the experiments with alkali feldspar in granite compositions, Tuttle and Bowen applied run times between 10 h (765 °C) and 984 h (725 °C); see Table 10 in Tuttle and Bowen (1958). Taking into consideration that the grains of the starting materials become smaller during melting and that some additional effects such as the catalytic action of water may increase the diffusivity of the alkalis, we can conclude that the results of Tuttle and Bowen represent equilibrium or at least near-equilibrium data. Further effects, such as the introduction of a high concentration of defects by stress or by the addition of water will also increase the reaction rate, and so favour a faster attainment of equilibrium composition.

Besides P–T values for wet (water-saturated) melting and quantitative data for cotectic and eutectic melt compositions, the solidus temperatures for water-undersaturated melting are of special importance for future research. The equilibrium P and T for constant X_{H_2O} and a_{H_2O} are shown in Figure 2.3.* The data in Figure 2.3 can help in estimating the water activities and minimum water contents prevailing during the formation of migmatites. In many migmatite areas, the maximum temperature of metamorphism was in the range 630° to 730 °C. Temperatures higher than 730 °C seem to be more the exception than the rule. The temperature estimates are based on Fe–Mg fractionation, oxygen isotope fractionation, and dehydration reactions (see e.g. Tracy, 1978). If we assume a relatively high migmatite-forming temperature of 730 °C and a rock pressure of 5 kbar, the water activity must be higher

*I am very grateful to John Holloway for calculating the values given in Figure 2.3.

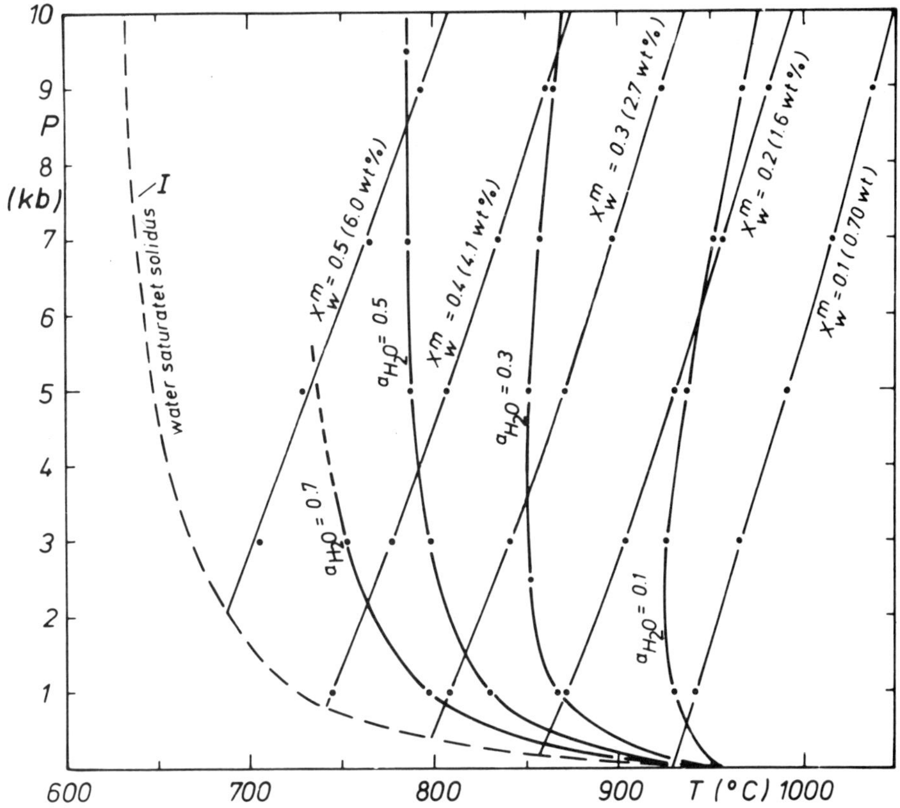

Figure 2.3 Melting equilibria in the system Qz–Or–Ab–H_2O calculated by John R. Holloway (pers. comm.) The a_{H_2O} curves represent the solidus for the specified H_2O activity. The X_w^m curves show the equilibrium P and T for the specified minimum water content. Calculations were made using equations and data from Burnham (1981) and a modified version of a program written by Hanna Nekvasil-Coraor, Pennsylvania State Univ. (1983).

than 0.7 (see extrapolated solidus curve for $a_{H_2O} = 0.7$ in Figure 2.3). It may be mentioned that at high temperature and high pressure, rock fluid is never constituted of pure water only, but always contains dissolved silicate components. In addition, appreciable amounts of salts such as NaCl (up to 10 or even 20 wt.%) could be dissolved in the supercritical gas phase. This implies that the activity of water in such a fluid would usually be less than 1, and the amounts of additional volatile components (such as CO_2, N_2, CH_4) responsible for further reducing the water activity must be low if melting is to occur. In other words, the fluids responsible for the formation of migmatites through partial melting will be water-rich and poor in CO_2.

Knowledge of the amount of water necessary to transform a metamorphic-looking rock into a migmatite is as important as the prevailing activity of water. According to viscosity measurements of van der Molen and Paterson

(1979), the critical amount of melt required to segregate and thus form a leucosome may be about 30 vol.%. If we consider a mesosome/leucosome ratio of 4:1, and assume incomplete segregation, so that mesosomes were molten up to 20 vol.% and leucosomes up to 40 vol.% at the peak of metamorphism (migmatite formation), then the total amount of melt would be as low as 24 vol.%. At 730 °C and 5 kbar, the minimum water content necessary to produce the 24 vol.% melt is 1.4 wt.%. If we believe that melts were present only in the leucosomes, then the amount of water needed (to transform 20% of a rock into a partially molten state with 40 vol.% melt) is only 0.48 wt.% (see position of curve $X_w^m = 0.5$ (6.0 wt.%) in Figure 2.3). So it may be concluded that the amount of free water required to transform gneisses or schists into migmatites is very low (around or even below 1 wt.%).

2.3 Experimental studies in the tonalite system Qz–Ab–An–H_2O

Investigations in the tonalite system are important for various reasons: there are migmatites in which both the metamorphic portions (mesosomes) and the magmatic-looking parts (leucosomes) are of tonalitic composition (examples: Mehnert, 1951, 1968; Mehnert and Büsch, 1982; Misch, 1968; Brown, 1983). The tonalite system, when compared with the granite one, is less complicated because of the lack of alkali feldspars in the former, and consequently the melting behaviour of plagioclase can easily be studied in this system, and there is a good chance to investigate the kinetics of melting reactions with plagioclases of intermediate compositions. Moreover, the results obtained in the tonalite system can be used to evaluate phenomena observed in the granite system and in migmatites of granitic composition.

It is often believed that melting temperatures of tonalitic rocks are very high and that metamorphic conditions are not usually severe enough to form migmatites of tonalitic composition. Complete data sets of solidus temperatures for different pressures, with different plagioclase compositions, can help in elucidating this problem. Unfortunately such data sets are not available. But there are solidus temperatures for the Qz–Ab–H_2O system (Tuttle and Bowen, 1958; Luth, Jahns and Tuttle, 1964) and for the system Qz–An–H_2O (Stewart, 1967). There is also a 5 kbar section for a series of quartz–plagioclase mixtures with plagioclases of different compositions (Johannes, 1978). Phase relations of natural tonalites were determined at 2 and 3 kbar water pressure by Piwinskii and Wyllie (1968). Temperatures at the beginning of melting were investigated for a quartz diorite by Kilinc (1979).

2.3.1 *The tonalite system: studies with natural rocks*

Kilinc (1979) determined beginning-of-melting curves in the quartz diorite–H_2O and quartz diorite–H_2O–CO_2 systems. He used a natural quartz diorite rock from the Henry Mountains (Utah) but did not state its mineralogical composition, only its chemical analysis. Perhaps the mineralogical composition of the starting material was the same as that of the glass-free 'run products' which consisted of plagioclase and hornblende with small amounts of quartz and magnetite. The chemical analysis shows almost 2% K_2O, 4% Na_2O and 5.7% CaO. It is well known that solidus temperatures depend upon the structural state and the chemical composition of the minerals, especially in the case of feldspars. For example, in the tonalite system Qz–Ab–An–H_2O, beginning-of-melting temperatures change with changing An/Ab ratio of the plagioclase involved (see below). But such data are not given by Kilinc (1979), making it impossible to interpret his results fully.

Winkler (1976) performed melting experiments on alkali feldspar-free biotite–quartz–plagioclase gneiss and showed that there is a considerable increase in the initial melting temperature with (a) increasing anorthite content of the plagioclase and (b) increasing Mg/Fe ratio of the biotite. According to him the solidus temperature is increased by 25 °C if the An content of plagioclase is increased from An_{21} to An_{25}. If Winkler's data could be considered correct, the beginning-of-melting temperature would be reduced by 30 °C merely by replacing Mg-rich biotite with Fe-rich biotite. This is very unlikely, and the findings of Winkler do not agree with the results presented in the next section. Winkler's minimum beginning-of-melting temperature of a K-feldspar-free tonalite gneiss is 25 °C lower than that of a K-feldspar-bearing metagreywacke ($P_{H_2O} = 2$ kbar, plagioclase: An_{13} and An_{18}, compare tables on pages 313 and 316 in Winkler, 1976). This is simply improbable because addition of alkali feldspar to a biotite–quartz–plagioclase gneiss can decrease the solidus temperature but would not increase it. So Winkler's data on the beginning of melting in tonalite compositions are very puzzling and internally inconsistent.

Piwinskii and Wyllie (1968) studied the phase relations of granodiorites and tonalites from the Needle Point Pluton (Oregon). Pressure–temperature curves were plotted for the beginning of melting and for the disappearance of the main rock-forming minerals (K-feldspar, quartz, plagioclase, biotite, and hornblende). Representative results are: granodiorite melting begins at 705 °C ($P_{H_2O} = 2$ kbar), K-feldspar disappears at 710 and quartz at 735 °C; tonalite melting begins at 725 °C and the trace constituent K-feldspar also disappears at this same temperature, whereas quartz disappears at 755 °C. Hornblende and plagioclase are still present at 900 °C. The approximate amounts and compositions of the partial melts are also given by Piwinskii and Wyllie (1968). The striking feature of these experimental results is that the granodiorite and tonalite consist of two groups of minerals with rather different characteristics. The assemblage potassium feldspar, sodic feldspar, and quartz fuses to a granitic melt at low temperature, whereas the assemblage hornblende, calcic plagioclase, and biotite remains rather insoluble in the granitic liquid in the lower part of the melting interval because of the refractory nature of these minerals.

Piwinskii and Wyllie obtained their experimental results under conditions where $P_{H_2O} = P_{total}$. Their data also provide guidelines for the phase relations under the condition $P_{H_2O} < P_{total}$ (for details see Piwinskii and Wyllie, 1968), but it must be noted that most of the data of Piwinskii and Wyllie (1968) are not precise equilibrium data because of unstable melting of plagioclase of intermediate compositions (see below, and Johannes, 1978, 1980). It seems that they do provide almost correct relative phase relations and sometimes achieved equilibrium, and the work of Piwinskii and Wyllie (1968) is nevertheless a good example for a critical and careful interpretation of experimental findings. The experimental results (solidus temperatures, dissolution of minerals in the melt, amount and composition of partial melts at various P and T, as determined by Piwinskii and Wyllie) may be used to

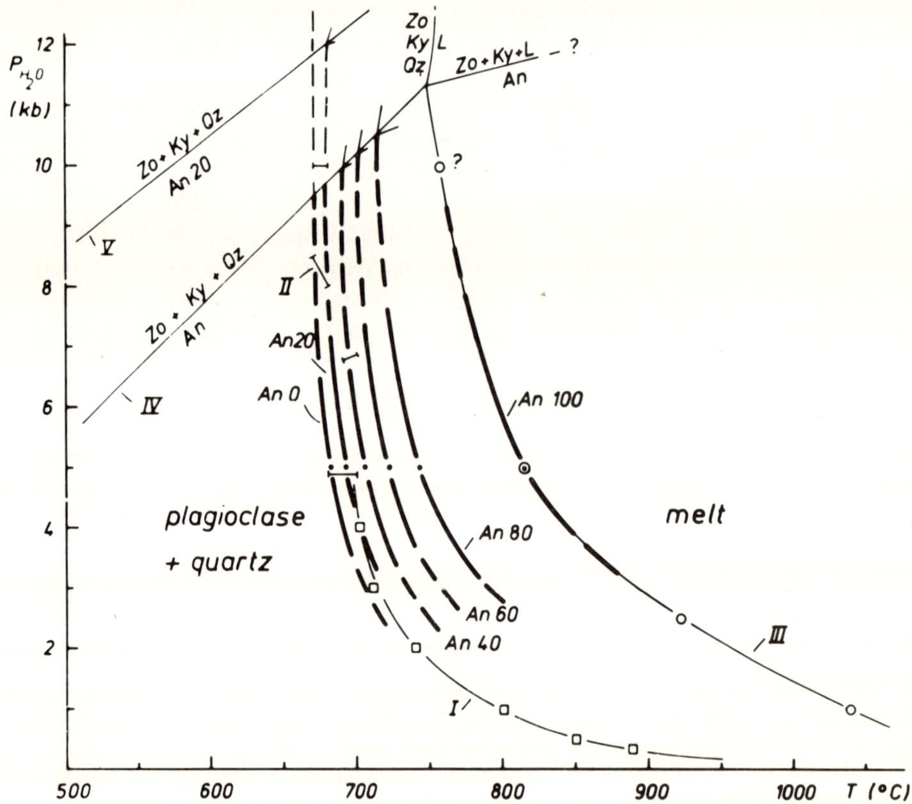

Figure 2.4 Phase relations in the system Na_2O–CaO–Al_2O_3–SiO_2–H_2O. Data sources: I, Tuttle and Bowen (1958); II, Luth *et al.* (1964); III, Stewart (1967); IV, Johannes (1984); V, Goldsmith (1982). Dots at 5 kbar: Johannes (1978). The solidus temperatures for plagioclase–quartz mixtures as determined by Johannes (1978) for 5 kbar and shown in Figure 2.5 have been used to construct the preliminary P–T solidus curves for plagioclase–quartz mixtures.

reconstruct the development of anatectic migmatites or magmatic rocks. Of course, a critical evaluation of experimental data and a careful application of such data to natural processes are absolutely necessary (as has also been emphasized by Piwinskii and Wyllie) to avoid a misuse of the experimental findings.

The easiest experimental procedure seems to be the determination of solidus temperatures. Piwinskii and Wyllie (1968) report a solidus temperature of 705 °C for a granodiorite and 725 °C for a tonalite ($P_{H_2O} = 2$ kbar). The solidus temperature appears to be controlled by the assemblages of the minerals in the rocks as well as by the compositions (and not the amounts) of the minerals dissolved in the first melt at a given fluid pressure. Both the rocks (i.e. granodiorite and tonalite) contained quartz, plagioclase, and K-feldspar

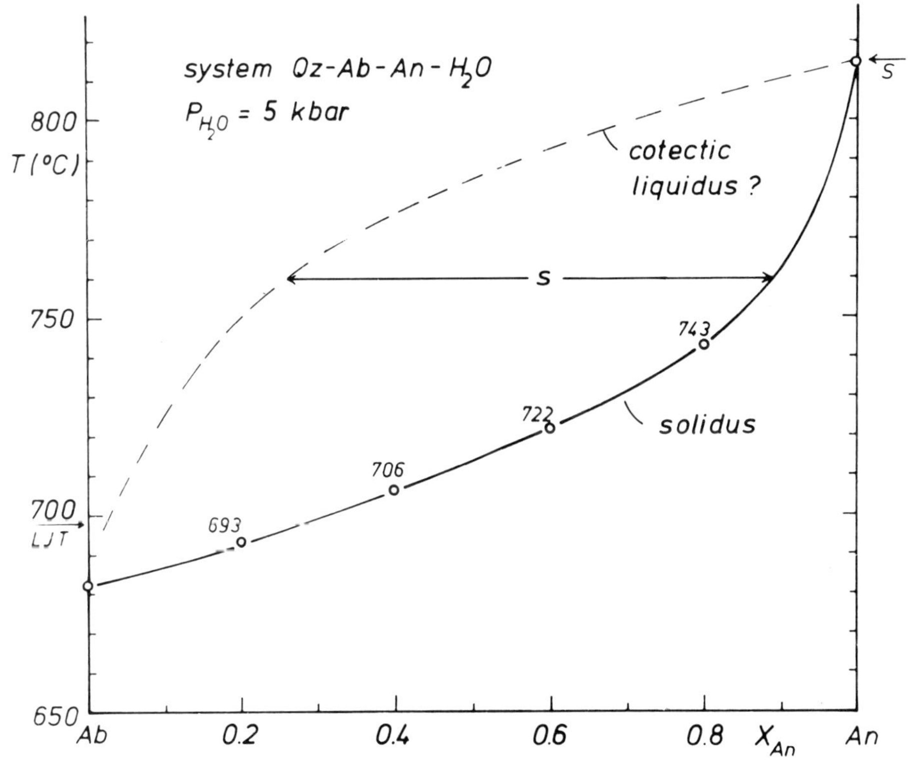

Figure 2.5 Projection of the solidus surface and liquidus boundary curve of the system Qz–Ab–An–H_2O for $P_{H_2O} = 5$ kbar. Dashed curve: liquidus temperatures for cotectic compositions from Yoder (1968). The data points for $X_{An} = 0$, 0.2, 0.4, 0.6, 0.8, and 1.0 have been used to construct the P–T curves of Figure 2.4. LJT, result of Luth *et al.* (1964). S, Stewart (1967). The arrows, at 760 °C, indicate the fractionation of plagioclase components at this temperature. If the given data are correct, the reaction of a plagioclase An_{60} with quartz (and H_2O) would yield a new plagioclase An_{89} and a melt having X_{An} of 0.26 (modified after Figure 3 of Johannes, 1978).

(besides biotite, hornblende, opaques and others). In the granodiorite the average core composition of the plagioclase was An_{37} and the composition of the rim was An_{29}, whereas in the tonalite corresponding values were An_{46} and An_{35}. According to my findings there should be a difference of approximately 2 °C between the solidus temperatures of the above-mentioned rocks (see Figures 2.4, 2.5 and Johannes, 1984), but Piwinskii and Wyllie (1968) reported 20 °C difference, which may be due either (1) to the difficulty in detecting small amounts of initial melt or (2) to heterogeneities of starting material.

Experimental studies such as that published by Piwinskii and Wyllie (1968) (other papers of Wyllie and co-workers are discussed below) can provide arguments for or against an anatectic origin of migmatites, granites, or other plutonic-looking rocks. But they cannot prove or disprove the origin of a migmatite by partial melting. This will become obvious in the present discussion and perhaps also through the other chapters in this book.

2.3.2 *The system Qz–Ab–An–H_2O: solidus and liquidus temperatures*

In quartz–plagioclase mixtures, the beginning of melting is determined at a given fluid pressure by the composition of the plagioclase. According to the plagioclase system Ab–An (Bowen, 1913) albite–quartz mixtures should have the lowest and anorthite–quartz the highest temperatures of beginning of melting. Equilibrium phase relations in the system Qz–Ab–H_2O were studied at atmospheric pressure and up to 4 kbar water pressure by Tuttle and Bowen (1958). The eutectic temperatures for the pressure range 500–4000 bars are included in Figure 2.4. At higher water pressure (4.9–10.0 kbar) the beginning-of-melting curve was determined by Luth *et al.* (1964); see Figure 2.4. The eutectic melt compositions and temperatures of quartz–anorthite assemblages were studied by Stewart (1967) at total water pressures of 1, 2, 5 and 10 kbar. The beginning-of-melting temperatures for this subsystem are also shown in Figure 2.4.

Liquidus temperatures in the system Qz–Ab–An–H_2O were determined by Yoder (1968) for a total water pressure of 5 kbar. Most of Yoder's data are liquidus temperatures near the liquidus boundary curve along which quartz + plagioclase + melt coexist at equilibrium, thus fixing this cotectic line of the system at the particular P_{H_2O} (5 kbar). The cotectic line and the liquidus temperatures are projected on to the Ab–An join in Figure 2.5. It is evident that the temperature difference between the solidus and liquidus is about 70 °C when we consider the plagioclase composition range An_{30}–An_{70}. This in turn shows that even if the initial solid assemblage has a cotectic composition, the melting interval is about 70 °C. If the quartz–plagioclase ratio deviates from the cotectic composition then the difference between the liquidus and solidus must be relatively large (>70 °C) as demonstrated in Figure 2.6.

The solidus for plagioclase–quartz mixtures (Qz + Ab, An_{20}, An_{40}, An_{50}, An_{70}, An_{80}, An_{100}) was determined by Johannes (1978) for $P_{H_2O} = 5$ kbar.

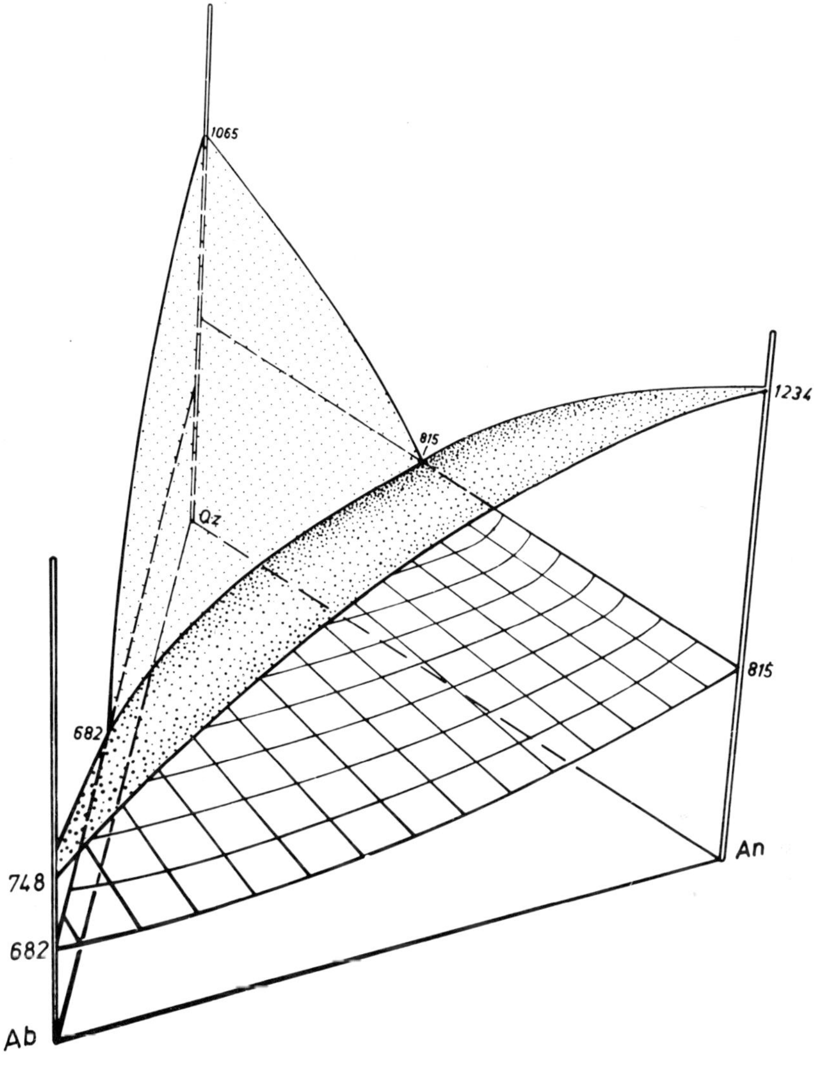

Figure 2.6 Three-dimensional model of the tonalite system Qz–Ab–An–H_2O. The stippled areas are the quartz and plagioclase liquidus surfaces; the plane with a grid pattern is the solidus surface. The solidus surface and the liquidus boundary curve (cotectic line connecting the two eutectic points at 682° and 815 °C) are projected on to the *T–X* plane in Figure 2.5. The melting temperatures are from Yoder (1968). The solidus surface is based on data from Johannes (1978).

Only a small amount of quartz was added to the feldspars, and the disappearance and reappearance of the quartz was used to bracket the solidus curve. In order to dissolve the quartz only small quantities of melts are required to be formed, and it is likely that only the outermost rims of the very fine-grained synthetic plagioclases take part in the melting reaction. The change in the chemical composition of these outermost rims remains small if the solidus temperature is exceeded by only a few degrees. Equilibrium or near-equilibrium melting seems to be likely in this type of experiment, even if the interdiffusion coefficient in plagioclase is as low as $10^{-18}\,cm^2/sec$, for example (see next section). A result favouring the conclusion that the solidus temperatures investigated at 5 kbar represent equilibrium or near-equilibrium temperatures is the fact that the melting reaction with quartz, An_{50}, and water could be reversed within a bracket of 8 °C (Johannes, 1978). The 5 kbar solidus data shown in Figure 2.5 have been used to construct the preliminary pressure–temperature curves in Figure 2.4. In the case of the melting temperature of albite (An_0), quartz, and water there is a difference of approximately 10 °C at 5 kbar from the data of Luth *et al.* (1964). This difference needs further investigation. The 5 kbar result of the anorthite–quartz–water melting reaction is identical with that of Stewart (1967).

The 5 kbar portion of Figure 2.4 is the only precise part. The extrapolation of the *P–T* curves for beginning of melting from lower to higher temperatures was done with the help of melting temperatures for the respective An-free and Ab-free subsystems. The melting curves for An_{40}, An_{60}, An_{80} and An_{100} end at the high-pressure stability boundary for anorthite (An + V = Zo + Ky + Qz, see Figure 2.4), because all plagioclase compositions from An_{40} to An_{100} break down along the same *P–T* curve as pure anorthite (Goldsmith, 1982). The intersection of the upper pressure stability curve of $An + H_2O$ with the beginning of melting of An + Qz + V is shown to be at 11.3 kbar and 750 °C. This is inconsistent with the data compiled by Boettcher (1970). The 10 kbar value of Stewart (1967) needs to be checked. A 9 kbar section and a low-pressure section are under investigation.

The data in Figure 2.4 clearly show that there are no unique solidus temperatures for tonalitic compositions. The beginning-of-melting temperature increases with increasing An content of the plagioclase. This increase is small in the case of Ab-rich and large in the case of An-rich plagioclases. There are the following temperature differences between the different curves: An_0–An_{20} = 11 °C, An_{20}–An_{40} = 13 °C, An_{40}–An_{60} = 16 °C, An_{60}–An_{80} = 21 °C, An_{80}–An_{100} = 72 °C. These differences decrease slightly with increasing water pressure.

The solidus temperatures in Figure 2.4 are for the pure system Qz–Ab–An–H_2O if the given water pressure equals total rock pressure. Reduced water activities will increase these temperatures, and additional components will decrease the solidus temperatures provided they are dissolved in the first melts.

2.3.3 *The tonalite system Qz–Ab–An–H_2O: kinetic studies*

The main mineralogical difference between the haplogranites on the one hand and tonalites, normal granites, and granodiorites on the other hand is the occurrence of intermediate plagioclases in the latter instead of albite or Ab-rich alkali feldspars in the former. In plagioclases Al replaces 1/4 to 1/2 of the Si in the tetrahedra of the network structure. Changing the composition of plagioclases (by coupled substitution of Na + Si by Ca + Al) means that the network structure has to be rearranged. The kinetics of reactions in which plagioclases of intermediate composition take part are obviously controlled by the behaviour of the plagioclases. There are different kinds of plagioclase reactions, e.g. melting reactions, dry subsolidus reactions and 'hydrothermal' reactions. The chemical diffusion in the plagioclases should be the rate-determining factor in the case of melting reactions and dry solid-solid reactions.

Meagre data on oxygen diffusion in plagioclases (Muehlenbachs and Kushiro, 1974; Giletti *et al.*, 1978; Matthews *et al.*, 1983), and a few estimates of interdiffusion data (Grove *et al.*, 1983) are available in the literature. Grove *et al.* (1983) calculated an interdiffusion coefficient $\tilde{D}$ of $10^{-17}\,cm^2\,sec^{-1}$ for 1240–1300 °C from heating experiments of Nord *et al.* (1974), who homogenized the lamellae of the Huttenlocher intergrowth by annealing them. For $T \sim 650$ °C, Grove *et al.* (1983) also estimated $\tilde{D}$ of $\sim 10^{-26}\,cm^2\,sec^{-1}$ from the thickness of Huttenlocher lamellae grown under amphibolite-facies conditions. The data are included in Figure 2.7. Very recently, partial melting experiments on plagioclase feldspars have been carried out by Tsuchiyama and Takahashi (1983) to investigate textures and kinetics of melting. The authors found that the melt was homogeneous and the temperature dependence of liquidus composition agreed approximately with the liquidus as given by Bowen (1913). Compositional gradients were observed in the residual crystals after short periods of melting. The An content of the crystals increased with increasing time until equilibrium composition was reached. It was concluded that the rate of increase in amount of the melts is controlled by diffusion in the crystals. A mutual diffusion coefficient of CaAl–NaSi in plagioclase ($\tilde{D}$) is estimated to be $2.6 \times 10^{-11}\,cm^2\,sec^{-1}$ at 1450 °C and 7.5×10^{-13} at 1360 °C. The data are also shown in Figure 2.7.

I have performed experiments to homogenize synthetic plagioclases (starting compositions 50% An_{20} + 50% An_{60}, maximum grain size 2μm), and observed a new homogeneous run product An_{40} after run times of 4–5 days. The temperature was 900 °C and P_s (bone dry, i.e. H_2O absent) was 10 kbar. If a relation $\tilde{D} = \rho_o^2/4t$ (ρ_o = radius of starting materials in cm, t = time in seconds) is assumed for homogenization of spheres, a diffusion coefficient of $\tilde{D} = 10^{-14}\,cm^2\,sec^{-1}$ can be estimated (data presented in Figure 2.7).

Diffusion data for plagioclase feldspars may also be deduced from the results of melting experiments performed in the systems Ab–An–H_2O, Qz–Ab–An–H_2O (Johannes, 1978) and Qz–Or–Ab–An–H_2O (Johannes, 1980). The experimental results obtained in these systems demonstrate that reaction rates and attainment of equilibrium are extremely dependent on temperature. Time studies were performed at 1000 °C, 900 °C, and 800 °C (P_{H_2O} = 5 kbar) in

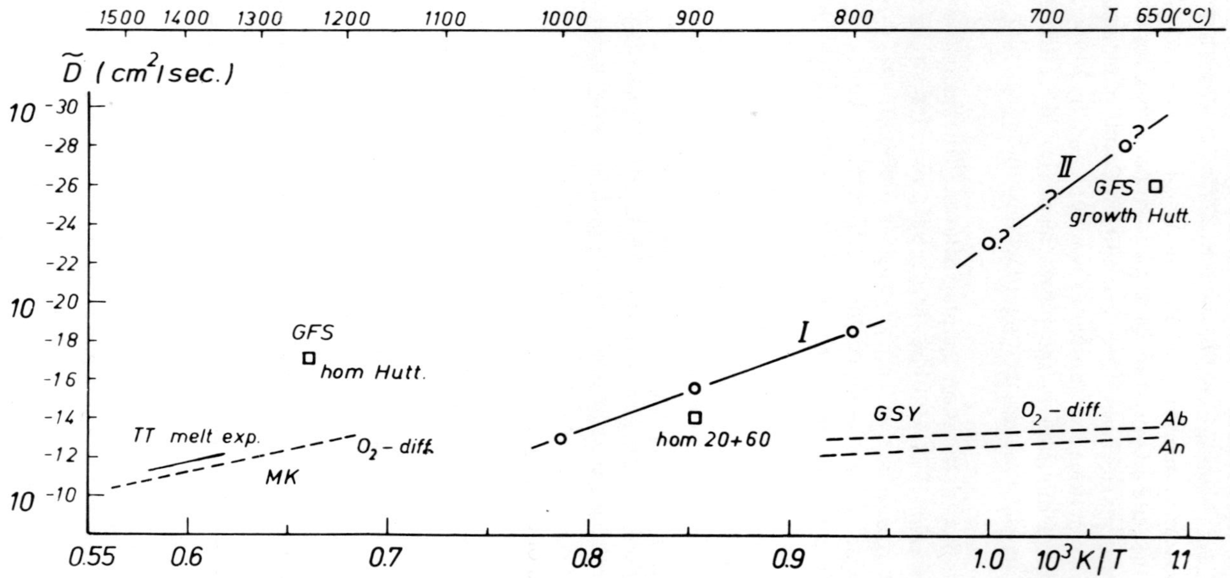

Figure 2.7 Estimated and calculated oxygen diffusion and cation interdiffusion data for plagioclase. MK: data from Muehlenbachs and Kushiro (1974) for dry oxygen diffusion. GSY: data from Giletti *et al.* (1978) for hydrothermal oxygen diffusion in albite and anorthite ($P_{H_2O} = 1$ kbar). TT melt exp.: interdiffusion coefficients calculated by Tsuchiyama and Takahashi (1983) from the results of plagioclase melting experiments. GFS: Grove *et al.* (1983). 'hom Hutt.' and 'growth Hutt.', estimations of interdiffusion coefficients based on experimental homogenization and natural growth of Huttenlocher lamellae. 'Hom 20 + 60', interdiffusion coefficients calculated from experimental homogenization of synthetic plagioclases (this chapter). I and II, estimated diffusion coefficients based on the results of melting experiments (Johannes 1978, 1980, 1983*a*).

the plagioclase–water system. At 1000 °C a run time of less than 1 hour is necessary to reach equilibrium composition in the run products. At 900 °C this run time is almost 200 hours and at 800 °C an equilibrium run time of 10^5 to 10^6 hours is estimated (Johannes, 1983*a*). The diffusion coefficient for interdiffusion ($\tilde{D}$) deduced from these experiments is 10^{-15}–10^{-16} $cm^2 sec^{-1}$ at 900 °C (Figure 2.7). (In the case of these melting experiments the diameter of starting materials was approximately 1 μm). In the tonalite system Qz–Ab–An–H_2O, time studies were accomplished at 730 °C ($P_{H_2O} = 5$ kbar). The starting composition was 50% plagioclase An_{40} + 50% quartz. The mean composition of the plagioclases in the run products was determined to be An_{42} after 10 h, An_{45} after 100 h, and An_{48} after 1000 h (Johannes, 1978). An extrapolation of these data showed that the probable equilibrium composition An_{70} (for $P_{H_2O} = 5$ kbar and 730 °C) would be reached after a run time of 10^{10} hours. This very sluggish reaction suggests an interdiffusion coefficient of 10^{-23} $cm^2 sec^{-1}$ (Figure 2.7).

For cotectic melting at the same water pressure (5 kbar) in the granite system, the solidus temperatures range from 650 to 700 °C (Johannes, 1980, 1983*a*). The time necessary to reach equilibrium composition becomes extremely long in this temperature range. For 665 °C this time is $\sim 10^{15}$ hours with an estimated interdiffusion coefficient of 10^{-28} cm^2/sec. Although this value differs by only two orders of magnitude from the data estimated by Grove *et al.* (1983) (see above), it appears to be very low, perhaps unacceptably so, and needs to be re-estimated. For comparison of the data under discussion, see Figure 2.7.

The compositions of the plagioclases in the run products were determined with the help of the K-exchange method of Viswanathan (1972). In this method, Na is exchanged for K from KCl melt and then several X-ray lines can be used to determine the An-content of the original plagioclase. The advantage of this method is that the chemical composition of the very fine-grained run products and synthetic plagioclases can be investigated, but weak zonation and other small differences in anorthite content cannot be detected.

In an earlier paper on melting of plagioclase (Johannes, 1978), I have reported plagioclases growing to considerable sizes, and continuously changing their composition while reacting together with quartz to form a (water-saturated) tonalitic melt. The changes of composition could be reproduced, but not the considerable increase of grain size. It may have been due to uncontrolled crystallization. Because of the homogeneity of the feldspars I concluded that the slow exchange of Na + Si and Ca + Al between melt and crystal may have been due to special surface processes and not due to slow diffusion. The large plagioclase crystals could be reproduced in runs where crystals grew at the expense of partial melts but not in the reverse case where partial melting occurred, when the temperature was kept constant without any observable fluctuation. My present interpretation is that the very sluggish melting of mixtures containing plagioclases of intermediate composition is controlled by the very slow interdiffusion of plagioclase components, and I agree with the interpretation of Tsuchiyama and Takahashi (1983). The reasons will be discussed elsewhere.

If diffusion controls the melting process, it is of special interest to see how melting plagioclases change their composition by diffusion. The first schematic profiles illustrating the melting of plagioclase with diffusion were produced by Tsuchiyama and Takahashi (1983) who calculated an interdiffusion coefficient for An-rich plagioclases of approximately $10^{-12}\,cm^2\,sec^{-1}$ at 1400 °C (Figure 2.7). This means that within a run time of 10 days the composition of plagioclase can be changed to a depth of approximately 10 μm. At 900 °C the interdiffusion coefficient may be as low as 10^{-16} cm^2 sec^{-1} (Figure 2.7). The relation $x = \sqrt{D.t}$ is responsible for the development of only a very thin reaction rim (10^{-1} μm) which cannot be detected with a microprobe. Although we cannot yet observe in detail how plagioclases melt in hydrothermal runs with tonalitic or granitic starting materials, we do have a rough idea of what would happen in such experiments. The following considerations are valid for quartz–plagioclase mixtures as well as for quartz–plagioclase–alkali feldspar assemblages. If a quartz–plagioclase mixture is, at a given water pressure, heated to a temperature between the solidus and the cotectic liquidus, there is a very unstable situation, resulting in the melting together of quartz and plagioclase. The driving force for melting is the difference between the composition of the starting plagioclase and the equilibrium solidus composition. (The estimated equilibrium composition of coexisting melts and crystals can be read, for $P_{H_2O} = 5$ kbar, from the projected solidus and liquidus boundary curves, shown in Figure 2.5). The mutual interdiffusion of Na + Si and Ca + Al is extremely slow within the plagioclase at low temperatures ($T < 1000$ °C) whereas the breakdown of crystalline structures by melting is a relatively fast process, so a direct and fast equilibrium fractionation of plagioclase components forming an Ab-rich melt and An-rich coexisting crystals cannot occur. So the hydrothermal melting of plagioclase–quartz mixtures may give rise to zoned crystals and melts having Ab/An ratios somewhere between the composition of the starting plagioclase and the (stable) Ab/An ratio of the cotectic liquidus. This melt obviously persists below cotectic liquidus temperatures like a supercooled melt.

The following experiment was performed to elucidate the problem of non-equilibrium melting: plates of quartz and plagioclase were kept in contact during a hydrothermal run. The contact surfaces of both the minerals were well polished. The contact surface of the plagioclase was (001) and that of the quartz was not known. The plates were surrounded by water. The water pressure was 5 kbar, temperature 760 °C and the run time 14 days. The plagioclase (from Lake View, Oregon) used for the experiment was of An_{60} composition, transparent, homogeneous and almost disordered. After the run time of two weeks, melt had developed between the two minerals, adhering in the form of glass to the plagioclase but not to the quartz. The glass selvage

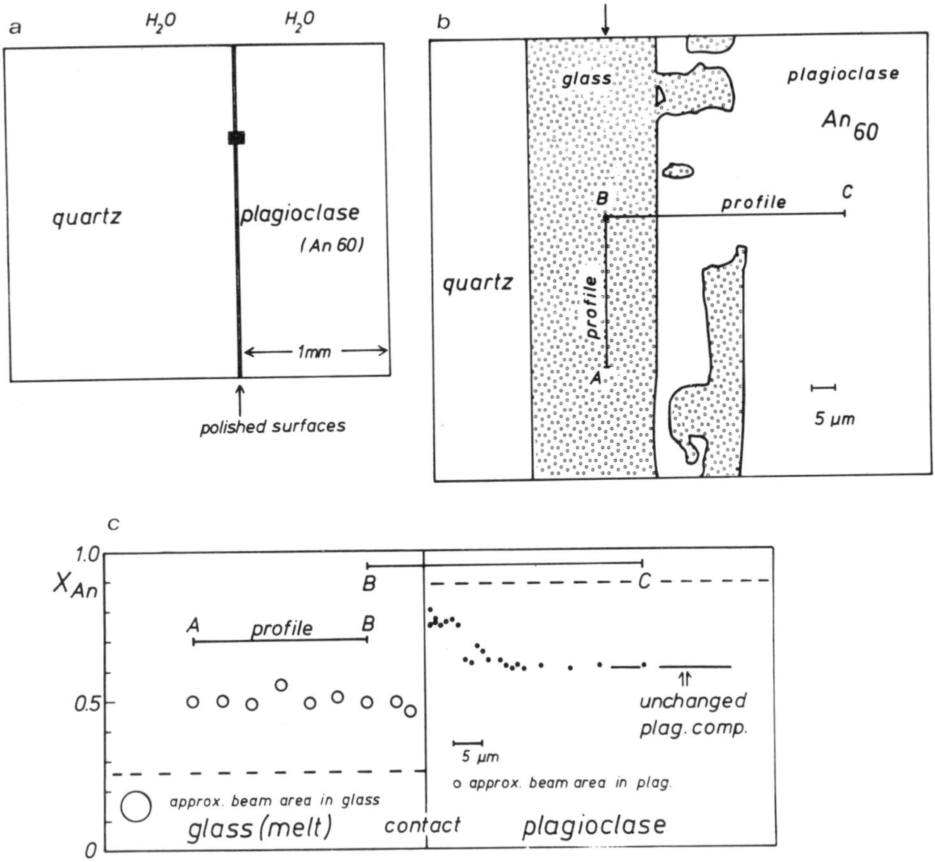

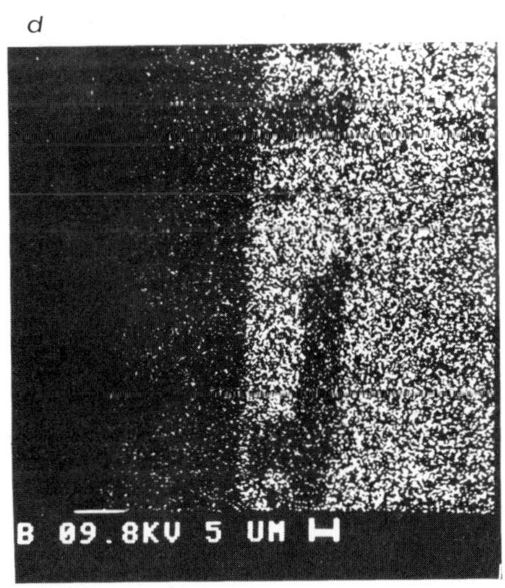

Figure 2.8 Sketches illustrating non-equilibrium melting in the tonalite system.

(*a*) Starting arrangement: polished quartz and plagioclase plates were kept in tight contact for 14 days at 760 °C and 5 kbar water pressure. The filled square marks the area enlarged in *b* and *d*.

(*b*) Results: the stippled areas are glass (frozen melt), the arrow at the top indicates the former contact of quartz and plagioclase.

(*c*) Results of microprobe analyses—the positions of the profiles A–B and B–C are shown in (*b*): the dashed curves represent the equilibrium compositions of coexisting quartz and plagioclase as indicated for 760 °C and 5 kbar water pressure in Figure 2.5.

(*d*) X-ray image of Ca Kα of the area shown in *b*.

covering the formerly polished surface of the plagioclase is approximately 30 μm thick (Figure 2.8*b*). However, glass is not only observed on the polished surface of the plagioclase but is also present inside the plagioclase (Figure 2.8*b,d*). Borders between the plagioclase and glass inside the plagioclase plate are almost parallel to (001) and (010) of the plagioclase. It is suggested on the basis of this experiment that silicon was dissolved in water which migrated into the plagioclase along fractures parallel to (010) and (001). Silicon, dissolved in water, reacted with the plagioclase and produced melt.

It is further observed that the composition of the glass on the (001) surface is relatively homogeneous in the investigated profile A–B. The Qz/Ab/An ratio is approximately 5/2/2 (the mole proportion (An)/(An + Ab) is shown in Figure 2.8*c*). However, the glass within the plagioclase is less homogeneous. It has approximately the same X_{An} as that of the surrounding plagioclase. The composition of the plagioclase has changed to a depth of 10 to 15 μm (Figure 2.8*c*). Within a rim 5 μm thick, X_{An} is 0.75–0.80. The composition of the unchanged plagioclase, An_{60}, is observed approximately 15 μm from the glass–plagioclase contact. The broken lines in Figure 2.8*c* show the equilibrium X_{An} of glass and coexisting plagioclase as it can be read from Figure 2.5 for 760 °C (the applied run temperature). The measured plagioclase and glass compositions do not agree with the solidus and liquidus values given in Figure 2.5. They seem to be the result of non-equilibrium melting. There is no clearly defined diffusion profile within the plagioclase. If the effects in the surface layer, 5–10 μm deep, are considered to be the result of volume diffusion, then the diffusion coefficient can be estimated as approximately $10^{-12}\,cm^2\,sec^{-1}$ (compare Figure 2.7), which is quite high for 760 °C and seems to be unrealistic. Perhaps equilibrium compositions are present at the immediate contact of melt and plagioclase, but could not be resolved with an electron microprobe.

2.4 Experimental studies in the granite system Qz–Or–Ab–An–H_2O

Granite differs from the haplogranite in containing enough An-component to form plagioclase of intermediate composition, though both the rock types are constituted of more than 90% normative Ab, Or, and Qz (see Tuttle and Bowen, 1958). Winkler *et al.* (1975) and Winkler (1979) emphasize that small contents of An-component may have a considerable effect on the solidus temperatures and on the compositions of cotectic melts. Discussing the formation of migmatites and granites, they consider it necessary to treat the component An in addition to Qz, Ab, Or and H_2O.

Experimental solidus temperatures determined in pure systems and in natural rocks will be reviewed in the next sections, followed by a discussion of the results of cotectic melting and their significance for the petrogenesis of migmatites.

2.4.1 *The system Qz–Or–Ab–An–H_2O: solidus temperatures*

In general it is accepted that the beginning of melting in rocks of granitic or granodioritic composition depends upon the plagioclase composition. According to experimental results of Winkler and von Platen (1961), the beginning-of-melting temperature increases enormously with increasing An of the plagioclase (see Figure 2 of Winkler and von Platen 1961, and Figure 2.9, this chapter). Recently Winkler (1979, pp 326 ff.) has claimed that different metagreywackes having qualitatively very similar compositions revealed very different solidus temperatures because of the different plagioclase compositions in these gneisses.

Johannes (1980) determined the solidus temperatures of quartz–plagioclase–alkali feldspar mixtures at $P_{H_2O} = 5$ kbar, using synthetic feldspars and natural quartz as the starting materials. The plagioclase compositions ranged from An_0 to An_{100} with five intermediate compositions in between. (For exact data and compositions of the coexisting alkali feldspars see Johannes, 1980). The experimental technique to determine the beginning of melting was the same as described above for the tonalite system.

Solidus temperatures of quartz–plagioclase–alkali feldspar mixtures at 2

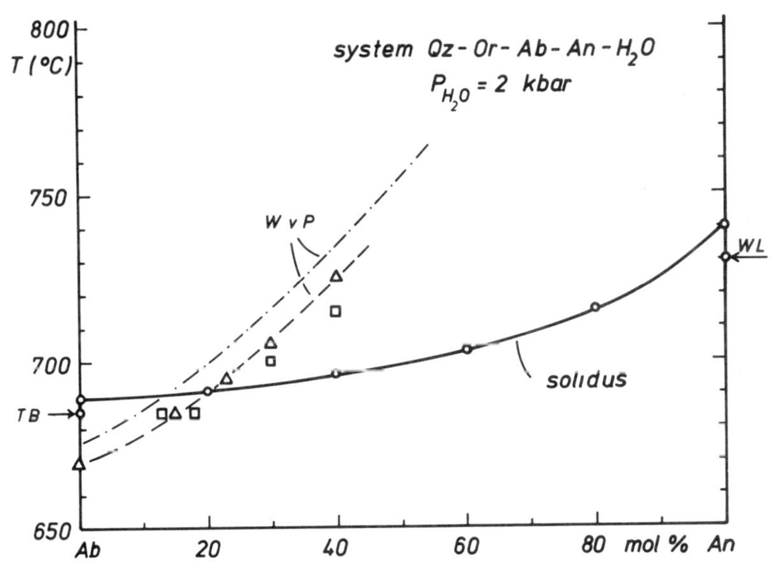

Figure 2.9 2 kbar section showing the compositions of plagioclase and beginning-of-melting temperatures for cotectic melting in the system Qz–Or–Ab–An–H_2O. Heavy line labelled 'solidus', Johannes (1984); TB, Tuttle and Bowen (1958); WL, Winkler and Lindemann (1972). Dashed-dotted curve: composition of plagioclase coexisting with cotectic melts as determined by Winkler and von Platen (1961). Dashed curve: beginning of melting temperatures as given by Winkler and von Platen (1961). Squares: data from Winkler (1979, Table 18–6).

and 3 kbar were recently investigated (Johannes, 1984). The beginning of melting was also established at higher pressures for mixtures with albite and Ab-rich plagioclases. The 2 kbar projection of Johannes (1984) is shown in Figure 2.9 together with solidus temperatures of Winkler (1979) and Winkler and von Platen (1961). The beginning of melting in the system Qz–Or–Ab–H_2O as determined by Tuttle and Bowen (1958), and the eutectic temperature of the system Qz–Or–An–H_2O (Winkler and Lindeman, 1972), are also given in Figure 2.9. The data of Winkler (1979) and of Winkler and von Platen (1961) indicate a considerable increase in the solidus temperatures with increasing An content of the plagioclases. This finding is quite different from that of Johannes (1984). As pointed out above, Winkler's data are internally inconsistent. An extrapolation of the 1961 and 1979 data to higher An/Ab ratios should fit the eutectic temperatures given by Winkler and Lindemann (1972), but this is not the case.

The 2 kbar curve of Johannes (1984) demonstrates a slow increase in the beginning-of-melting temperature with increasing An of the plagioclases in contrast to the steep increase shown by Winkler (1979) and by Winkler and von Platen (1961). The difference between the beginning of melting for An-free alkali feldspar–quartz mixtures and mixtures with plagioclase An_{40} is only 7 °C at $P_{H_2O} = 2$ kbar. (This difference increases only a little with increasing water pressure: see Figure 2.10). The total difference between the beginning of melting in the An-free and Ab-free subsystem is as small as 50 °C (Figure 2.9). Beginning-of-melting curves for constant An content of plagioclase have been drawn on the basis of various pressure sections and are presented in Figure 2.10 (see also Johannes, 1984).

At water pressures of approximately 8 kbar the assemblages anorthite + sanidine and An-rich plagioclase + K-rich alkali feldspar are replaced by zoisite + muscovite + quartz assemblages (Johannes, 1984). This means that at high water pressures the first melts of granitic composition are formed by the breakdown of zoisite + muscovite + alkali feldspar + quartz.

The beginning-of-melting curves compiled in Figure 2.10 clearly demonstrate that the beginning of melting of rocks having granitic or granodioritic composition does not very much depend on the composition of the plagioclases present in the rocks. In the compositional range 0–50% An, the increase in solidus temperature is only ~1 °C if the An content of the plagioclase is changed by 5% (see Figure 2.9 and 2.10). The An contents of plagioclases present in the layers of a migmatite and its country rock (parent rock) are usually not very different and vary only within a few % (Johannes and Gupta, 1982). In the light of the experimental data shown in Figures 2.9 and 2.10, it may be concluded that the relatively small differences in the compositions of gneiss and migmatite plagioclases do not support the idea that the composition of plagioclase could be an important factor in controlling partial melting and formation of migmatites. This is in contrast to the findings and conclusions of Winkler (1979, pp. 326 ff.).

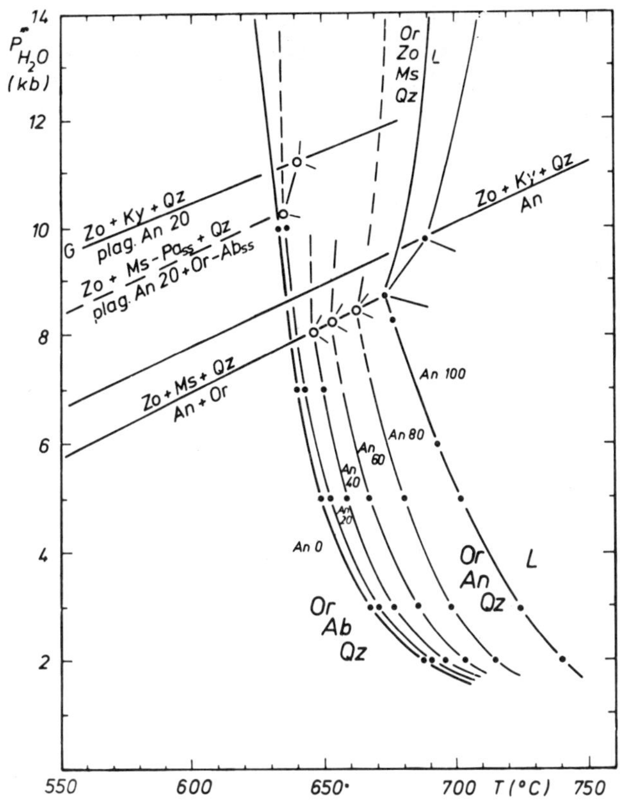

Figure 2.10 Solidus phase relations in the granite system Qz–Or–Ab–An–H_2O and subsolidus reactions of the system Na_2O–K_2O–CaO–Al_2O_3–SiO_2–H_2O. Data sources: Goldsmith (1982) and Johannes (1984). For details see text and Johannes (1984).

2.4.2 *Solidus temperatures: the importance of additional volatile components*

Additional components in the haplogranite system Qz–Or–Ab–H_2O or in the granite system Qz–Or–Ab–An–H_2O may change the solidus temperatures considerably and can increase or decrease a beginning-of-melting temperature. A component (like boron or lithium) soluble in granitic melts may decrease the solidus temperatures, while inert volatiles (like CO_2) concentrated in the fluid phase reduce the partial pressure of water, thus causing higher solidus temperatures at a given total rock pressure or total fluid pressure. Such a relationship is shown in Figure 2.3 for haplogranite. Variations in the beginning-of-melting temperature curves with changing X_{H_2O} in the tonalite and granite systems would look quite similar to that of the haplogranite. However, additional components which are present in the solid phases but do not take part in the melting reaction have no influence upon the solidus temperatures.

The effect of the presence of P_2O_5, HF, and LiOH in addition to H_2O was investigated by Wyllie and Tuttle (1964). According to these authors, 5% of P_2O_5, HF, or LiOH added to H_2O lowered the beginning of melting of a granite by 20 °C, 60 °C and 90 °C respectively. Von Platen (1965) found a lowering of solidus temperatures by 5–20 °C in melting experiments with granites of various compositions when 0.5 M HCl was used instead of pure H_2O. In the case of 0.5 M HF, the solidus was lowered by 35 °C. In all experiments the pressure of the fluid phase was 2 kbar.

Manning (1981) determined liquidus phase relations for the system Qz–Ab–Or with excess water and with 1, 2 and 4 wt% added fluorine at 1 kbar pressure. The position of the minimum melting composition changed progressively from Qz_{37} Ab_{34} Or_{29} for the fluorine-free system (Tuttle and Bowen, 1958) to Qz_{15} Ab_{58} Or_{27} when 4% fluorine was added to the starting materials. The minimum melting temperature in the system decreased by 100 °C as a consequence of adding 4% fluorine to the haplogranite (Figure 2.11). Exploratory experiments carried out by Manning below the liquidus indicate that, in certain bulk compositions, two alkali feldspars, quartz and vapour can coexist with melt at temperatures as low as 550 °C at 1 kbar.

Pichavant (1981) studied the effect of boron on water-saturated haplogra-

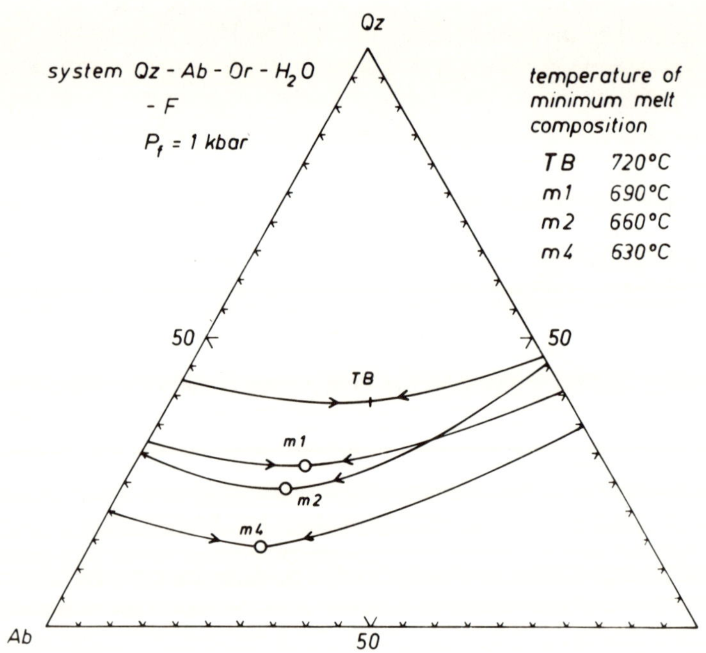

Figure 2.11 Liquidus phase relations for the system Qz–Ab–Or with excess H_2O and F added to the system. TB: data from Tuttle and Bowen for the fluorine-free system; m1, m2, m4: data from Manning (1981), system Qz–Ab–Or + 1, 2, 4% F added.

nite at 1 kbar vapour pressure. He found that the solidus temperatures of Qz–Or–Ab mixtures are lowered by 60 °C when 5% B_2O_3 is added to water and by 130 °C when 17 wt% is added (Figure 2.12). In other words, it may be said that boron as an additional component gives similar results to those for fluorine. These data clearly demonstrate that the composition of first melts can be very different from those in pure systems, and that the beginning-of-melting temperatures can be lowered considerably by additional components preferentially dissolved in the melt. It is important to see if the above experimental findings can help to elucidate the formation of migmatites by partial melting.

Additional and more precise data are needed to explain the formation of migmatites by magmatic processes (partial melting), in addition to those for granitic rocks in the pure systems. To trace the genesis of migmatitic rocks in the light of the experimental data discussed above, the following information and problems, concerning migmatites in general, need to be taken into consideration. Some parts of migmatites still have a metamorphic appearance (mesosomes) while others seem to have crystallized through a (partial) melt (leucosome). If it is believed that variation in the degree of partial melting is controlled by a restricted supply of water-rich fluids, then it needs to be explained why water is preferentially concentrated into some layers of the parent rocks. If on the other hand leucosomes are considered as concentrates of partial melt, then how do these melts separate from the unmolten portions and accumulate in the form of leucosome layers? How have restite minerals been separated from partial melts, if it is believed that melanosomes represent concentrated restite material (Mehnert, 1968)?

The leucosomes of common (layered) migmatites do not contain any special

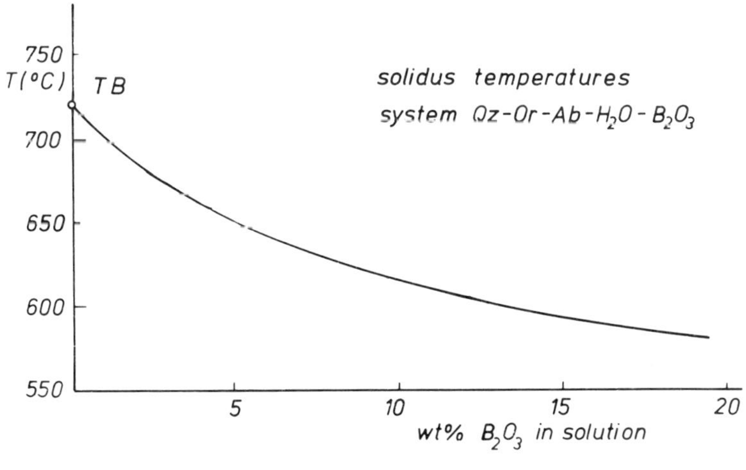

Figure 2.12 Beginning-of-melting temperatures in the system Qz–Or–Ab–H_2O–B_2O_3 for various B_2O_3 concentrations in solution; modification of Figure 1 of Pichavant (1981). TB: solidus of F-free cotectic minimum after Tuttle and Bowen (1958).

enrichment of components which could have lowered the melting temperatures, exclusively in these layers, to transform some of the country-rock layers into magmatic (plutonic) looking portions (called leucosomes) while leaving the others as mesosomes (see Johannes and Gupta, 1982; Gupta and Johannes, 1982). If it is supposed that the low-temperature melt-forming volatile components were concentrated in some particular layers undergoing migmatization then we ought to find a preferred enrichment of such volatiles in the leucosomes. This is not, however, supported by the available data. Moreover, the solidus-depressing volatiles are not needed to explain the formation of migmatites: rather a preferred concentration of water is sufficient to serve the purpose, as discussed below. Special volatiles (like fluorine and boron) may have caused low-temperature pegmatitic melts, which, however, seem to have no importance in the formation of migmatites.

2.4.3 *Solidus temperatures: the significance of additional and excess components (MgO, FeO, Al_2O_3, etc.)*

It may be worthwhile to consider to what extent (if at all) additional and excess components (MgO, FeO, MnO, TiO_2, Al_2O_3 etc.) may change the beginning-of-melting temperatures and equilibrium phase relations of granitic or tonalitic mineral assemblages. If there is a large decrease of solidus temperatures through the presence of additional components, this could be of significance for the genesis of migmatites by partial melting. Primary differences in the different portions of the parent rock of a migmatite may be responsible for different solidus temperatures and different amounts of melts at given temperatures and water pressures.

Thompson and Algor (1977) have emphasized that the system $KAlO_2$–$NaAlO_2$–Al_2O_3–SiO_2–H_2O, even without CaO, FeO and MgO, provides a much more realistic representation of peraluminous granites and migmatites than the simple system Qz–Or–Ab–H_2O. Thompson elucidated the significance of excess and additional components for the formation of migmatites and granites. For details the reader is referred to Thompson (1974), Thompson and Algor (1977), Thompson and Tracy (1979), and Huang and Wyllie (1974). For the purpose of this chapter it may be sufficient to discuss a few examples and to evaluate their significance for the genesis of migmatites.

The experimental findings of Schairer and Bowen (1955) indicated that silicate liquids saturated with mullite produced a ternary eutectic involving potassium feldspar + tridymite + mullite at 985 °C in the system K_2O–Al_2O_3–SiO_2 at 1 atm pressure. The ternary eutectic is 5 °C lower than the Or–Qz cotectic line. This lowering of temperature is caused by the additional Al_2O_3 in the liquid phase, which was only 1 wt.% in excess of that in the feldspar. At lower temperatures and in the presence of H_2O the peraluminous phase could be muscovite or sillimanite or andalusite instead of mullite.

According to Thompson and Algor (1977) the eutectic beginning-of-melting curves for the reactions Ms + Or + Qz + H_2O = L and Als + Or + Qz + H_2O = L occur at slightly lower temperatures in a P–T diagram than does the eutectic Or + Qz + H_2O, due to the assumed solubility of excess alumina in the silicate melt. As already noted by Thompson and Algor (1977), the solubility of excess alumina is so small that minimum melting of peraluminous compositions can rarely be distinguished from reactions involving alumina and alkali proportions as present in feldspars.

In order to provide information on the relative location of melting reactions with and without additional Al-rich and Ca-rich phases, Schliestedt and Johannes (1984) carried out some experiments at $P_{H_2O} = 6$ kbar involving (a) quartz + sanidine + anorthite + H_2O, (b) quartz + sanidine + anorthite + muscovite + H_2O, and (c) quartz + sanidine + anorthite + zoisite + H_2O. At this pressure, melting started at 695 °C in all three cases and no distinctions were noticed between the initial melting reactions in the three mineral assemblages. These findings agree with analogous experiments in the system K_2O–Al_2O_3–SiO_2–H_2O by Huang and Wyllie (1974) and in Na_2O–Al_2O_3–SiO_2–H_2O by Chatterjee (1974), and indicate only a small solubility of excess alumina in quartz–feldspar melts (Thompson and Algor, 1977).

After determining the beginning of melting for synthetic mixtures whose compositions closely approach those of the average granite, Tuttle and Bowen (1958) selected two granites for a similar set of experiments to test the possible effect of constituents other than the feldspar and quartz in lowering the temperature of beginning of melting. One granite was a plagioclase–microcline–quartz granite with biotite and muscovite and the other a perthite–quartz granite, carrying a sodium-iron amphibole as the principal dark mineral. It was found (within the experimental errors) that the two contrasting natural granites begin to melt at the same temperatures as synthetic mixtures having compositions near the ternary minimum. Thus the data of Tuttle and Bowen (1958) indicate a low solubility of additional components in quartz–feldspar melts and show no significant change in the beginning-of-melting temperatures.

Winkler's experimental findings are different from those of Tuttle and Bowen. Winkler (1976, p. 313) reported beginning-of-melting temperatures between 660° and 715 °C ($P_{H_2O} = 2$ kbar) for four different gneisses having similar mineralogical compositions but differing in the Ab/An ratios of plagioclases and FeO/MgO ratios of biotites. However, Winkler emphasized that two of the gneisses had the same amounts of biotite and plagioclase, with similar plagioclase compositions but differing Mg and Fe contents of biotites. He reported solidus temperatures of 660 °C and 690 °C for gneisses having biotite FeO/(FeO + MgO) ratios of 0.8 and 0.3 respectively. Thus Winkler's data suggest a high solubility of iron-rich biotite in the first melt, leaving Mg-rich biotite in the residue. Such a change in biotite compositions is not observed in nature and not yet reproduced by other experimentalists. On the contrary, biotites of migmatites showed higher Fe/Mg ratios than those of adjacent, lower-grade (amphibolite-facies) schists (White, 1966).

Hypothetical liquidus relations in the subsystem Al_2O_3–FeO–MgO projected from quartz, alkali feldspar and plagioclase for $a_{H_2O} \leqq 1.0$ are discussed by Abbott and Clarke (1979). All silicate liquids were considered to be saturated with respect to quartz, alkali feldspar, plagioclase and one or more of the mafic minerals biotite, garnet, cordierite and Al_2SiO_5 (andalusite or sillimanite). For temperatures higher than the breakdown of muscovite and an activity of H_2O of 1.0 and 0.6, equilibrium phase relations are constructed for different P–T regimes. It is shown that with increasing temperature (and decreasing pressure), first Fe-rich biotite + Al_2SiO_5, then biotite + cordierite, then biotite + garnet coexist with low-temperature granitic melts. The temperature difference for $a_{H_2O} = 1$ (at constant P) between the locus of the liquidus minimum in the system Qz–Or–Ab–H_2O and in the system saturated with MgO–FeO–Al_2O_3 is estimated to be 15–30 °C. These interesting findings need further experimental clarification.

Recently Naney (1983) determined phase equilibria of rock-forming ferromagnesian silicates in thc granite system at 2 and 8 kbar with coprecipitated gels of granitic and granodioritic composition as starting materials. The experiments were conducted at temperatures up to 1200 °C under anhydrous to 20 wt.% H_2O conditions. Naney compared his data determined in the FeO-bearing and MgO-bearing systems with those of Whitney (1975) achieved in the system with no FeO or MgO. A difference in solidus temperature of approximately 30 °C at 2 kbar pressure was estimated between the pure and the MgO–FeO-containing systems. At a pressure of 8 kbar, the solidus was found to be approximately 40 °C lower than in the pure (Mg–Fe-free) granodiorite system (Whitney, 1975). In comparison to the data of Tuttle and Bowen (1958) discussed above, the difference of 30–40 °C observed by Naney (1983) and Whitney (1975) between the solidus temperatures of FeO–MgO-bearing and pure systems seems to be large. Although the data are reported to be internally consistent, unstable reactions cannot be ruled out in the gel experiments of Naney and may have been responsible for the observed results. The kinetics of such complex reactions need further clarification.

Existing published data on beginning-of-melting temperatures and phase relations for pure systems and with additional and excess components are not without contradictions. A significant reduction of solidus temperatures and formation of low-temperature melts is unlikely but cannot be excluded. It is not possible that low-temperature melts were preferentially produced in the leucosomes through the presence of additional mafic components. If excess components were the reason for selectively low-temperature melting in the leucosomes they should have been concentrated in these layers. But in all migmatites investigated under the supervision of the author (two examples are published: Johannes and Gupta, 1982; Gupta and Johannes, 1982; five more are still unpublished) the leucosomes have the lowest contents of mafic components in comparison with mesosome and melanosome layers. Even if the leucosomes and adjacent melanosomes are regarded as a unity (this combination is the so-called 'neosome' of Mehnert, 1968) and compared with mesosomes, the latter still have higher FeO, MgO, MnO etc. in most cases. Obviously the additional mafic components do not have any influence in melting the leucosome layers preferentially. However, the mafic minerals could control the formation of low-temperature melts preferentially if they had different stabilities in the different layers of a migmatite. Fleming and White (1984) suggest that an OH-mica component of either muscovite or biotite (or both) could have broken down to provide water, leaving a fluorine-rich mica as residue. Some evidence is found for this in the selvage (melanosome) biotite which has high F (besides high TiO_2 and high Fe/Mg) in comparison with the biotite of the country rocks belonging to the lower-grade amphibolite facies (White, 1966).

2.4.4 *Solidus temperatures determined with natural rocks as starting materials*

Many solidus curves have been determined with natural rocks or mixtures of natural minerals as starting materials. What is the significance of such data for the interpretation of the formation of migmatites and granites? Do they supply more useful information than those achieved with synthetic materials? Does a solidus determined for a natural rock represent unique, stable solidus conditions?

Any reader not very familiar with petrology should be reminded that the melting temperature of a mineral (at a given pressure) depends not only on its chemical composition but also its structural state. In the case of a granite (or migmatite or gneiss) the feldspars usually do not have the chemical and structural composition of the minerals which initially crystallized from a melt or via a metamorphic reaction. Chemical and structural compositions are frozen in at a (much) lower temperature during slow cooling of crustal rocks. If natural minerals are used as starting materials for melting experiments they should be brought to their chemical and structural equilibrium compositions immediately below the solidus temperature. It may be impossible to attain this equilibrium, or it may take a very long time (Eberhard, 1967). Inappropriate structural state of feldspars alone may change a beginning-of-melting temperature by more than 10° (Tuttle and Bowen, 1958; Johannes, 1978). In my experience and opinion, natural minerals are even more problematic as starting materials for melting experiments than are synthetic ones. It requires much experience to perform worthwhile melting experiments with natural rocks.

Solidus curves and solidus P–T data determined with natural granites and granodiorites are compiled in Figure 2.13. For comparison, the beginning-of-

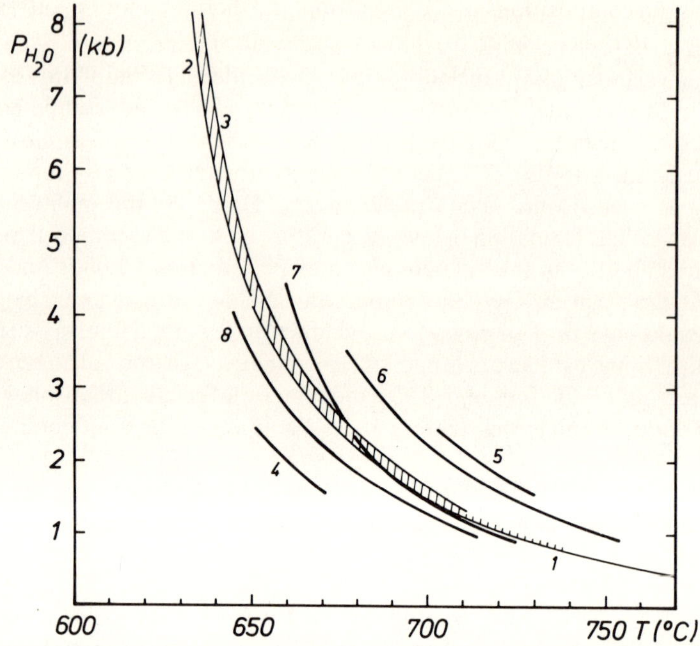

Figure 2.13 Solidus P–T data determined with natural rocks as starting materials in comparison with the solidus for the pure system Qz–Or–Ab–H_2O. Data sources: (1) Tuttle and Bowen (1958); (2) Luth *et al.* (1964); (3) Johannes (1984); (4) Winkler (1976), starting composition (st co) alkali feldspar-free paragneiss; (5) Winkler (1976), st co metagreywacke: (6) Piwinskii and Wyllie (1968), st co granodiorite 779; (7) Piwinskii (1973), st co quartz monzonite MO-1B; (8) Piwinskii (1968), st co granite 104.
Except (4) all rocks contain alkali feldspar (besides quartz, Ab-rich plagioclase, and mafic minerals). The solidus curves of these rocks should almost coincide. The differences may be mainly due to experimental difficulties. The solidus of (4) (tonalitic composition) should plot on the high-temperature side of the 'granite' curves and not on the left side.

melting values for alkali feldspar granite as determined by Tuttle and Bowen (1958), Luth *et al.* (1964) and Johannes (1984) are also shown. The difference between the data shown in Figure 2.13 should be much smaller, because all rocks but one have quartz, Ab-rich plagioclase, Or-rich alkali feldspar and mafic minerals. The quantity of these minerals may be quite different in the different rocks but this is not important because solidus data are controlled by the chemical and structural compositions of the minerals taking part in the melting reaction and not their quantity. This must not be forgotten when discussing the formation of migmatites by partial melting.

2.4.5 *The equilibrium problem in experimental studies with starting materials of granitic composition*

In the section on experimental studies in the haplogranite system it was shown that run times of approximately 30 hours at 700 °C are sufficient to reach equilibrium compositions in the run products, whereas in the case of tonalitic starting materials (= quartz–plagioclase mixtures) it was observed that 10^{10} hours are necessary to reach equilibrium compositions in melting experiments at $T = 730\,°C$ ($P_{H_2O} = 5$ kbar). An interdiffusion coefficient of 10^{-23} cm^2 sec^{-1} was estimated for intermediate plagioclase at this temperature (Figure 2.7).

Granitic rocks, melting and crystallizing under crustal pressure and water-saturated conditions, show relatively low melting temperatures and need extremely long run times in order to reach equilibrium compositions. An estimate for the value of the interdiffusion coefficient in the plagioclase is extremely low in these rocks. An extrapolation of the data determined by Tsuchiyama and Takahashi (1983) with the help of the melting experiments on plagioclases gives a coefficient of 10^{-22} or 10^{-23} at 670 °C. If the diffusion data deduced from melting experiments on plagioclases at $P_{H_2O} = 5$ kbar (Johannes, 1978) are extrapolated down to 670 °C the interdiffusion coefficient is 10^{-24} cm^{-2} sec^{-1}. Calculations made on the rocks of the tonalite system may be extended to the granitic rocks. The latter mainly differ in having lower melting and crystallization temperatures, and the plagioclase has a very much slower mutual interdiffusion exchange rate of Ca + Al and Na + Si at this lower temperature.

An interdiffusion coefficient ($\tilde{D}$) of 10^{-22} cm^2 sec^{-1} would mean that within an experimental run time of 10 days (= almost 10^6 sec) only a thin rim, i.e. 10^{-7} mm (10^{-1} nm), of the plagioclase grains would be changed by diffusion. Now, if the maximum radius of the plagioclase grains in the starting material is 0.1 μm (for synthetic and natural starting materials), then the run time necessary to reach equilibrium compositions is 2×10^6 days or approximately 6000 years. This may be the minimum run time to reach equilibrium or near-equilibrium results in melting experiments with granitic or granodioritic compositions.

The most widely-read papers dealing with low-temperature granitic melts are those of Winkler *et al.* (1975) and of Winkler and Breitbart (1978). These findings are also incorporated in Winkler's book on metamorphic rocks (1976, 1979). Because of the popularity of Winkler's data and the obvious significance of these results for cotectic melting, it is important to investigate whether they really represent equilibrium data as claimed. Winkler *et al.* (1975) do not give evidence for having reached equilibrium compositions in their paper on low-temperature granitic melts, but state (p. 248): 'The duration of the runs was 60 hours which was a much longer time than necessary for the finely ground crystal powder to reach crystal/liquid equilibrium in the presence of water during the melting experiments'. (The starting minerals were synthesized hydrothermally from specpure oxides—SiO_2, Al_2O_3—and carbonates—K_2CO_3, Na_2CO_3). In 1978 Winkler and Breitbart performed melting experiments with several leucosomes from *in-situ* migmatites. In order to achieve equilibrium conditions, run times of one week were applied (p. 467): 'The duration of runs for the melting experiments was seven days; it thus was very much longer than the minimum required time (Piwinskii and Martin, 1970).' So onus of proof of having reached equilibrium is transferred to Piwinskii and Martin.

Piwinskii and Martin (1970) investigated rates of reaction in granite–water systems at 10 kbar water pressure. They report (p. 1): 'Composition and structural state of coexisting alkali feldspars from reversal runs on the hypersolvus granite further suggest that 18 days are insufficient to attain a tentative equilibrium state.' The authors summarize their experiences (page 9):

> Experimental results have shown that 18 days are not long enough to satisfy all prerequisites of equilibrium in hypersolvus granite BH7450. Considering that this granite is alumina undersaturated, that the fluid pressure in the experiments was 10 kb, and that an aqueous vapor phase was present in the runs, we suggest that extremely favorable kinetics prevailed in our experiment. Furthermore, the low calcium content of the hypersolvus granite BH7450 probably favors more rapid equilibration rates because kinetics of ordering in compositionally intermediate plagioclases are far more sluggish than in albite (Eberhard, 1967). Future investigators of granite–water systems, or of rock–water systems simulating anatectic processes should be cognizant of, and rigorously evaluate, all factors influencing reaction kinetics.

If Piwinskii and Martin had been able to determine the chemical composition of the product plagioclases, they would have detected that their run products were much further away from the supposed equilibrium composition.

As has been shown for tonalitic compositions (see above), unstable melting of plagioclases controls phase relations of granitic mineral assemblages too. In case of experimental melting in the system Qz–Or–Ab–An–H_2O, the Ab–An ratio of cotectic melts is too low in comparison to stable compositions, and the amounts of An and Or are too high in relation to Ab (and Qz?). The melting temperatures determined in hydrothermal melting experiments are lower than they should be if the melting feldspars continuously achieved equilibrium compositions (Johannes, 1980). For these reasons, Winkler's data on cotectic melting in the granite system Qz–Or–Ab–An–H_2O (Winkler *et al.* 1975; Winkler 1976, 1979) should not be used for petrogenetic interpretations. The data are results of unstable melting, obviously caused by slow mutual diffusion of Ca + Al and Na + Si in the plagioclases, and may deviate considerably from stable equilibrium compositions. The projections of cotectic lines and isotherms shown in Figures 2–10 of Winkler *et al.* (1975) should not be taken very seriously. The gross features are of course correct: the amount of Or increases, Ab decreases (and Qz also increases slightly) in low-temperature cotectic melts of granitic compositions when albite is replaced by intermediate plagioclase. The precise cotectic data between the well-known eutectics of the subsystems Qz–Or–Ab–H_2O and Qz–Or–An–H_2O are—in the author's belief—still unknown.

Further papers deal with the kinetics of plagioclase reactions. Piwinskii

(1967) performed experiments to test equilibration in hydrothermal melting with natural rocks. The compositions of plagioclases in the rocks and in the run products were determined with a microprobe. At 800 °C an increase of more than 6 wt.% An over the original starting composition could be observed after a run duration of 11 days, while a zoned plagioclase remained almost unchanged after 33 days at 690 °C. Piwinskii assumes attainment of equilibrium in the 800 °C run and no equilibration at 690 °C. Had the solidus of the granite system been known in 1967 it would have become obvious that equilibrium was also not achieved at 800 °C.

Further intensive tests were conducted by Piwinskii (1968) and by Piwinskii and Wyllie (1970) to determine whether the results obtained in melting experiments with granitic compositions represent stable equilibrium conditions. It was found, in runs at low temperature (below or slightly above the solidus) and at moderate hypersolidus temperatures, that the feldspars did not reach their stable equilibrium condition with regard to their compositions and structural states. However, the positions of the phase boundaries were believed to be correct or to represent a close approach to reversible equilibrium boundaries. The problem of equilibrium in melting experiments is discussed in several publications of Wyllie and co-workers (Piwinskii and Wyllie, 1968, 1970; Boettcher and Wyllie, 1968; Robertson and Wyllie, 1971; Maaløe and Wyllie, 1975). The problem was also critically reviewed and summarized by Wyllie (1977). The experimental findings of Piwinskii and of Wyllie and co-workers may help to reconstruct the history of crystallizing magmas and to elucidate the genesis of plutons, but they are of less importance for the interpretation of migmatite formation. In the case of migmatite research there is still a lack of equilibrium data on cotectic melting in the granite system Qz–Or–Ab–An–H_2O.

2.4.6 *Is there a criterion indicative of magmatic origin of granitic rocks?*

Winkler and Breitbart (1978) presented a criterion to ascertain a magmatic origin of rocks. This is based on the experimental findings of Winkler *et al.* (1975) discussed above. According to Winkler and Breitbart (1978, p. 471)

> ... it must be determined whether or not a very large part of a granitic rock, i.e., about 80% or more, can be liquid at relatively low temperatures in the presence of water vapour. If this is possible the probability is great that the rock is of magmatic origin.

Winkler and Breitbart (1978) performed melting experiments with seven granites and gneisses covering the compositional range from alkali feldspar granite to tonalite. Winkler's interpretation and conclusions are clear from the following quotation (p. 479):

> The investigation of seven representative granitic rock compositions shows that they are not always located on a cotectic surface or close to it; they may as well be located relatively far away from such a surface. However, even if located relatively far away

from a cotectic surface, the rock composition in terms of Qz: Ab: Or: An is such that a very large amount of melt is present at relatively low temperatures. Thus all seven rocks investigated are of magmatic origin, although some magmas will have had some amount of solids suspended in the melt. In the rocks investigated, it is several per cent of plagioclase which has to be regarded as solid "resisters" suspended in the magma (together with some biotite).

So far we have not found a granitic rock which does not yield a very large amount of melt at relatively low temperatures; in other words, the large number of granitic rocks which have been investigated in our laboratory are all of magmatic origin.

In order to make Winkler's simple criterion very clear, one example of the investigated rocks (an Archaean Nuk-gneiss from Greenland) may be quoted (p. 478):

> At 710 °C the liquid part of the magma amounts to about ... 70%. At 730 °C ... about 75–80% .... In the case of this Archean tonalite ... we may assume that 730 °C has been reached or even has been exceeded yielding an even larger liquid part of the tonalite magma. Consequently, we conclude that also this tonalite, like the other rocks discussed, is of magmatic origin.

Lack of water, or water-undersaturated melting, due to dilution of water by inert components (e.g. CO_2), were not taken into consideration. If we follow Winkler's logic we will find hardly any felsic high-grade metamorphic rock in the Earth's crust which is not of magmatic origin.

Winkler's 'criterion' is tested by Winkler and Breitbart (1978) by experimental partial melting of nine leucosomes of *in-situ* migmatites. According to Winkler and Breitbart (1978, pp. 465–466):

> Leucosomes of in-situ migmatites (not of injection migmatites) constitute the only easily accessible place where crystallized granitic, granodioritic, or tonalitic magma is still found at its site of generation. In these migmatites, the leucosomes have formed by partial melting (anatexis) of gneisses and certain schists under high-grade metamorphic conditions. By means of melting experiments at small temperature intervals, it can be determined whether such leucosomes have been completely liquid under anatectic high-grade metamorphic conditions or not.

Winkler does not doubt that leucosomes are formed by partial or complete melting. With the background of the results of melting experiments Winkler and Breitbart (1978) distinguished four different cases: (a) leucosomes which have been completely liquid at low temperatures; (b) and (c) leucosomes of granitic and tonalitic composition but giving rise to melts with *small* amounts of solids present even at rather high temperature; and (d) leucosomes giving rise to melts with rather *large* amounts of solid (or solids) at high temperatures. According to Winkler and Breitbart (1978, p. 469):

> The leucosomes of category (d) show most clearly that an anatectically generated magma may not have been completely liquid but may have suspended crystals of plagioclase or/and quartz that amount to 15, 25, and rarely to 40 per cent, at 730 °C and 5 kb water pressure. Of course, the amount of solids will be even greater at lower

temperatures. On the other hand, leucosomes are known which, at relatively low temperature, have only a few per cent of suspended leucocratic minerals or none at all in the melt.

It must be remembered that Winkler regarded partial melting as the *only* rock-forming process in case of *in-situ* migmatites, and he believed in water-saturated melting during anatexis of high-grade metamorphic rocks. Winkler attempted to corroborate his theory by experimental hydrothermal studies using only leucosomes as starting material. The insignificance of this type of experimental study would have become obvious if Winkler had also melted the mesosomes ('palaeosomes' in the terminology of Mehnert, 1968) adjacent to the investigated leucosomes. Mesosomes and leucosomes of a migmatite are of rather similar composition (Mehnert, 1968; Gupta and Johannes, 1982; Johannes, 1983*b*). The phase relations and the amounts of melts present in melting mesosomes are very similar to those of leucosomes as will be shown below. If Winkler had performed melting experiments with mesosome compositions he would have proved (according to his criterion for the magmatic origin of rocks) the magmatic origin of mesosomes (palaeosomes)! But in Winkler's interpretation, mesosomes are unmolten (or almost unmolten) restite materials or parent rocks.

This discussion of Winkler's papers (Winkler *et al.* 1975, 1978) may show the limits of the significance of hydrothermal melting experiments. It also demonstrates that we need more information than the results of water-saturated melting experiments in order to prove or disprove a magmatic origin of a rock.

2.4.7 *Melting at grain boundaries and distribution of partial melts*

There is no doubt that the formation of migmatites is a complex problem and even more difficult to explain than the genesis of anatectic granites. If partial melts did separate from the unmolten restite (Winkler, 1961) or if restite minerals separated from partial melts (Mehnert, 1968) then such processes should be related to differences in the surface energies of the melts and minerals involved.

Recently some further studies have been made in this direction. Mehnert *et al.* (1973) and Büsch *et al.* (1974) tried to elucidate the formation of initial melts and related petrogenetic processes in hydrothermal melting experiments, with cylinders of natural rocks as starting materials. Jurewicz and Watson (1984) examined the distribution of partial melts in a felsic system with quartz–albite melt in equilibrium with a quartzite. Mehnert *et al.* (1973) studied initial melting at grain boundaries of silica-rich rocks (granulites, gneisses and feldspar-bearing quartzites). The beginning of melting was observed at polyphase contacts of mineral aggregates (consisting of quartz, alkali feldspar and plagioclase) and at quartz/feldspar grain boundaries. This occurred at or near minimum melting temperatures. From these experimental results, it was suggested that melting at the grain boundaries and polyphase contacts plays an important role in the formation of anatectic granites on a regional scale.

Büsch *et al.* (1974) performed melting experiments with rock cylinders of intermediate composition and found that the melts are of leucogranodiorite to leucotonalite compositions. They also observed the decomposition of biotite, and enrichment of potassium in the melt at quite high temperatures. Plagioclase–quartz assemblages had developed rims of melts which were strictly confined to the grain contacts of quartz and plagioclase, while the contacts between quartz/quartz and plagioclase/plagioclase remained free of melt. Even quartz crystals included in plagioclase (and thus without any obvious contact with the gas phase circulating along the grain boundaries) showed rims of melt of rather the same thickness as could be observed at normal quartz/plagioclase contacts. The quartz–plagioclase melt appeared 15–20 °C below the temperatures of beginning of biotite decomposition, which was accompanied by the appearance of tiny hornblende crystals.

In comparison with the commonly performed hydrothermal melting experiments with finely ground rock powder or synthetic mineral assemblages, the experimental studies of Mehnert and Büsch have some advantages. They clearly show (i) which minerals are reacting to form a melt, (ii) where the melts are formed and (iii) if the melts are dispersing, or gathering or remaining in place. They may also give information under favourable conditions, about congruent or incongruent melting of minerals. Of course there are obvious disadvantages of melting in rock cylinders, e.g. (i) the results may be even further away from equilibrium compositions than the results of experiments with fine-grained rock powder, and (ii) spatial conditions prevailing in nature, such as distribution and diffusion of volatiles, migration of elements, and very slow variation of stress or fluid pressure, cannot be simulated in the laboratory.

The mineralogical composition in the experiments of Jurewicz and Watson (1984) was much simpler than that of the starting materials of Mehnert *et al.* (1973) and Büsch *et al.* (1974). Jurewicz and Watson (1984) used a quartz–albite melt and a synthetic, texturally equilibrated quartzite for their distribution and melting experiments, performed at 1250 °C and 8 kbar piston-cylinder pressure. They demonstrated that a 50% Ab, 50% Qz melt in chemical equilibrium with the quartzite will not penetrate the dry grain boundaries, and that the stable configuration of this partially-molten system is one in which the melt remains at the sites of generation, either in small pockets at grain corners or at the intersections of grain edges. The initial melt was found to be stable in this dispersed form, but not as a film wetting all grain boundaries. It thus appears that 'external' driving forces such as gravity or stress are required to induce collection of melt into larger bodies capable of migrating towards the surface (see also Ashworth and McLellan, this volume). Jurewicz and Watson emphasized that melt distribution and infiltration tendency, in other systems and in nature, may vary with rock type, water content, impurities, stress, oxygen fugacity, and the temperature–pressure regime involved.

The experiments of Jurewicz and Watson (1984) are quite informative, especially bearing on the interaction of minerals responsible for melt generation, but their experiments are confined to synthetic material, and more such experiments are needed with synthetic as well as natural rocks, to achieve better understanding of the external forces responsible for the collection and segregation of melt. The latter processes are widespread in the natural anatectic migmatites on the one hand, and poorly understood on the other. Perhaps such experiments may also throw some light on the quantitative significance of the mineral proportions in a parent rock for the formation of leucosomes.

2.4.8 *Experimental studies with leucosome and mesosome compositions*

The melting interval (difference between solidus and liquidus temperatures) is small in the case of minimum or near-minimum compositions in the system Qz–Or–Ab–An–H_2O (it is called a 'minimum' because there is no eutectic composition in the granite system with two coexisting ternary feldspars). This means that the amount of melt changes considerably with small changes of the run temperature in hydrothermal melting experiments. (The influence of pressure on the amount of melt is not significant at high water pressure).

In order to compare the amounts and composition of melts formed in various hydrothermal melting experiments with different rocks as starting materials, the run temperatures have to be known very precisely. The best practical way to achieve the same run temperature (and pressure) in two sample containers is to place the samples side by side (parallel) at the same temperature position of an autoclave. Such parallel runs were performed with layers of a migmatite from Nelaug, Southern Norway (for description of this migmatite see Gupta and Johannes, 1982). The mesosomes investigated (nos. 2 and 3, see Figure 2.14) are of granodioritic composition and the leucosomes (nos. 16 and 17, Figure 2.14) are of granitic composition. The melting experiments were performed at $P_{H_2O} = 5\,\text{kbar}$.

Mineralogical and chemical compositions within the mesosomes and leucosomes are very similar and show almost identical Qz/Or/Ab ratios within each of the two rock types (Figure 2.14*b*). The same minerals are present in all the four layers investigated (Figure 2.14*a*). At the present time, it is believed by many geologists that granites and granodiorites having different quantitative mineral compositions should have different solidus temperatures. But the solidus temperature of a rock depends upon its qualitative mineral composition (the assemblage present) rather than quantitative (modal) proportions, as discussed above. Since leucosomes and mesosomes of the Nelaug migma-

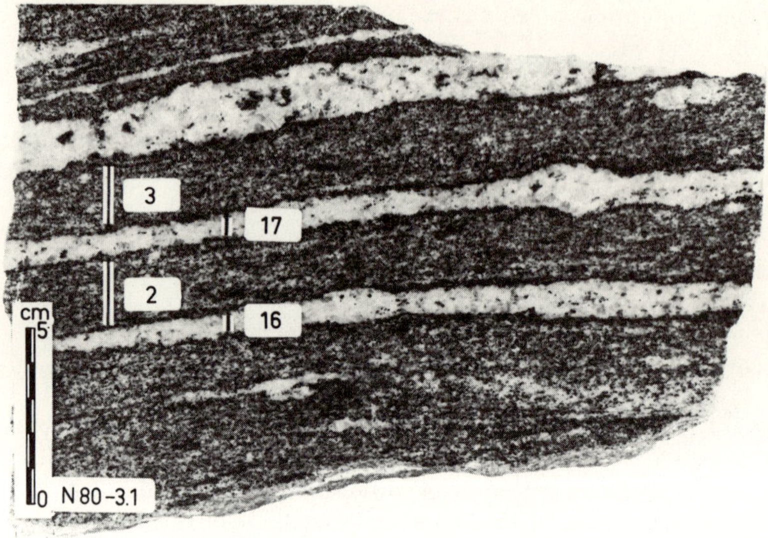

(*a*)

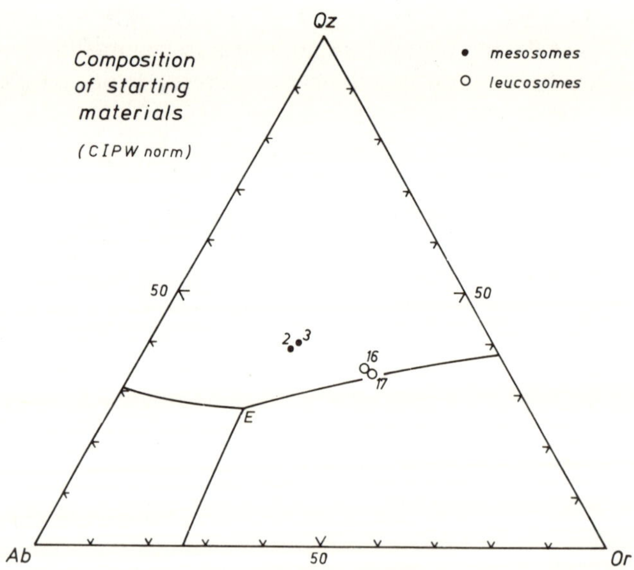

(*b*)

Figure 2.14 (*a*) Hand-specimen of a layered (stromatic) migmatite from the Nelaug area (Southern Norway). For description of this migmatite see Gupta and Johannes (1982). The leucosomes (16) and (17) and the mesosomes (2) and (3) are used as starting materials for hydrothermal melting experiments. (*b*) CIPW-norm data of leucosome and mesosome starting materials. The eutectic point E and the cotectic lines are 5kbar data from Luth *et al.* (1964).

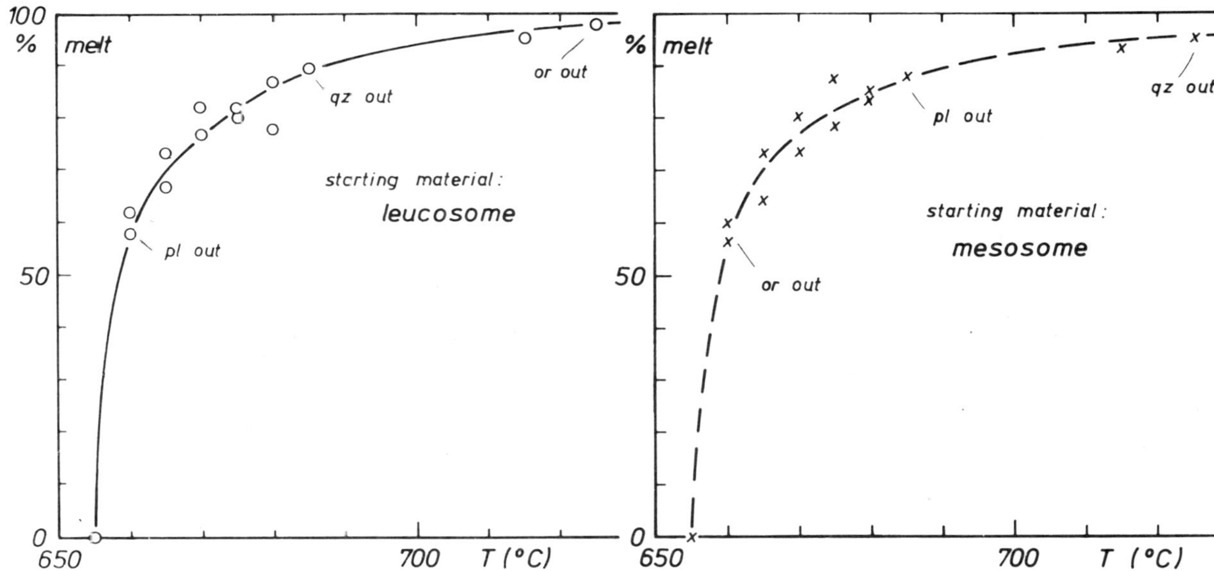

Figure 2.15 Results of melting experiments performed at $P_{H_2O} = 5$ kbar with leucosomes (16) and (17) and mesosomes (2) and (3) as starting materials.

qz = quartz, or = alkali feldspar, pl = plagioclase

tites have similar qualitative mineral compositions, they should have the same solidus temperatures. This theoretical prediction was confirmed in melting experiments performed at the water pressure of 5 kbar (Figure 2.15).

In nature, different migmatite layers, e.g. mesosome and leucosome (see Gupta and Johannes, 1982), with similar qualitative but different quantitative mineralogical composition should have had different amounts of melt, which fact is responsible for the formation of a layered migmatite. But the experimental results show the same amount of melt up to 670°C ($P_{H_2O} = 5$kbar) in both types of layers (Figure 2.15). There is no noticeable difference between the amounts of melt present in leucosomes and mesosomes until 80% melting is achieved. Therefore it may be concluded that the common quantitative differences between the mineralogical composition of migmatite layers (more precisely: between the layers of the parent rock) cannot be the only reason for development of leucosomes and mesosomes by partial melting. If it is true that different amounts of melt changed one suite of layers of a parent rock into leucosomes leaving the other unchanged (mesosomes) then the difference in the proportion of melts should be much more pronounced than observed in the experiments. According to experimental findings of van der Molen and Paterson (1979) this difference should become apparent when the amount of melt is still relatively low (30 or 40 vol%?), which is experimentally shown not to be the case (Figure 2.15).

Are the experimental results shown in Figure 2.15 reasonable? Because of unstable melting of plagioclase (see above), they do not represent equilibrium compositions but they do show correct relative information. If all kinds of felsic minerals are in contact via a fluid phase, melting should occur until one of the felsic phases taking part in the melting reaction is exhausted. In the case of the mesosomes this is alkali feldspar, and in the case of the leucosomes, plagioclase. The amount of plagioclase in the leucosomes is approximately 20–25%, and of alkali feldspar in the mesosomes, also 20–25%. The exact stable compositions of the cotectic melts formed in the leucosomes and mesosomes are not yet known (Winkler's data are the result of unstable melting, as shown above). Such compositions should be very poor in An and the Ab/Qr/Qz ratio may be around 1/1/1 at $P_{H_2O} = 5$ kbar. If these assumptions are correct, there will be cotectic melting producing (at a given temperature and water pressure), the same proportion and the same composition of melt in both the leucosome and mesosome layers until at least 60% of melt is formed.

The quintessence of the experimental results is that the formation of migmatites cannot be explained by water-saturated hydrothermal melting experiments. There seems to be a close relationship between the layers of a migmatite and the structure of its parent rock (Johannes and Gupta, 1982). Differences in the composition of the layers of a parent rock are obviously responsible for the development of leucosomes and mesosomes by different degrees of partial melting in these layers. But at present the formation of (layered) migmatites cannot be related to the results of the simple water-saturated experiments. There must be some additional factors controlling and causing the evolution of leucosomes (+ melanosomes) and thus giving rise to migmatites.

2.5 Plagioclase compositions: observations in nature, experimental findings and conclusions

Conclusions based upon the anorthite content of the plagioclase in various migmatites are not free from contradictions! It is usual to make some hypothetical assumptions, e.g. that the system was closed or open, water was available or not, especially when certain observations are inexplicable otherwise. As already pointed out, leucosome compositions vary greatly, from granitic to tonalitic in various migmatites, but the composition of the minimum initial melt as determined experimentally is always more or less the same in the system Qz–Or–Ab(–H_2O) (Bowen, 1913; Tuttle and Bowen, 1958; Luth *et al.*, 1964).

Similarity of plagioclase anorthite contents in the layers of a migmatite has been interpreted as evidence for partial melting on the one hand (Johannes and Gupta, 1982), and also has been cited in favour of metamorphic derivation of leucosomes on the other (Misch, 1968). In contrast, dissimilarity of plagioclase anorthite contents between leucosome and mesosome has also been interpreted as favouring partial melting, of course with certain assumptions (Ashworth, 1976).

In general, the composition of plagioclases is very similar in the various layers of a migmatite belonging to a regionally metamorphic terrain (Mehnert, 1968; Johannes and Gupta, 1982). In places, the An content of a leucosome plagioclase may even be more calcic than that of the adjacent mesosome (Gupta and Johannes, 1982), whereas in some instances the plagioclases of mesosomes are more calcic than those of the leucosomes (Ashworth, 1976).

Plagioclase anorthite contents in contact metamorphic anatectic rocks show different trends from those of deep-seated migmatites. Plagioclases grown in a contact aureole usually have calcic cores and gradually become more sodic towards the margins (Gehle, pers. comm.). Calcic cores, with fluctuating anorthite contents towards margins, have also been observed in the initial stages of migmatization by Mehnert and Büsch (1982).

The observations enumerated above are directly or indirectly related to the problem of migmatite petrogenesis. It is not possible to provide a satisfactory answer to all these questions but I intend to discuss some facts observed by various workers and their interpretations, and also present my own ideas concerning the history of migmatite plagioclases.

Misch (1968) studied plagioclase compositions from the migmatitic gneisses in the Northern Cascade Mountains of Washington State. These migmatites are of trondhjemitic and quartz dioritic composition. Misch observed that there is a close relationship between the leucocratic migmatitic gneisses and the associated schists, paragneisses, amphibolites, etc. According to Misch (1968) this indicates that the migmatitic gneisses were directly derived from their metamorphic protoliths by metamorphic recrystallization, with a small shift towards more sodic plagioclase compositions in most, but not all, of the gneisses. Plagioclase relationships (and other data) seem to rule out both igneous injection and an anatectic origin of the leucocratic migmatitic gneisses. Besides metamorphic differentiation, metasomatism has contributed towards the development of leucosomes, causing the migmatization (Misch, 1968). Why metasomatism and metamorphic differentiation were active only in the leucosomes and not the mesosomes is not explained by Misch (1968).

Ashworth (1976) described migmatites from the sillimanite-potash feldspar zone of a contact aureole in the Huntly-Portsoy area, northeast Scotland. Within this zone, he described two types of lower-grade migmatites represented by trondhjemitoid and granitoid, and two types of high-grade migmatites represented by cordierite-granitoid and noritoid migmatites. According to Ashworth (1976), textural immaturity and consistently cotectic modal compositions indicate that (at least in case of the trondhjemitoids) the leucosomes originated as melts. Scatter of plagioclase compositions between leucosomes suggests that the partial melting occurred in small closed systems. Varying fugacities of water are believed to be responsible for the formation of melt in the layers which were converted into leucosomes. The plagioclase compositions determined by Ashworth (1976) in the Huntly-Portsoy migmatites vary within the range An_{16} to An_{54}. The range of plagioclase compositions in the mesosomes (palaeosomes in the terminology used by Ashworth, 1976) is An_{27} to An_{54}, and that of the leucosomes plus melanosomes together is An_{16}–An_{39}. The tendency of mesosomes to have more calcic plagioclases than adjacent leucosomes plus melanosomes is (according to Ashworth) not a differentiation effect; rather the mesosomes were more calcic from the outset. Their persistence as mesosomes is the result of inhibited migmatization. Ashworth (1976) emphasizes that plagioclase composition data rule out imposition of mineral compositions by metasomatism. If minor, local metasomatism was active it failed to eliminate zoning in plagioclases. The survival of the zoning rules out (Ashworth, 1976, p. 669) 'any interpretation of the lack of differentiation between leucosome and melanosome plagioclase in terms of thorough retrograde reaction between these bodies' (cf. Mehnert, 1968). According to Ashworth, melting occurred under water-undersaturated conditions. Rocks with plagioclase more sodic than An_{39} incipiently melted. But only those layers which had initially high fugacities of water could generate enough melt to produce leucosomes (and the adjacent melanosomes).

Harris (1974) described migmatites from the Barousse Massif, Central Pyrenees, determining anorthite contents from leucosomes and melanosomes, and attempting to explain the origin of the migmatites by various processes. In most examples, leucosome and melanosome plagioclases were found to be identical or almost identical. In only two specimens, there is a strong (10–20%) albite enrichment in leucosome plagioclases. On the contrary there are other examples which show anorthite enrichment in leucosome plagioclases. On the basis of distribution of plagioclase compositions, structure, fabrics and mineral assemblages, it is concluded by Harris that most of these migmatites have formed by recrystallization of sedimentary horizons, accompanied by metamorphic differentiation, although some migmatites appear to have formed by anatexis or magmatic injection.

Kenah and Hollister (1983) investigated migmatites in the Central Gneiss Complex in the Coast Mountains of British Columbia. Plagioclase compositions were determined from mesosome, leucosome and melanosome layers. The An contents of the plagioclases are An_{31}–An_{36} in the leucosomes and melanosomes, and relatively high (An_{38}–An_{44}) in the mesosomes. The low An in both leucosomes and melanosomes rules out the possibility that the mesosomes could have been the protoliths of leucosomes plus melanosomes (but see Mehnert, 1968). Kenah and Hollister (1983) emphasized that the leucosome cannot be added to the melanosome to make the mesosome because the SiO_2 content of leucosome is greater than that of the melanosome which is essentially the same as that of the mesosome (see Table 2 of Kenah and Hollister, 1983).

Mehnert and Büsch (1982) observed almost identical plagioclase compositions in leucosomes and mesosomes at the initial stage of migmatite formation. The authors state that changes in the An content can occur only if melt is clearly separated from its starting material, and conclude that the whole specimen reacted as a closed system with regard to plagioclase.

The equilibrium fractionation of plagioclase components between melts (of tonalitic and granitic composition) and restite plagioclases is not yet known precisely (see above). It may be as large, as indicated in Figure 2.5. The liquidus and solidus in Figure 2.5 may set a maximum for the effect in the case of the tonalite system. The preliminary results shown in Figure 2.8 represent stages on the way to equilibrium compositions. If melts and restite minerals are separated in nature, there should be pronounced differences in the compositions of the resulting plagioclases. In my opinion, all observations regarding the composition of plagioclase fit best into the model involving initial compositional differences in a layered protolith, as described by Johannes and Gupta (1982). This model is described as layer-by-layer transformation and is best shown by the authors in their Figure 23. The different layers (or portions of these) have to be regarded as closed systems at least with regard to plagioclases (see also Mehnert and Büsch, 1982; Ashworth, 1976). The word 'isochemical' should not be taken too literally. High potassium feldspar contents in the lowest-grade leucosomes may indicate (restricted) operation of alkali metasomatism.

In most deep-seated migmatites with very homogeneous plagioclases, the system seems to be completely re-equilibrated during crystallization of partial melts. In contact anatexites quenched rapidly, the partly-molten plagioclases are still zoned (Gehle, pers. comm.). Also in regional metamorphic migmatites the unmolten restite plagioclases can sometimes be recognized. An interesting example of a zoned migmatite plagioclase showing an obviously old core is given by Mehnert and Büsch (1982, Figure 8*a,b*). To my mind, the degree of re-equilibration is not only a question of the quenching time, but also depends on deformation, and the presence of a fluid phase and its composition.

2.6 Suggestions for future research

There is no doubt that the formation of migmatites is a complicated matter and a great deal of research is still needed to elucidate the unsolved questions. It seems clear that neither petrographic observations nor experimental investigations alone can supply satisfactory answers. In order to encourage colleagues, especially younger colleagues and co-workers, I should like to point out some topics for future research.

2.6.1 *The protoliths of migmatites*

Mehnert's model of isochemical partial melting is generally accepted. According to Mehnert (1968) the palaeosome ('mesosome' in the terminology of Johannes and Gupta, 1982) represents the parent rock of migmatites, and the palaeosome splits into pegmatoid leucosomes and biotite-rich melanosomes forming selvages around the leucosomes (see also Mehnert and Büsch, 1982). The model is based on the Urenkopf migmatite (Black Forest). This migmatite was re-investigated by Mehnert and Büsch (1982). In the meantime seven migmatites have been investigated by myself and co-workers (Schauinsland, Black Forest, Germany; Arivika, western Sweden; Graestorp, western Sweden; Nelaug, southern Norway; Rantasalmi, southern Finland; Mullardoch, Scotland; Port Navalo, Brittany, France). The mean compositions of these migmatites range from granitic to tonalitic and quartzitic. The compositions of mesosome and leucosome + melanosome layers have been analysed in all these examples. None of the data from the above-mentioned areas fits into the model of Mehnert (1968). The (layered) migmatites seem to have been derived from layered paragneisses or schists. We know the protoliths of three of the migmatite areas mentioned above, i.e. Arvika, Nelaug and Rantasalmi areas. The layer-by-layer transformation model developed by Johannes and Gupta (1982) is based on the composition of Arvika migmatites and gneisses.

In my opinion, a very important task of migmatite research is the discovery or (if that is not possible) the reconstruction of the protoliths. A comparison of parent rock and migmatite will give at least a general idea of what happened during migmatization of a protolith.

2.6.2 *Metasomatism*

The role of metasomatism during formation of migmatites is still not clear. Very different models have been developed explaining the genesis of migmatites by metasomatism on the one hand or by isochemical reactions on the other. If water had free access to the rocks undergoing migmatization then there must have been some changes in the bulk compositions. Future research work may decide whether the chemical changes were on a regional or local scale and what was the most prominent rock-forming process. Here again it is important to know the composition of the protolith as precisely as possible. Contributions of Olsen (1984 and this volume) on mass balance and mass transfer may show a way to elucidate this problem.

2.6.3 *Mineral compositions*

The amount of melt responsible for the formation of leucosomes is not yet known even to those petrologists who believe in partial melting as the most important migmatite-forming process. This amount may have been as low as 40 or 30%, which is probably enough to let the unmolten minerals swim in a liquid phase, or this melt may be present in the form of a film covering the mineral grains and serving as a transport medium. More precise knowledge about the mineral compositions, in the newly crystallized minerals and the restites, may help to elucidate this problem. Especially, old plagioclase cores having new plagioclase rims of different composition, or rounded inclusions of quartz and feldspars (in feldspars or in quartz) may serve as good indicators of restite materials.

Biotite is also an important constituent, next to quartzofeldspathic minerals, and it can help in reconstructing the petrogenesis of migmatites. Melanosomes are often observed at the margins of leucosomes, and the main mineral in them is biotite. It has been seen that during the process of transformation of the parent layers into leucosomes, biotite either breaks down into K-feldspar, cordierite, sillimanite etc. (Gupta and Johannes, in preparation), reacts with plagioclase to give rise to hornblende (Hoschek, 1976; Henkes and Johannes, 1981), or is recrystallized into larger flakes, which are transported and concentrated at the margins to form melanosomes or remain within the leucosomes as 'relict' melanosomes (see true and relict melanosomes, Johannes and Gupta, 1982). The questions arise: why does the biotite recrystallize into larger flakes, especially in the melanosomes? And why is it transported to the margins in some cases, but remains as relict melanosomes in others? Perhaps it may be possible to find the reasons if we discover:

(i) differences between the original biotite and recrystallized biotite and their stability fields,
(ii) the distribution coefficients of Mg and Fe between biotite (original and new), garnet, cordierite, hornblende etc.

2.6.4 *Fluid inclusions*

The most important answers to still unanswered questions of migmatite formation may come from fluid inclusion studies. According to preliminary observations, there may be different types of fluid inclusions in leucosomes and mesosomes, and water-rich fluid inclusions may be present only in leucosomes (Touret, 1981 and pers. comm.).

As shown above, the activity of water must be high in vapours able to produce granitic melts at relatively low temperatures. Such fluids should be detected in leucosomes. High contents of Na, K-chlorides or sulphates would indicate a high exchange capacity and allow metasomatism. There is definitely a need for fluid inclusion studies in migmatite research.

2.6.5 *Deformation*

There are relatively few investigations dealing with deformation in migmatites. Examples demonstrating the importance of such investigations are provided by Ashworth (1979), McLellan (1983) and Fleming and White (1984). If partial melting is controlled by the amount of excess water, and if this water is introduced into some selected layers or lenses of a gneiss or a schist, then it is extremely important to study the textural evolution of the protoliths during migmatization. From the relationship between thermal activity, partial melting and rock deformation (Fleming and White, 1984), non-anatectic leucosomes may be distinguished from leucosomes formed through partial melting (McLellan, 1983). In order to investigate the role of partial melt during structural developments of crustal rocks, small and large-scale tectonic structures of migmatites and their country rocks should be compared.

2.6.6 *Geothermometry*

It is generally believed that migmatites are the products of high-grade regional or contact metamorphism. This means that migmatites are formed at higher temperatures than their unmigmatized parent rocks. But the exact temperature of the initiation of the migmatization process, and the temperature interval prevailing during the genesis of migmatites in a particular belt, are not known.

Only a few hypersolidus temperature determinations on migmatites are available at present (e.g. Tracy, 1978; Kenah and Hollister, 1983). Subsolidus temperatures derived from the classical geothermometers (Mg–Fe distribution between femic minerals; oxygen isotope distribution between magnetite, quartz and feldspars; the two-feldspar geothermometer) generally do not support the partial melting theory. But if the same techniques are applied to a leucosome and its adjacent mesosome layers separately, we may get different subsolidus temperatures in both the cases (e.g. 30–50 °C lower temperatures in the leucosomes than mesosomes by oxygen isotope distribution: Hoernes, pers. comm.).

To my mind, the above-mentioned geothermometers often do not yield the maximum temperature prevailing during migmatization: rather, they represent the temperatures either at which recrystallization took place, or of exchange reactions occurring during cooling of the rocks or of superimposed lower-grade metamorphism. It would be even more interesting if these techniques are applied to estimate the extent of chemical exchange between solids, or between solids and fluids occurring at subsolidus temperatures.

Data on element or isotope fractionation are very useful and may help to elucidate the formation of migmatites. It would be rewarding to apply similar techniques to the various layers of a migmatite. It is to be hoped that more isotope studies and investigations of the distribution of elements (especially Fe and Mg) will be undertaken in future.

2.6.7 *Experimental investigations*

In order to evaluate the great number of experimental investigations concerning the formation of migmatites, it seems to be necessary to investigate the kinetics of the appropriate reactions. This is especially true for melting reactions including plagioclase of intermediate compositions, as has been shown above. It is obvious that we need more information about chemical (volume) diffusion in feldspars. But also the kinetics controlling the exchange of oxygen isotopes, or the distribution of Fe and Mg and other elements of interest (e.g. Rb and Ba; Fleming and White, 1984) need to be clarified by hydrothermal investigations.

We do not yet know the amount and role of water or the amount and role of melt involved in migmatite formation, but the experiments performed with natural or synthetic rock specimens (Mehnert *et al.*, 1973; Büsch *et al.*, 1974; Jurewicz and Watson, 1984) appear to be quite important. The type of experiment described by Jurewicz and Watson offers an opportunity to clarify the tendency of melts to infiltrate into rocks or to collect into larger bodies in synthetic systems, and perhaps also in natural rocks under crustal conditions.

Very helpful model systems for anatexis of pelitic rocks have already been developed (Thompson and Tracy, 1979). Such models should encourage experimentalists to work out the details as precisely as possible, e.g. the solubility of mafic components in granitic melts, or the interconnected breakdown of biotite and formation of hornblende (Hoschek, 1976).

In order to elucidate the importance and extent of metasomatism during genesis of migmatites there is a need for quantitative information about the composition of fluids coexisting with natural mineral assemblages. There are data on alkali exchange between vapour and alkali feldspar phases (Orville, 1963) and on plagioclase cation exchange equilibria with aqueous chloride solutions (Orville, 1972), but no information on fluid compositions coexisting with ternary feldspars is available. Since it has become possible to determine the compositions of very fine-grained feldspars (plagioclase besides alkali feldspar) in hydrothermal run products (Johannes, 1978) the equilibrium compositions of more complex (Na, K and Ca-bearing) metasomatically active solutions can be determined (and are under investigation). The results of such studies in combination with petrographic observations, chemical investigations and fluid inclusion studies may help to clarify the role of metasomatism during the formation of migmatites. There is still a lot to do, for experimentalists as well as other workers, in order to elucidate the formation of migmatites.

References

Abbott, R.N. and Clarke, D.B. (1979) Hypothetical liquidus relationships in the subsystem Al_2O_3–FeO–MgO projected from quartz, alkali feldspar and plagioclase for a $(H_2O) \leq 1$. *Can. Miner.* **17**, 549–560.

Ashworth, J.R. (1976) Petrogenesis of migmatites in the Huntly-Portsoy area, north-east Scotland. *Mineral. Mag.* **40**, 661–682.

Ashworth, J.R. (1979) Textural and mineralogical evolution of migmatites. In *The Caledonides of the British Isles—Reviewed*, eds. Harris, A.L., Holland, C.H. and Leake, B.E., *Spec. Publ. Geol. Soc. Lond.* **8**, 357–361.

Boettcher, A.L. (1970) The system CaO–Al_2O_3–SiO_2–H_2O at high pressures and temperatures. *J. Petrol.* **11**, 337–379.

Boettcher, A.L. and Wyllie, P.J. (1968) Jadeite stability measured in the presence of silicate liquids in the system $NaAlSiO_4$–SiO_2–H_2O. *Geochim. Cosmochim. Acta* **32**, 999–1012.

Bowen, N.L. (1913) The melting phenomena of the plagioclase feldspars. *Am. J. Sci.* **35**, 577–599.

Brady, J.H. and Yund, R.A. (1983) Interdiffusion of K and Na in alkali feldspars: Homogenization experiments. *Am. Miner.* **68**, 106–111.

Brown, M. (1983) The petrogenesis of some migmatites from the Presqu'ile de Rhuys, southern Brittany, France. In *Migmatites, Melting and Metamorphism*, eds. Atherton, M.P. and Gribble, C.D., Shiva, Nantwich, 174–200.

Burnham, C.W. (1981) The nature of multicomponent aluminosilicate melts. *Phys. Chem. Earth* **13 + 14**, 197–226.

Büsch, W., Schneider, G. and Mehnert, K.R. (1974) Initial melting at grain boundaries part II: Melting in rocks of granodioritic, quartzdioritic and tonalitic composition. *Neues Jb. Miner. Mh.* 345–370.

Chatterjee, N.D. (1974) Crystal-liquid-vapour equilibria involving paragonite in the system $NaAlSi_3O_8$–Al_2O_3–SiO_2–H_2O. *Indian J. Earth Sci.* **1**, 3–11.

Christoffersen, R., Yund, R.A. and Tullis, J. (1983) Inter-diffusion of K and Na in alkali feldspars: diffusion couple experiments. *Am. Miner.* **68**, 1126–1133.

Eberhard, E. (1967) Zur Synthese der Plagioklase. *Schweiz. Miner. Petrogr. Mitt.* **47**, 385–398.

Fleming, P.D. and White, A.J.R. (1984) Relationships between deformation and partial melting in the Palmer migmatites, South Australia. *Aust. J. Earth Sci.* (in press).

Giletti, B.J., Semet, M.P. and Yund, R.A. (1978) Studies in diffusion–III. Oxygen in feldspars: an ion microprobe determination. *Geochim. Cosmochim. Acta* **42**, 45–57.

Goldsmith, J.R. (1982) Plagioclase stability at elevated temperatures and water pressures. *Am. Miner.* **67**, 653–675.

Grove, T.L, Ferry, J.M. and Spear, F.S. (1983) Phase transitions and decomposition relations in calcic plagioclase. *Am. Miner.* **68**, 41–59.

Gupta, L.N. and Johannes, W. (1982) Petrogenesis of a stromatic migmatite (Nelaug, Southern Norway). *J. Petrol.* **23**, 548–567.

Harris, N.B.W. (1974) Some migmatite types and their origins, from the Barousse Massif, Central Pyrenees. *Geol. Mag.* **111**, 319–328.

Henkes, L. and Johannes, W. (1981) The petrology of a migmatite (Arvika, Värmland, western Sweden). *Neues Jb. Miner. Abh.* **141**, 113–133.

Hoschek, G. (1976) Melting relations of biotite + plagioclase + quartz. *Neues Jb. Miner. Mh.* 79–83.

Huang, W.L. and Wyllie, P.J. (1974) Melting relations of muscovite with quartz and sanidine in the K_2O–Al_2O_3–SiO_2–H_2O system to 30 kilobars and an outline of paragonite melting relations. *Am. J. Sci.* **274**, 378–395.

Huang, W.L. and Wyllie, P.J. (1975) Melting reactions in the system $NaAlSi_3O_8$–$KAlSi_3O_8$–SiO_2 to 35 kilobars, dry and with excess water. *J. Geol.* **83**, 737–748.

Johannes, W. (1978) Melting of plagioclase in the system Ab–An–H_2O and Qz–Ab–An–H_2O at $P_{H_2O} = 5$ kbars, an equilibrium problem. *Contrib. Miner. Petrol.* **66**, 295–303.

Johannes, W. (1979) Ternary feldspars: Kinetics and possible equilibria at 800 °C. *Contrib. Miner. Petrol.* **68**, 221–230.

Johannes, W. (1980) Metastable melting in the granite system Qz–Or–Ab–An–H_2O. *Contrib. Miner. Petrol.* **72**, 73–80.

Johannes, W. (1983*a*) Metastable melting in the granite and related systems. In *Migmatites, Melting and Metamorphism*, eds. Atherton, M.P. and Gribble, C.D., Shiva, Nantwich, 27–36.

Johannes, W. (1983*b*) On the origin of stromatic (layered) migmatites. In *Migmatites, Melting and Metamorphism*, eds. Atherton, M.P. and Gribble, C.D., Shiva, Nantwich, 234–248.

Johannes, W. (1984) Beginning of melting in the granite system Qz–Or–Ab–An–H_2O. *Contrib. Miner. Petrol.* **86**, 264–273.

Johannes, W. and Gupta, L.N. (1982) Origin and evolution of a migmatite. *Contrib. Miner. Petrol.* **79**, 114–123.

Jurewicz, S.R. and Watson, E.B. (1984) Distribution of partial melt in a felsic system: the importance of surface energy. *Contrib. Miner. Petrol.* **85**, 25–29.

Kenah, C. and Hollister, L.S. (1983) Anatexis in the Central Gneiss Complex, British Columbia. In *Migmatites, Melting and Metamorphism*, eds. Atherton, M.P. and Gribble, C.D., Shiva, Nantwich, 142-162.

Kilinc, I.A. (1979) Melting relations in the quartz diorite–H_2O and quartz diorite–H_2O–CO_2 systems. *Neues Jb. Miner. Mh.* 62–72.

Luth, W.C., Jahns, R.H. and Tuttle, O.F. (1964) The granite system at pressures of 4 to 10 kilobars. *J. Geophys. Res.* **69**, 759–773.

Maaløe, S. and Wyllie, P.J. (1975) Water content of a granite magma deduced from the sequence of crystallisation determined experimentally with water-undersaturated conditions. *Contrib. Miner. Petrol.* **52**, 175–191.

Manning, D.A.C. (1981) The effect of fluorine on liquidus phase relationships in the system Qz–Ab–Or with excess water at 1 kb. *Contrib. Miner. Petrol.* **76**, 206–215.

Matthews A., Goldsmith, J.R. and Clayton, R.N. (1983) On the mechanisms and kinetics of oxygen isotope exchange in quartz and feldspars at elevated temperatures and pressures. *Geol. Soc. Am. Bull.* **94**, 396–412.

McLellan, E.L. (1983) Contrasting textures in metamorphic and anatectic migmatites: an example from the Scottish Caledonides. *J. Metamorphic Geol.* **1**, 241–262.

Mehnert, K.R. (1951) Zur Frage des Stoffhaushalts anatektischer Gesteine. *Neues Jb. Miner. Abh.* **82**, 155–198.

Mehnert, K.R. (1968) *Migmatites and the Origin of Granitic Rocks.* Elsevier, Amsterdam.

Mehnert, K.R. and Büsch, W. (1982) The initial stage of migmatite formation. *Neues Jb. Miner. Abh.* **145**, 211–238.

Mehnert, K.R. Büsch, W. and Schneider, G. (1973) Initial melting at grain boundaries of quartz and feldspars in gneisses and granulites. *Neues Jb. Miner. Mh.* 165–183

Merrill, R.B., Robertson, J.K. and Wyllie, P.J. (1970) Melting reactions in the system $NaAlSi_3O_8$–$KAlSi_3O_8$–SiO_2–H_2O to 20 kilobars compared with results for other feldspar–quartz–H_2O and rock–H_2O systems. *J. Geol.* **78**, 558–569.

Misch, P. (1968) Plagioclase compositions and non-anatectic origin of migmatitic gneisses in Northern Cascade Mountains of Washington State. *Contrib. Miner. Petrol.* **17**, 1–70.

Muehlenbachs, K. and Kushiro, I. (1974) Oxygen isotope exchange and equilibrium of silicates with CO_2 or O_2. *Yb. Carnegie Instn Wash.* **73**, 232–236.

Naney, M.T. (1983) Phase equilibria of rock-forming ferromagnesian silicates in granitic systems. *Am. J. Sci.* **283**, 993–1033.

Nord, G.L., Heuer, A.H. and Lally, J.S. (1974) Transmission electron microscopy of substructures in Stillwater bytownites. In *The Feldspars*, eds. MacKenzie, W.S. and Zussman, J., Manchester University Press, Manchester, 522–535.

Olsen, S.N. (1984) Mass-balance and mass-transfer in migmatites from the Colorado Front Range. *Contrib. Miner. Petrol.* **85**, 30–44.

Orville, P.M. (1963) Alkali ion exchange between vapor and feldspar phases. *Am. J. Sci.* **261**, 201–237.

Orville, P.M. (1972) Plagioclase cation exchange equilibria with aqueous chloride solution: Results at 700 °C and 2000 bars in the presence of quartz. *Am. J. Sci.* **272**, 234–272.

Pichavant, M. (1981) An experimental study of the effect of boron on a water saturated haplogranite at 1 kbar vapour pressure. *Contrib. Miner. Petrol.* **76**, 430–439.

Piwinskii, A.J. (1967) The attainment of equilibrium in hydrothermal experiments with granitic rocks. *Earth Planet. Sci. Lett.* **2**, 161–162.

Piwinskii, A.J. (1968) Experimental studies of igneous rock series: Central Sierra Nevada batholith, California. *J. Geol.* **76**, 548–570.

Piwinskii, A.J. (1973) Experimental studies of granitoids from the Central and Southern Coast Ranges, California. *Tschermaks Miner. Petrogr. Mitt.* **20**, 107–130.

Piwinskii, A.J. and Martin, R.F. (1970) An experimental study of equilibrium with granitic rocks at 10 kb. *Contrib. Miner. Petrol.* **29**, 1–10.

Piwinskii, A.J. and Wyllie, P.J. (1968) Experimental studies of igneous rock series: a zoned pluton in the Wallowa batholith. Oregon. *J. Geol.* **76**, 205–234.

Piwinskii, A.J. and Wyllie, P.J. (1970) Experimental studies of igneous rock series: 'Felsic Body Suite' from the Needle Point pluton, Wallowa batholith, Oregon. *J. Geol.* **78**, 52–76.

Robertson, J.K. and Wyllie, P.J. (1971) Experimental studies on rocks from the Deboullie stock, northern Maine, including melting relations in the water-deficient environment. *J. Geol.* **79**, 549–571.

Schairer, J.F. and Bowen, N.L. (1955) The system K_2O–Al_2O_3–SiO_2. *Am. J. Sci.* **253**, 681–746.

Schliestedt, M. and Johannes, W. (1984) Melting and subsolidus reactions in the system K_2O–CaO–Al_2O_3–SiO_2–H_2O: corrections and additional experimental data. *Contrib. Miner. Petrol.* (in press).

Steiner, J.C., Jahns, R.H. and Luth, W.C. (1975) Crystallization of alkali feldspar and quartz in the haplogranite system $NaAlSi_3O_8$–$KAlSi_3O_8$–SiO_2–H_2O at 4 kb. *Geol. Soc. Am. Bull.* **86**, 83–98.

Stewart, D.B. (1967) Four phase curve in the system $CaAl_2$–Si_2O_8–SiO_2–H_2O between 1 and 10 kilobars. *Schweiz. Miner. Petrogr. Mitt.* **47**, 35–39.

Thompson, A.B. (1974) The instability of feldspar in metamorphism. In *The Feldspars*, eds. MacKenzie, W.S. and Zussman, J., Manchester University Press, Manchester, 645–672.

Thompson, A.B. and Algor, J.R. (1977) Model systems for anatexis of pelitic rocks I. Theory of melting reactions in the system $KAlO_2$–$NaAlO_2$–Al_2O_3–SiO_2–H_2O. *Contrib. Miner. Petrol.* **63**, 247–269.

Thompson, A.B. and Tracy, R.J. (1979) Model systems for anatexis of pelitic rocks II. Facies series melting and reactions in the system CaO–$KAlO_2$–$NaAlO_2$–Al_2O_3–SiO_2–H_2O. *Contrib. Miner. Petrol.* **70**, 429–438.

Touret, J. (1981) Fluid inclusions in high grade metamorphic rocks. *Min. Soc. Canada Short Course Handbook.* **6**, 182–208.

Tracy, R.J. (1978) High grade metamorphic reactions and partial melting in pelitic schist, West-Central Massachusetts. *Am. J. Sci.* **278**, 150–178.

Tsuchiyama, A. and Takahashi, E. (1983) Melting kinetics of a plagioclase feldspar. *Contrib, Miner. Petrol.* **84**, 345–354.

Tuttle, O.F. and Bowen, N.L. (1958) Origin of granite in the light of experimental studies in the system $NaAlSi_3O_8$–$KAlSi_3O_8$–SiO_2–H_2O. *Geol. Soc. Am. Mem.* **74**.

van der Molen, I. and Paterson, M.S. (1979) Experimental deformation of partially-melted granite. *Contrib. Miner. Petrol.* **70**, 299–318.

Viswanathan, K. (1972) Kationenaustausch an Plagioklasen. *Contrib. Miner. Petrol.* **37**, 277–290.

von Platen, H. (1965) Kristallisation granitischer Schmelzen. *Beitr. Miner. Petrologie* **11**, 334–381.

White, A.J.R. (1966) Genesis of migmatites from the Palmer region of South Australia. *Chem. Geol.* **1**, 165–200.

Whitney, J.A. (1975) The effects of pressure, temperature and X_{H_2O} on phase assemblage in four synthetic rock compositions. *J. Geol.* **83**, 1–31.

Winkler, H.G.F. (1961) Die Genese von Graniten und Migmatiten auf Grund neuer Experimente. *Geol. Rdsch.* **61**, 347–364.

Winkler, H.G.F. (1976) *Petrogenesis of Metamorphic Rocks.* 4th edn., Springer-Verlag, New York etc.

Winkler, H.G.F. (1979) *Petrogenesis of Metamorphic Rocks.* 5th edn., Springer-Verlag, New York etc.

Winkler, H.G.F. and Breitbart, R. (1978) New aspects of granitic magmas. *Neues Jb. Miner. Mh.* 463–480.

Winkler, H.G.F. and Lindemann, W. (1972) The system Qz–Or–An–H_2O within the granitic system Qz–Or–Ab–An–H_2O. Application to granitic magma formation. *Neues Jb Miner. Mh.* 49–61.

Winkler, H.G.F. and von Platen, H. (1961) Experimentelle Gesteinsmetamorphose IV. Bildung anatektischer Schmelzen aus metamorphisierten Grauwacken. *Geochim. Cosmochim. Acta* **24**, 48–69.

Winkler, H.G.F., Boese, M. and Marcopoulos, T. (1975) Low temperature granitic melts. *Neues Jb. Miner. Mh.* 245–268.

Wyllie, P.J. (1977) From crucibles through subduction to batholiths. In *Energetics of Geological Processes*, eds. Saxena, S.K. and Bhattacharji, S., Springer-Verlag, New York, etc., 389–433.

Wyllie, P.J. and Tuttle, O.F. (1964) Experimental investigation of silicate systems containing two volatile components. Part III. The effects of SO_3, P_2O_5, HCl, and Li_2O, in addition to H_2O on the melting temperatures of albite and granite. *Am. J. Sci.* **262**, 930–939.

Yoder, H.S. (1968) Albite–anorthite–quartz–water at 5 kb. *Yb. Carnegie Instn Wash.* **66**, 477–478.

3 Phase equilibria in partial melting of pelitic rocks

JAMES A. GRANT*

3.1 Introduction

Melting of rocks is a logical extension of normal high-grade metamorphism and there is no question that it occurs: magma exists. Many petrologists ascribe the origins of many migmatites to partial melting with segregation of solids from liquid. However, it is by no means axiomatic that all migmatites are formed essentially by this process. As we have been recently reminded by Olsen (1983), partial melting, metamorphic differentiation, injection or metasomatism are all still processes possible in the formation of migmatites, and more than one of these may be involved in the formation of a given migmatite (e.g. Johannes, 1983).

In order to better understand the many facets of partial melting in the crust, it is desirable to compare predictions from experimental and theoretical studies with observations from natural migmatites, using systems sensitive to changing metamorphic conditions.

Pelitic compositions have been used as sensitive indicators of metamorphic conditions, at least since the pioneering work of Barrow (1893), and this sensitivity extends into the realm of partial melting (Grant, 1973). Pelitic compositions not only are more sensitive to change in metamorphic conditions than quartzofeldspathic compositions (which correspond to the most common migmatites), but also have another major advantage. Pelites commonly contain abundant mica, and thus have their own 'built-in' source of an H_2O-component, which, even in the absence of an aqueous phase, results in melting being possible at quite accessible temperatures.

Of course, migmatites are not the only products of partial melting of pelitic rocks. If separation of solids from liquid is effective, two distinct products can form. On the one hand, we may find residual refractory 'restites', often rich in alumina and poor in silica and alkalis, for which a partial melting origin seems

* Department of Geology, University of Minnesota, Duluth, MN 55812, USA.

plausible (e.g. Grant, 1968, 1985; Lal *et al.*, 1978). This is also a possible mode of origin of some granulites (e.g. Powell, 1983*a*). On the other hand, channelling and uprise of derivative melts may give rise to peraluminous instrusions and extrusions (for examples, see Chappell and White, 1974, and many of the papers in Clarke, 1981).

This review will cover some of the aspects of phase equilibria in partial melting of compositions that are pelitic in the broad sense. Particularly we shall be concerned with equilibria involving the common minerals quartz, potassium feldspar, muscovite, aluminosilicates, biotite, cordierite, garnet, orthopyroxene and spinel, as well as liquid and vapour. Discussion will be confined mainly to relatively simple systems, K_2O–FeO and/or MgO–Al_2O_3–SiO_2–H_2O (abbreviated as KF/MASH, Table 3.1), with or without CO_2, for in these one finds most of the major kinds of relationships, possibilities and uncertainties that are still fundamental when one goes to the more complex systems. We shall start with consideration of possible equilibria in the absence of liquid, and proceed to melting with P_{H_2O} equal to P_{total}, then with P_{H_2O} less than P_{total} and finally under vapour-absent conditions. We shall conclude with discussions of internal and external control of intensive variables and of paths through pressure–temperature–time space.

This review will not deal to any significant extent with several topics of importance. Geobarometry and geothermometry from chemistries of coexisting phases are well covered in the recent reviews by Essene (1982), Newton (1983) and Bohlen *et al.* (1983*a*). The interaction of silicates and oxides is considered in any detail only for hercynitic spinel, and the reader is referred particularly to papers presented at the Mineralogical Society of America symposium in 1983, including Bohlen and Dollase (1983), Robinson and Tracy (1983), Haggerty (1983), and also to Kars *et al.* (1980), Grew (1981) and again to Bohlen *et al.* (1983*a*). Nor does this review attempt to relate many of the recent papers on natural occurrences to the equilibria presented (see Grant, 1973, for an earlier review of this kind).

Table 3.1 Abbreviations

Components		*Phases*	
C	CaO	Qtz	quartz
N	Na_2O	Kfs	potassic feldspar
K	K_2O	Pfs	sodic feldspar
F	FeO	Mus	muscovite
M	MgO	Als	aluminosilicate (Al_2SiO_5)
A	Al_2O_3	Bio	biotite
S	SiO_2	Cdt	cordierite
H	H_2O	Gar	garnet
		Opx	orthopyroxene
		Cor	corundum
		Spl	spinel
		L	liquid
		V	vapour

Square brackets [ ] denote phases not participating in a particular equilibrium.

A final note on nomenclature: in this review 'liquid' means a silicate-rich magma, and 'vapour' and 'fluid' are used interchangeably to mean a supercritical fluid. It may be remarked that Grecian terminology for migmatites is kept to a bare minimum, and the comments of Johannes (1983, p. 234–235) should be borne in mind, as should the prologue to Zen (1967).

3.2 Subsolidus phase equilibria and experimental data

Let us start with the phase relations in KFASH and KMASH and then proceed to KFMASH. There is good agreement that for both systems, KFASH and KMASH, in the absence of orthopyroxene, orthoamphibole and olivine, equilibria between Qtz–Kfs–Mus–Als–Bio–Cdt–Gar–V are as shown schematically in Figure 3.1. (Abbreviations are given in Table 3.1.) Figure 3.1 is compatible with the grids of Thompson (1976, Figure 4 and 1982, Figure 1, which are essentially the same except for incorporation of Tschermak's $Al_2Fm_{-1}Si_{-1}$ substitution in the latter), and of Vielzeuf and Boivin (1984,

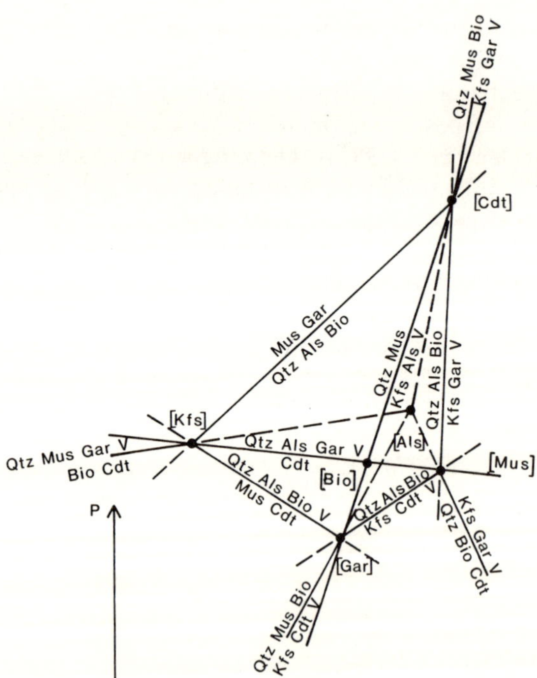

Figure 3.1 Schematic pressure-temperature grid of invariant and univariant equilibria in KFASH or KMASH for quartz-bearing, vapour–present and orthopyroxene-absent assemblages among Qtz–Kfs–Mus–Als–Bio–Cdt–Gar–V. Metastable reactions are shown by dashed lines. Qtz + Mus = Kfs + Als + V is assumed degenerate. [Kfs], [Mus], [Cdt] and [Gar] are stable, [Als] and [Bio] metastable.

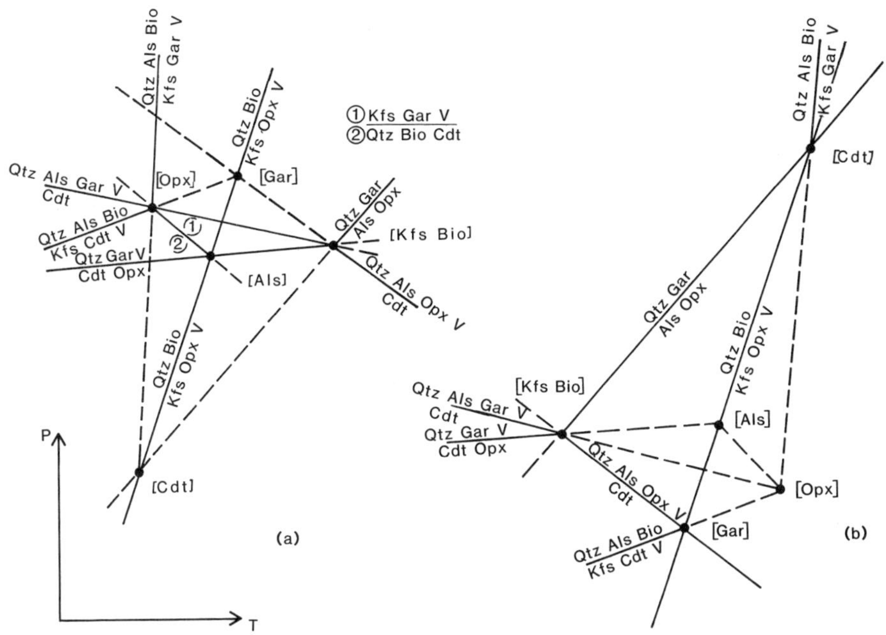

Figure 3.2 Schematic pressure-temperature grids of possible invariant and univariant equilibria in KFASH or KMASH for quartz-bearing, vapour-present and muscovite-absent assemblages among Qtz–Kfs–Als–Bio–Cdt–Gar–Opx–V. Metastable reactions are shown by dashed lines. Qtz + Bio = Kfs + Opx + V is assumed degenerate. (After Vielzeuf and Boivin, 1984, Figure 7*A* and 7*B*.) In (*a*), [Als], [Bio] and [Opx] are stable, [Kfs], [Cdt] and [Gar] are metastable, [Kfs] coincides with [Bio] and [Cdt] is metastable at *low* pressure. In (*b*), [Kfs], [Cdt] and [Gar] are stable, [Als], [Bio] and [Opx] metastable, again [Kfs] coincides with [Bio], but [Cdt] is stable (as far as these phases are concerned) at *high* pressure.

Figure 3B) i.e. the invariant assemblage [Mus, Opx] is stable at a temperature above Qtz + Mus = Kfs + Als + V.

Assemblages bearing orthopyroxene (or orthoamphibole or olivine) may be analysed similarly in the absence of muscovite, which yields relations as shown in Grant (1983*a*, 1985) and Vielzeuf and Boivin (1984, Figure 7). Of course, the relations about [Mus, Opx] are identical in both analyses. The major question is whether the relations between Qtz–Kfs–Als–Bio–Cdt–Gar–Opx–V are similar for both KFASH and KMASH (both conforming to Figure 3.2*a*) as in Thompson (1982), or different (with KFASH conforming to Figure 3.2*a* and KMASH to Figure 3.2*b*) as in Vielzeuf and Boivin (1984). To answer this we must consider the sparse available experimental data.

3.2.1 *KMASH*

Experimental data in KFASH and KMASH are shown in Figure 3.3, including reactions involving melting, which are most conveniently discussed

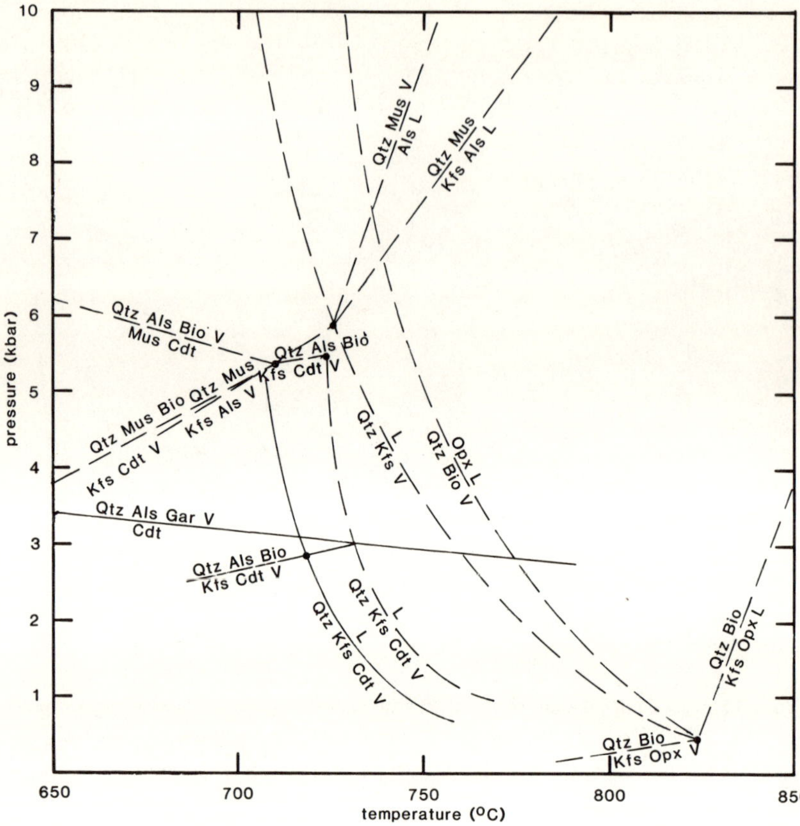

Figure 3.3 Experimentally determined reactions: dashed lines refer to reactions in KMASH and the subsystem KASH, solid lines to reactions in KFASH. See text for sources and discussion.

in this section. Equilibria involving Qtz–Kfs–Mus–Als–L–V are common to both KFASH and KMASH, and in this figure, equilibria with pure muscovite are shown (see Thompson, 1982, and the discussion of Figure 3.9 in this chapter). Qtz + Mus = Kfs + Als + V is from Chatterjee and Johannes (1974) (see also Evans, 1965; Althaus *et al.*, 1970; Day, 1973). This ties in well with the determinations of Qtz + Mus + V = Als + L and Qtz + Mus = Kfs + Als + L (Storre and Karotke, 1971, 1972; Huang and Wyllie, 1974). The solidi Qtz + Kfs + Mus + V = L and Qtz + Kfs + Als + V = L occur immediately below the solidus Qtz + Kfs + V = L (Lambert *et al.*, 1969; Huang and Wyllie, 1974), and this last reaction is after Shaw (1963) and Bohlen *et al.* (1983*b*).

In KMASH, Qtz + Mus + Bio = Kfs + Cdt + V (Seifert, 1976) is essentially indistinguishable from Qtz + Mus = Kfs + Als + V, as determined by Chatterjee and Johannes (1974) above 3 kbar. These must intersect at the invariant point where Qtz + Als + Bio + V = Mus + Cdt and Qtz + Als + Bio = Kfs + Cdt + V are generated (IP2 of Seifert, 1976). The position of Qtz + Als + Bio + V = Mus + Cdt in Figure 3.3 is compatible with the experimental data of both Seifert (1970) and Bird and Fawcett (1973). Qtz + Als + Bio = Kfs + Cdt + V terminates in melting within about 20° of this point, and is shown with a slope of 10 bars deg^{-1} after Schreyer and Seifert (1969). The solidus Qtz + Kfs + Cdt + V = L (with an implied degeneracy) is from Seifert (1976), with the minor change that his determination of the beginning of melting at 1 kbar has been discounted to yield curvature more in keeping with the solidus Qtz + Kfs + V = L and with the results of Grant and Hoffer (1977) on melting of Qtz + Kfs + Cdt + V = L in KFMASH. The relations shown are not significantly different from those depicted by Schreyer (1976, p. 312) and by Seifert (1976). As both these writers note, the relationships with respect to the beginning of melting are complex and not well known, largely due to uncertainty in the compositions of liquids involved, as will be discussed later.

Qtz + Bio = Kfs + Opx + V (Luth, 1967; Wendlandt, 1981; Bohlen *et al.*, 1983*b*) terminates in melting, presumably generating Qtz + Kfs + Opx + V = L and Qtz + Kfs + Bio + V = L, just below Qtz + Kfs + V = L, and the relations shown are from Wendlandt (1981) and Bohlen *et al.* (1983*b*). Note that 'Bio' here is phlogopite, and that this is a simplification over the aluminous compositions found in natural high-grade biotites (see below, KFMASH).

Cordierite breakdown reactions pose problems because of the variable and important water content (Newton, 1972; Newton and Wood, 1979; Lonker, 1981; Martignole and Sisi, 1981) and occasionally because of use of an erroneous structural formula for cordierite (*all* the Al is in tetrahedral coordination, see Gibbs, 1966). The most important point in the present context is that Mg-garnet and quartz are only stable at pressures greater than about 15 kbar (Boyd and England, 1959; Schreyer, 1968; Hensen and Essene, 1971, Figure 3) whereas hydrous Mg-cordierite is not stable at pressures greater than about 11 kbar and anhydrous Mg-cordierite is not stable at pressures greater than about 8 kbar (Newton, 1972). Thus the invariant point [Kfs, Mus, Bio] involving the assemblage Qtz–Als–Cdt–Gar–Opx–V in KMASH is metastable. It lies at the intersection of Qtz + Gar = Als + Opx and Cdt = Qtz + Als + Gar + V. If we use the data for the former reaction from Hensen and Essene (1971, Figure 1, data for 'Als' = sillimanite) and the calculated position for the latter reaction from Newton and Wood (1979, Figure 5, again for sillimanite) this point is at approximately 600 °C and 7.5 kbar—very approximately. Further, Qtz + Gar = Als + Opx (Hensen and Essene, 1971) and the extension of Qtz + Bio = Kfs + Opx + V surely intersect in an invariant point ([CD] of Vielzeuf and Boivin, 1984) at *high pressure*. Thus the liquid-absent phase relations in KMASH correspond to the grid of Figure 7B of Vielzeuf and Boivin (1984), from which the present Figure 3.2*b* is taken.

3.2.2 *KFASH*

The phase relations between Qtz–Kfs–Als–Bio–Cdt–Gar–Opx–V in KFASH are discussed in some detail in Grant (1985), from which much of the following is taken. Equilibria involving orthopyroxene, even though metastable with respect to quartz–fayalite in KFASH, are the more pertinent to a review of phase relations affecting pelitic compositions. To a first approximation, the transition between orthopyroxene- and olivine-bearing assemblages can be considered as if polymorphic.

The experimental data of Richardson (1968) and Holdaway and Lee (1977) on Cdt = Qtz + Als + Gar + V, and of Day (1971) and Holdaway and Lee (1977) on Qtz + Als + Bio = Kfs + Cdt + V constrain these two reactions well. The invariant point at the intersection of the two reactions ([Opx] in Figure 3.2) was placed by Holdaway and Lee (1977) at 740 °C and about 3 kbar, whereas in the writer's estimation it may be as low as 720 °C, and compatible with all the experimental data cited. To be sure, this is not a great difference, and leaves the major conclusions of Holdaway and Lee unchanged. In Figure 3.3, the invariant point is placed at 730 °C as a compromise. The slopes of the other two reactions generated at the invariant point were calculated to be conformable with Holdaway and Lee's work, using data from Newton and Wood (1979), Burnham *et al.* (1969), Helgeson *et al.* (1978), Robie *et al.* (1967, 1978) in addition to those of Holdaway and Lee (1977). The calculated slopes are + 340 bars/deg for Qtz + Als + Bio = Kfs + Gar + V, and − 100 bars/deg for Qtz + Bio + Cdt = Kfs + Gar + V (Grant, 1983*a*, 1985). The former reaction, Qtz + Als + Bio = Kfs + Gar + V, and Qtz + Mus = Kfs + Als + V intersect at the invariant point [Cdt] of Figure 3.1. Thus this point must lie at about 6 ± 0.5 kbar and 730 ± 10 °C. (And it is moot whether this point is metastable with respect to melting: this is a busy location in pressure–temperature space!) There are no experimental data on the two reactions (Figure 3.1) generated on the low temperature side of Qtz + Mus = Kfs + Als + V at this point, nor on the others generated at pressures lower than those of Cdt = Qtz + Als + Gar + V. The latter, however, are really outside the realm of this review.

We turn once again to temperatures above Qtz + Mus = Kfs + Als + V, and to the invariant point [Mus, Opx]. From this invariant point arises Qtz + Bio + Cdt = Kfs + Gar + V, and the invariant point [Mus, Als] lies at the intersection of this reaction with the stability limit of quartz + aluminous Fe-biotite (cf. Qtz + Bio = Kfs + Opx + V, Figure 3.2*a*). At 2 kbar, this reaction would be at about 720 °C, following Eugster and Wones (1962) and Rutherford (1969). Thus [Mus, Als] would lie at about 735 °C and 2.5 kbar, given the location of [Mus, Opx] as in Figure 3.3. Recall that in KFASH the reaction involves olivine rather than orthopyroxene, and (because of the aluminous composition of the biotite) cordierite. Thus this point is at a maximum pressure and minimum temperature for the metastable equivalent involving orthopyroxene. Unfortunately there are no experimental data on the degenerate reaction Cdt + Opx = Qtz + Gar + V which also passes through this point, [Mus, Als]. However, it has a very shallow slope (Grant, 1985), and will apparently intersect Cdt = Qtz + Als + Gar + V at about 2.5 kbar in the vicinity of 850–900 °C. This point, [Kfs, Mus, Bio], is important because it is one termination of the reaction Cdt + Gar = Qtz + Als + Opx + V in the system KFMASH. (It is metastable with respect to quartz–spinel, to be discussed below.) Clearly the invariant point lies at temperatures above the breakdown of quartz–biotite in KFASH, in contrast to the situation in KMASH, and the pertinent phase equilibria for muscovite-absent assemblages in KFASH are shown in Figure 3.2*a*.

As far as melting reactions in KFASH are concerned, Holdaway and Lee (1977) found that at about 3 kbar, melting ensued in Qtz–Kfs–Als–Bio–Cdt–V between 710–722 °C. This is just above the beginning of melting found for Qtz–Kfs–Cdt–V in KFMASH by Grant and Hoffer (1977), and the degenerate solidus shown in Figure 3.3 is compatible with these results. (As will be discussed below, we are embarrassingly ignorant of the compositions of the first melts formed, never mind their changes with increasing degree of melting.)

In concluding this section, let us return to the main point of this discussion. As far as can be judged, the orthopyroxene- and liquid-absent relations for KMASH and KFASH are essentially as shown in Thompson (1976, Figure 4 and 1982, Figure 1), Vielzeuf and Boivin (1984, Figure 3A), Grant (1983*a*, 1984*a*) and here, in Figure 3.1. However, the muscovite- and liquid-absent relations for KFASH and KMASH are different, and correspond to Figure 3.2*a* and 3.2*b*, respectively, which are after Vielzeuf and Boivin (1984, Figure 7A and 7B). The difference is largely due to the drastic difference in the stability fields of quartz-pyrope and quartz-almandine.

3.2.3 *KFMASH*

Equilibria in the two limiting system KFASH and KMASH are integrated with those in the system KFMASH essentially as shown in Vielzeuf and Boivin (1984, Figure 9), on which Figure 3.4 is based. The net result is a simple grid of reactions in KFMASH which is similar to that shown in Grant (1973, Figure 4) with two corrections. The invariant point [Mus, Opx] is now known to be metastable (Holdaway and Lee, 1977) and Cdt + Gar = Qtz + Als + Opx + V (in KFMASH) should have a negative slope. In the two previous sections, the positions of the metastable end-points for this reaction, the invariant points [Kfs, Mus, Bio] in KMASH and KFASH, were estimated, albeit with considerable uncertainty. However, from data on natural assemblages (Grant, 1973), the pertinent equivalent of this reaction should lie dominantly in the sillimanite field (and in the vicinity of the aluminosilicate triple point if orthoamphibole rather than orthopyroxene is involved). This is compatible

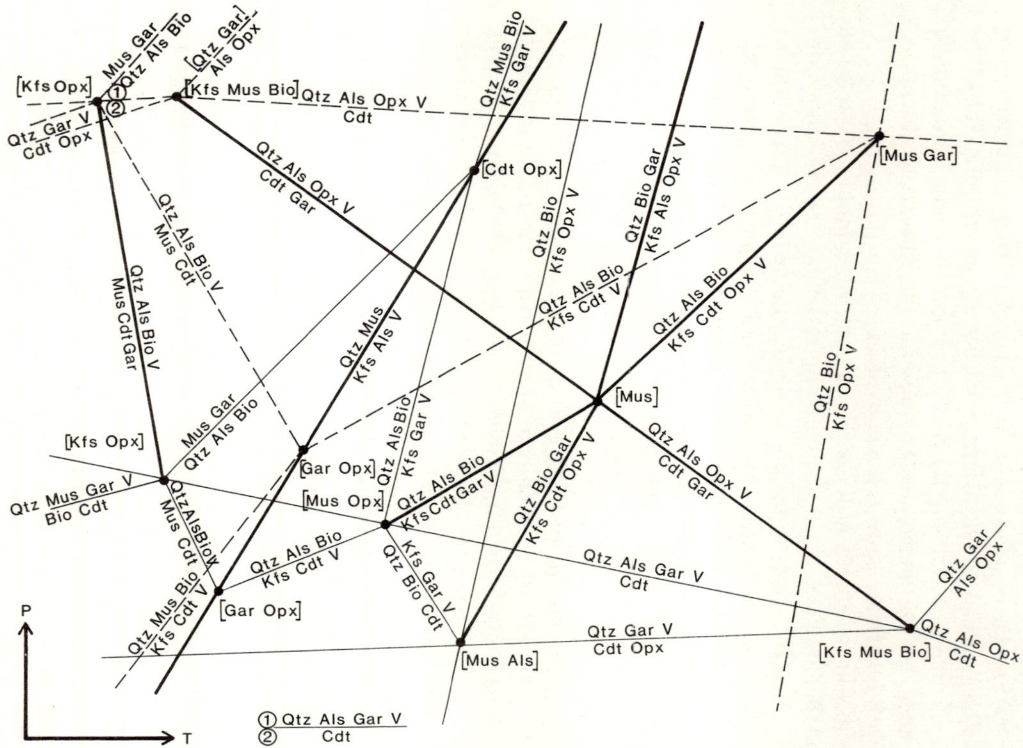

Figure 3.4 Schematic pressure–temperature grid of invariant and univariant equilibria for quartz-bearing and vapour-present equilibria among Qtz–Kfs–Mus–Als–Bio–Cdt–Gar–Opx–V in KFASH (light lines), KMASH (dashed lines) and KFMASH (heavy lines). (After Vielzeuf and Boivin, 1984, Figure 9.) Only invariant points that are stable in this model are labelled. Note that the two points labelled [Kfs, Mus, Bio] are stable but degenerate (not involving a potassic phase), and that in KFMASH, [Bio] is metastable at the intersection of two degenerate univariant reactions.

with the present positioning of the reaction involving hydrous cordierite. Indeed, none of the reactions in KFMASH have been determined by direct experiment, but given the constraints of Holdaway and Lee on the slope of Qtz + Als + Bio = Kfs + Cdt + Gar + V, approximately 25 bars/deg, it is probable that [Mus] (Figure 3.4; equivalent to [MU] of Vielzeuf and Boivin, 1984, Figure 9), lies metastably at roughly 790 °C and 4.5 kbar.

3.2.4 *Additional solid phases*

In addition to the phases considered above, even quartz-bearing assemblages may bear staurolite (Fe-rich compositions at lower temperatures), spinel (Fe-rich compositions at higher temperatures), osumilite, sapphirine and kornerupine (Mg-rich compositions at high temperatures).

Staurolite–quartz in FASH is limited to temperatures below about 680 °C according to Richardson (1968), Ganguly (1972) and Rao and Johannes (1979), although Zn may stabilize it to higher temperatures (Spry, 1982). This is discussed, for example, in Grant (1973) and Ashworth (1975). Phase relations between Qtz–Als–Cdt–Gar–Opx–spinel–sapphirine and Qtz–Kfs–Cdt–Gar–Opx–spinel–sapphirine–osumilite were discussed by Hensen (1971 and 1977, respectively). According to Hensen (1977) osumilite is only likely to occur at low a_{H_2O}, being occluded by melting reactions at $P_{H_2O} = P_{total}$ (see also Berg and Wheeler, 1976). Sapphirine–quartz apparently only occurs in low a_{H_2O} environments at high pressure and temperature (Seifert, 1974, p. 196) although sapphirine has a large stability field in Mg-rich and Si-poor compositions (see also Ackermand *et al.*, 1975; Lal *et al.*, 1978; Ellis *et al.*, 1980). Kornerupine is well discussed by Seifert (1975) and Schreyer (1976). Although these phases may surely be involved in melting, this review concentrates on the relationships between the more common minerals, and of these phases, only spinel will be considered further here.

In FASH, quartz–spinel assemblages are stable at high temperature (Richardson, 1968, Figure 6; see also Keesmann *et al.*, 1971). Bohlen and Dollase (1983) show quartz–spinel stable at temperatures above an invariant point at 775 °C and 2.9 kbar, and note that zinc and ferric iron expand the stability field of quartz–spinel. This results in [FeMuBi] of Vielzeuf and Boivin (1984) being metastable, and the phase relations involved are discussed in Vielzeuf (1983). However, there he preferred a model with x_{Mg} in spinel greater than that in garnet. To the data noted in Vielzeuf (1983), one may add the analytical results of Berg (1977), Harris (1981), Harris *et al.* (1982) and Grant (1983*b*), where, in quartz-bearing assemblages, spinels have x_{Mg} less than or equal to that in coexisting garnet. In this case, the relations in FASH and FMASH are as shown in Figure 3.5 for assemblages involving Qtz–Als–Cdt–Gar–Opx–Spl–V, with Qtz–Cdt–Spl collinear (see Richardson, 1968 and Schreyer, 1976 for comments on magnetite solid solution in hercynitic spinel). Thus the termination of Cdt + Gar = Qtz + Als + Opx + V in FMASH will

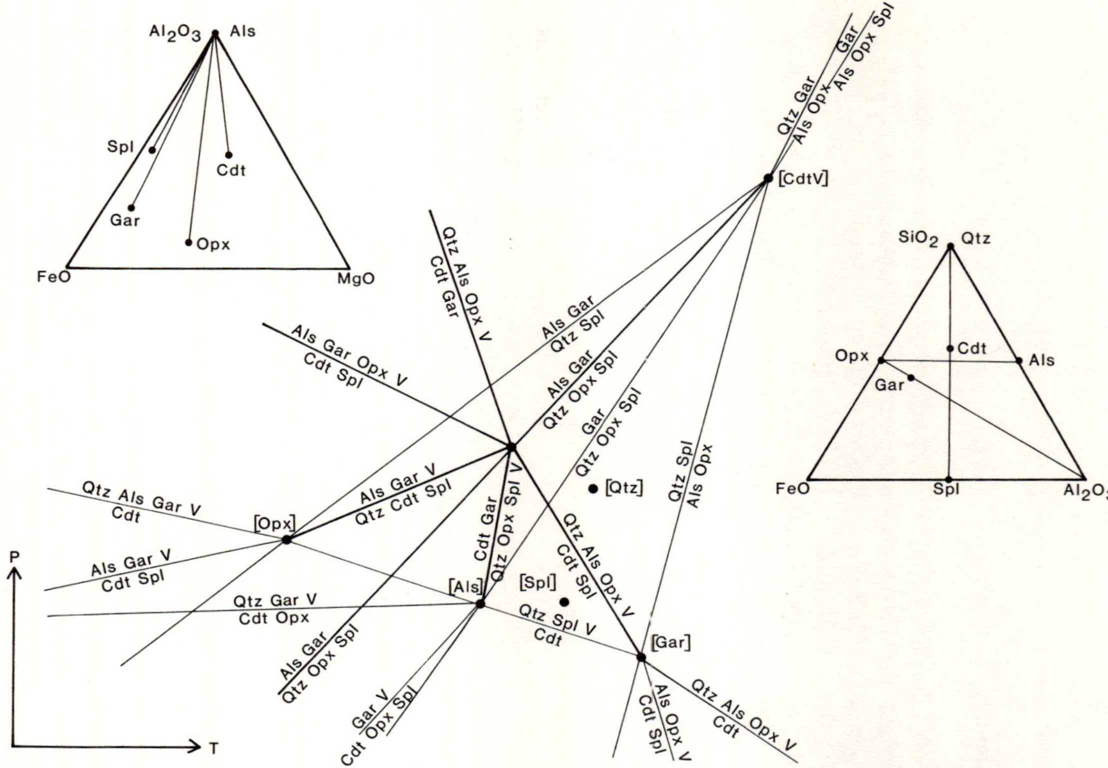

Figure 3.5 Schematic pressure–temperature grid of invariant and univariant equilibria in the vicinity of [Kfs, Mus, Bio] of KFASH in Figure 3.4. Equilibria in FASH are shown in light lines, and the corresponding chemographic relations are shown in projection from H_2O in the Al_2O_3–FeO–SiO_2 triangle on the right. [Qtz], [Spl] and [V] are metastable. Equilibria in FMASH are shown in heavy lines, and the corresponding chemographic relations are shown in projection from SiO_2 and H_2O in the Al_2O_3–FeO–MgO triangle on the left.

correspond to Vielzeuf's (1983) Figure 5, location # 3. If the anhydrous systems are considered, probably the only significant change would be due to withdrawal of the stability field of cordierite to lower pressures. Here it will suffice to note the modification of Figure 3.4 required by relations shown in Figure 3.5.

3.3 Partial melting

Several general considerations arise repeatedly in considering partial melting, so, with the forbearance of readers to whom these are self-evident, let us raise them now.

3.3.1 *Composition of melts*

Probably the most frustrating—or challenging—aspect of dealing with melting phenomena lies in our uncertainty of the chemistries and changes in chemistry of the silicate liquids that are involved in different systems and under different conditions.

We know in general that the low-melting fractions from pelitic compositions are likely to be 'granitic', that is, involving high normative quartz and feldspars (e.g. Winkler, 1979; Grant, 1973; Wyllie, 1977*a,b*; Fyfe *et al.*, 1978, pp. 177ff; Abbott and Clarke, 1979, among many). We know that in the absence of K-feldspar in the subsolidus, a 'K-feldspar component' can be provided through incongruent melting of micas (e.g. Winkler, 1979; Shaw, 1963; Luth, 1967; Lambert *et al.*, 1969; Brown and Fyfe, 1970; Kerrick, 1972; Huang and Wyllie, 1974, and many others).

What we do not know well is the extent to which 'excess' Al_2O_3, MgO and FeO, to name only major additions, may be dissolved in pertinent silicate melts (e.g. Luth, 1967; Lambert *et al.*, 1969; Abbott and Clarke, 1979). Generally, we would anticipate excess alumina over that in the feldspars, and relatively high Fe/Mg in lower temperature liquids. However, it is *not* axiomatic that the lowest temperature liquids must be in the iron-end member system: the role of almandine-rich garnet (and hercynite and Fe-oxides) is important here (compare Green, 1976; Abbott and Clarke, 1979 and Clemens and Wall, 1981).

It is worthwhile to emphasize here the role of cordierite in this context. As shown by Schairer (1954) and by Seifert (1976), Qtz–Kfs–Cdt–V in the system KMASH melt at temperatures below the Qtz + Kfs + V = L solidus. Seifert concluded that Qtz–Kfs–Cdt–V melts congruently and several tens of degrees below that solidus. With the addition of iron to the system, the beginning of melting in Qtz–Kfs–Cdt–V, using cordierite of intermediate Fe/Mg, was found to be lowered to near coincidence with the Qtz + Pfs + V = L solidus (Grant and Hoffer, 1977). In the system KFASH, Holdaway and Lee (1977, p. 181) found that melting started in Qtz–Kfs–Als–Bio–Cdt–V between 710 and 722 °C at 3 kbar—not significantly different from the result of Grant and

Hoffer. Thus even in KFMASH, whether or not melting starts in the system KFASH, there would seem to be a low-temperature liquid field extending at least to intermediate values of Fe/Mg over a very small temperature interval. Moreover, even in KMASH it is not now a good approximation to assume that low-temperature liquids lie in Qtz–Kfs–V (see Grant, 1973 and the necessary revision in Grant, 1983*a*, 1985).

A major part of the problem is that of determining the composition of small amounts of quenched liquid. In experimental melting, microprobe analysis of quenched melt near the relevant solidus is commonly plagued by interference from unmelted solid grains, round which melt tends to occur (e.g. Hoffer and Grant, 1980, p. 19), possible loss of iron to the enclosing capsule (Green, 1976) and correction for loss of alkalis during analysis. In nature, the analogous problem also lies in the separation of melt from solid, especially where the ratio melt/solid is small. This is true both in actuality (e.g. Arzi, 1978) because of the apparent threshold of perhaps 20% melting which must occur before significant segregation can proceed, and in the eye of the beholder trying to interpret as previously coexisting melt and solid what is now all solid. Consider that a migma close to the solidus may contain material that was never molten and material that crystallized from melt (even in the same grain) as well as residual melt.

3.3.2 *Thermal barriers*

The possibility of waxing and waning of thermal barriers is of concern with changing conditions, especially due to the effect of pressure on the relative stability of a hydrate (Wyllie and Tuttle, 1960; Eggler, 1973). Since the data just mentioned raise the possibility of a thermal barrier involving Qtz–Kfs–Cdt–V, at least in the system KMASH, let us extend this as in the example given by Grant and Hoffer (1977).

Consider the hypothetical relations in quartz-bearing assemblages involving Qtz–Kfs–Als–Bio–Cdt–L–V in the system KMASH, shown in the schematic diagrams of Figure 3.6. The simplest interpretation of the congruent melting in Qtz–Kfs–Cdt–V is that a thermal barrier exists, separating two eutectic systems on the low pressure side of the invariant point (where Qtz–Kfs–Cdt–V can be stable). It would be fortuitous if the projected composition of liquid at this invariant point lay on the Kfs–Cdt join, hence two possibilities arise. Either the liquid is more aluminous than Kfs–Cdt, generating Qtz + Kfs + Als + Cdt + V = L, or less aluminous than Kfs–Cdt, generating Qtz + Kfs + Bio + Cdt + V = L, as in this figure. Close to the invariant point, an incongruent melting reaction is generated (Qtz + Kfs + Cdt + V = Als + L) and, with emergence of the thermal barrier, this gives way to two congruent melting reactions, Qtz + Kfs + Als + Cdt + V = L and Qtz + Kfs + Cdt + V = L, the latter, higher-temperature, reaction being the thermal barrier. Note that the composition of 'L' is by no means constant

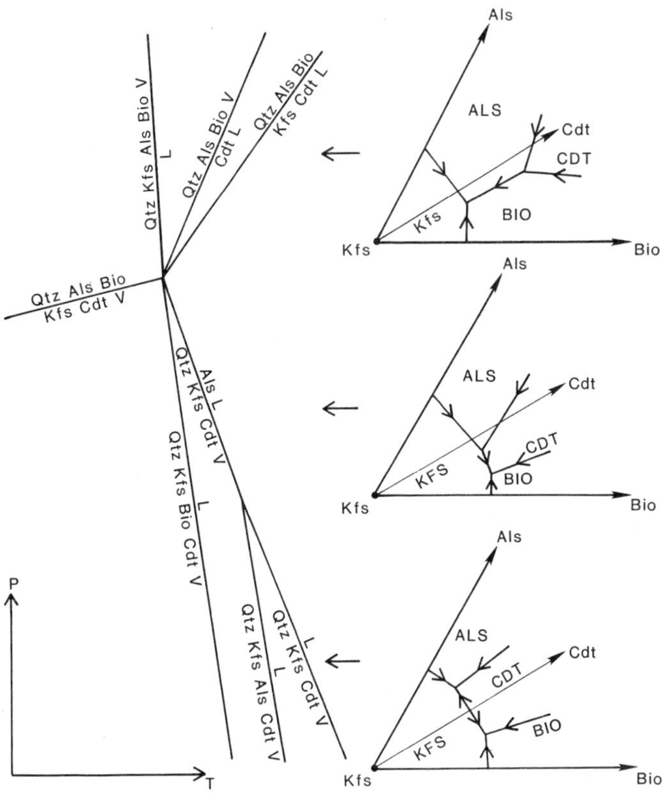

Figure 3.6 Schematic pressure–temperature grid of univariant equilibria around the invariant point involving quartz-bearing and vapour-present assemblages among Qtz–Kfs–Als–Bio–Cdt–L–V in the system KMASH, illustrating the possible appearance of a thermal barrier Qtz + Kfs + Cdt + V = L. Three liquidus diagrams at different pressures are shown for potassic compositions projected from SiO_2 and H_2O onto $KAlO_2$–Al_2O_3–MgO.

over this range of conditions, nor should it be expected to be so. Further, the degeneracy Qtz + Kfs + Cdt + V = L, used in Figure 3.3, is an approximation to reactions such as those extending to the low-pressure side of this invariant point.

3.3.3 *Melting involving a hydrous mineral*

Melting of pelites commonly involves mica, muscovite or biotite or both, and it is appropriate to review the general features of this process. This need be but brief to begin with, because much of the topic has been well covered by others, notably Brown and Fyfe (1970), Fyfe *et al.* (1978) and Wyllie (1977 *a,b*). Usually the melting of quartz–muscovite is used as an example, as this is simple and relatively well known. Here, we shall use quartz–biotite in the system

KMASH as the example, not just for the sake of equal opportunity for biotite, but to make some points that may not be relevant in the case of quartz–muscovite.

Consider, then, possible equilibria among Qtz–Kfs–Bio–Opx–L–V, where 'Bio' is phlogopite and 'Opx' is enstatite. A schematic pressure-temperature grid for the relations in the vicinity of melting is given in Figure 3.7 (cf. Luth, 1967; Bohlen *et al.*, 1983*b*). The dehydration reaction Qtz + Bio = Kfs + Opx + V terminates in melting slightly below the Qtz + Kfs + V = L solidus, at an invariant point from which extend five reactions involving melt. At low

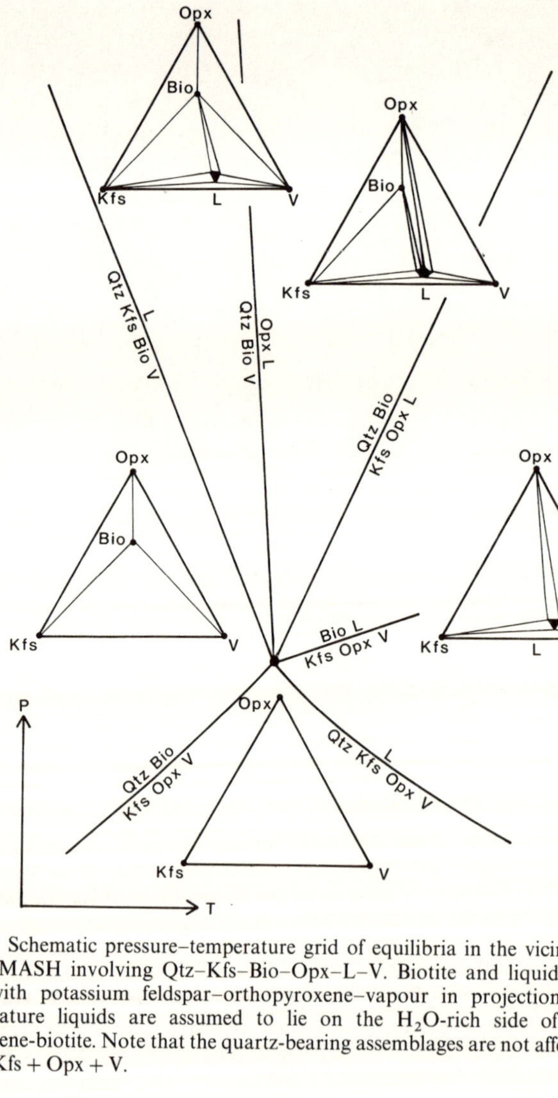

Figure 3.7 Schematic pressure–temperature grid of equilibria in the vicinity of the invariant point in KMASH involving Qtz–Kfs–Bio–Opx–L–V. Biotite and liquid are assumed to be coplanar with potassium feldspar–orthopyroxene–vapour in projection from quartz. The low-temperature liquids are assumed to lie on the H_2O-rich side of the extended join orthopyroxene-biotite. Note that the quartz-bearing assemblages are not affected by the reaction Bio + L = Kfs + Opx + V.

pressure, dehydration sets in before congruent vapour-saturated melting on Qtz + Kfs + Opx + V = L. In such an environment, anhydrous solid-phase assemblages can coexist with water vapour. If vapour is not present, melting will not occur until the dry-melting reaction Qtz + Kfs + Opx = L (not shown). At pressures greater than at the invariant point, however, there are three possible univariant melting reactions involving quartz-bearing assemblages. Qtz + Kfs + Bio + V = L is a typical water-saturated congruent melting reaction on which melting will start as long as water vapour is available to an assemblage of the three solid phases. Although water-undersaturated melts can form as soon as this solidus is passed, the next reaction, Qtz + Bio + V = Opx + L, also produces a water-saturated melt, but is incongruent, with the formation of orthopyroxene. The form of the reaction implies that the liquid is more water-rich than the extended join biotite–orthopyroxene. This is certainly true for the analogous reaction involving muscovite (Shaw, 1963) and largely true for phlogopite (Wones and Dodge, 1966; Luth, 1967; Wendlandt, 1981; Bohlen *et al.*, 1983*b*), except possibly close to the invariant point (see below). As temperature rises above this reaction, the orthopyroxene–liquid field expands at the expense of biotite–liquid, and the stability of quartz–biotite is eventually limited by the vapour-absent reaction Qtz + Bio = Kfs + Opx + L, in which the liquid is water-undersaturated. Note again that the form of the reaction implies that the liquid is still more water-rich than the extended join biotite-orthopyroxene in the chemographic projections. As is shown, for example, in Fyfe *et al.* (1978, p. 180), a melt formed close to the solidus can rise but a short distance through the crust before crystallizing, whereas a liquid formed by melting of quartz–biotite in the absence of vapour may rise significantly through the crust, to pressures lower than at the invariant point, before crystallization is complete. The final reaction, Bio + L = Kfs + Opx + V, is again different from the others in that the hydrate and liquid lie on the high-pressure side of the shallow reaction. This gives rise to a vapour phase, which would tend to escape to still lower pressures, leaving behind an assemblage impoverished in water, and possibly triggering melting at shallow depths.

It is worthwhile to consider what may occur on crystallization of a water-undersaturated melt coexisting with quartz, potassium feldspar and orthopyroxene at high temperature. If this melt were isolated from the solid phases, it could eventually form Qtz–Kfs–Bio–V or Qtz–Kfs–Opx–V, depending on the path. However, if separation were not complete, and the migma cooled at high pressure, some quartz and biotite would form on regaining Kfs + Opx + L = Qtz + Bio, the extent of reaction being limited by the liquid/solid ratio and possibly by armouring of the anhydrous assemblage by the reaction products (which could produce a selvage between restite and melt). Were this so, any remaining liquid would attain the vapour-saturated solidus Qtz + Kfs + Bio + V = L before crystallizing completely.

Let us return to the question of the water contents of liquids in the above scenario. It is *not* axiomatic that the water contents of liquids are always greater than the extended join orthopyroxene–biotite. Water-undersaturated liquids may well extend to the opposite side of this join, yielding a thermal barrier, and two regions for coexistence of Qtz–Bio–Opx–L, where 'L' may be relatively water-rich (as above) or water-poor (cf. Naney, 1983, pp. 1022–1023). This is illustrated in the schematic diagram of Figure 3.8. This is a liquidus diagram for quartz–bearing assemblages in the system $KAlSi_3O_8$–$MgSiO_3$–SiO_2–H_2O, involving potassium feldspar, biotite, orthopyroxene and water vapour. Note that the join orthopyroxene–biotite extends to the potassium feldspar-vapour sideline at $Kfs_{0.5}V_{0.5}$ whereas, for example, a water-saturated granite with about 5 wt% H_2O, corresponding to 767 °C and 2 kbar (Shaw, 1963), would project at $Kfs_{0.4}V_{0.6}$. Since water-undersaturated melts are axiomatically impoverished in H_2O, it would not be unlikely that they extend to the H_2O-poor side of the extended join orthopyroxene–biotite, while still in equilibrium with biotite. In fact, near the invariant point of Figure 3.7, at 0.5 kbar, according to Shaw (1963) even the *water-saturated* liquid in Qtz + Kfs + V = L lies to the *left* of the extended join orthopyroxene–biotite. In this case, at low pressure, the reaction involving Qtz–Kfs–Bio–Opx–L would change from Qtz + Bio =

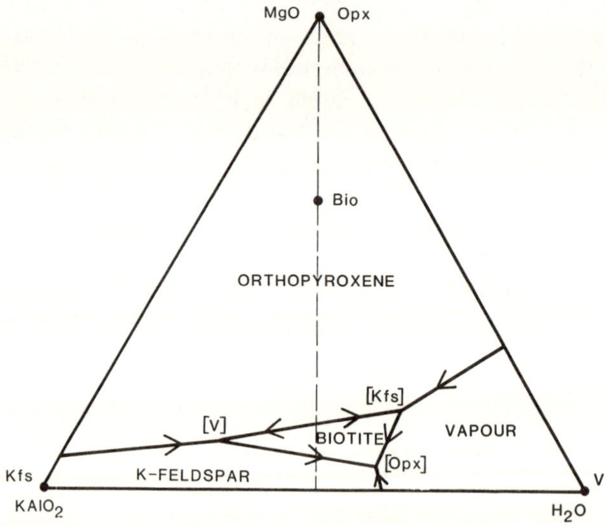

Figure 3.8 Schematic isobaric 'liquidus' diagram for quartz-bearing assemblages in part of the system KMASH involving Qtz–Kfs–Bio–Opx–L–V, where again biotite and liquid are coplanar with potassium feldspar–orthopyroxene–vapour in projection from quartz. Phase compositions are plotted in mole percent, and the MgO contents of liquids are exaggerated for clarity. Bulk compositions in the somewhat unorthodox, but valid field for vapour (water-saturated and vapour-present compositions) will consist of liquid and vapour, the liquids being *less* water-rich than water-saturated liquids on the true liquidus. Only when the liquids become saturated in both vapour and solid will the liquid compositions (lower right, pseudo-ternary boundary curves) achieve the liquidus. There are no silicate-rich liquid compositions within the vapour field at any temperature. (After Grant, 1984).

Kfs + Opx + L (as above) to Qtz + Kfs + Bio = Opx + L, with generation of thermal barrier, Qtz + Bio = Opx + L (cf. Qtz + Kfs + Cdt + V = L in Figure 3.6). An isothermal section at a temperature just below the maximum would be similar in form to that shown in Yoder and Kushiro (1969, Figure 5) for the silica-undersaturated system. The recent contribution by Naney (1983) is tantalizing in this regard. He investigated the phase relations in granitic systems with ferromagnesian silicates, including biotite and orthopyroxene. Fully realizing that his system is much more complex than the simple one modelled here, and in particular that quartz is not in excess for most of his experiments, it is irresistible to compare his results with the model of Figure 3.8. The closest comparison is with his results at 200 MPa (2 kbar), shown in his Table 4. If we consider just the relations among Kfs–Bio–Opx–L–V, we note that the liquidus phase is orthopyroxene at all water contents, and that this is eventually resorbed in all cases. At water contents (in weight percent) of 0.5% and 1.5%, the sequence is orthopyroxene, potassium feldspar, biotite, orthopyroxene resorbed, as would be the case in Figure 3.8 for compositions lying in Kfs–Bio–V and to the left of the join Opx–[V]. At a water content of 3.5%, Naney found the sequence orthopyroxene, biotite, potassium feldspar, orthopyroxene resorbed, similar to that for compositions in Kfs–Bio–V lying between Opx–[V] and Opx–Bio extended. At a water content of 5%, the sequence is orthopyroxene, biotite, orthopyroxene resorbed, potassium feldspar. This is analogous to crystallization of a composition in Kfs–Bio–V lying between the extended join Opx–Bio and the join Opx–[Kfs]. The sequence for his vapour-saturated experiments is the same, except for the presence of V throughout. This is analogous to crystallization of a composition in Kfs–Bio–V lying to the right of Opx–[Kfs] in Figure 3.8. The major point to be made is that if this comparison has some validity, the reaction at [V] seems to be Qtz + Kfs + Bio = Opx + L, as in Figure 3.8, rather than Qtz + Bio = Kfs + Opx + L, as in Figure 3.7. The sequences at low and high water contents would be the same in either case, but the sequence at 3.5% H_2O does not seem to be compatible with the latter reaction. Perhaps this underlies Naney's comment (1983, pp. 1026–1027) that for H_2O contents greater than about 3% (and for higher temperatures) this reaction is not appropriate, but rather one of the form orthopyroxene + liquid + H_2O = biotite. This implies a liquid lying to the *left* of the extended join orthopyroxene–biotite, as well as the higher water contents and higher temperatures, which are all required for the segment of Qtz–Bio–Opx–L lying between the maximum in Figure 3.8 and the invariant point [V].

Although this discussion is intended mainly to point out the possibility of biotite coexisting with liquids of low a_{H_2O}, it does lead to the much broader topic of the relations between water-saturated, vapour-saturated and vapour-absent melting, to which we shall return later.

Suffice it to say at present that low a_{H_2O} may be achieved by dilution of a vapour phase by CO_2 or some other component. The major results of such dilution are, using the grid of Figure 3.7 for examples, (a) withdrawal of a dehydration reaction such as Qtz + Bio = Kfs + Opx + V to lower temperatures, (b) withdrawal of a hydrous melting reaction such as Qtz + Kfs + Bio + V = L to higher temperatures, (c) concomitant shift in the position of an invariant point to higher pressures along the relevant vapour-absent reaction, such as Qtz + Bio = Kfs + Opx + L. These kinds of relations are well displayed in Kerrick (1972), Fyfe *et al.* (1978) and Wyllie (1977*b*). Of course, the fact that a vapour phase is present can have major effects on the physical properties of the system and on the potential for metasomatism (e.g. Holloway, 1976), but we shall not go into such topics here.

3.4 Melting at $P_{H_2O} = P_{total}$

3.4.1 *Preliminary remarks*

As in the first section, we shall consider equilibria in the limiting systems KFASH and KMASH. These constrain the reactions possible in KFMASH, which will follow. Finally we shall introduce an albitic component, and extend the discussion to NKFMASH. As noted by Greenwood (1975), variation in natural assemblages is often less than might be expected from the numerous reactions on grids such as those shown in this chapter. In part this is a function of the extent of solution in most of the phases involved, so that only a small range in composition may be affected by a given univariant reaction. Much of

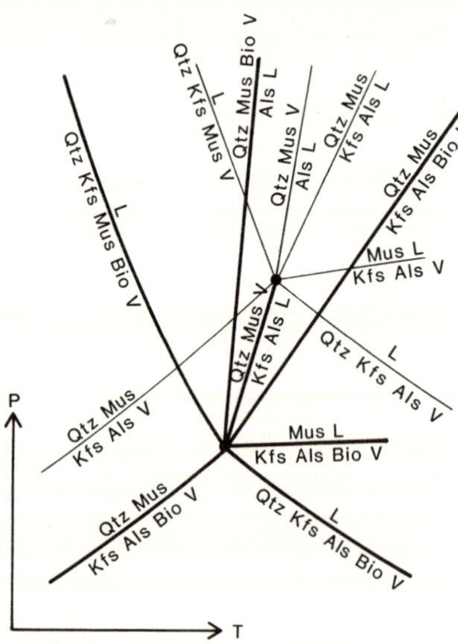

Figure 3.9 Schematic pressure–temperature grid of relations between Qtz–Kfs–Mus–Als–Bio–L–V in KFASH (or KMASH), shown in heavy lines, and Qtz–Kfs–Mus–Als–L–V in KASH, shown in light lines. After Thompson (1982, Figure 2).

the compositional range involved may be taken up by relatively high variance assemblages, and one usually has to hunt for low variance assemblages. (Figure 3.22 is a particularly good illustration of the situation.) This is in marked contrast to the case of assemblages of phases of fixed compositions. This is also why grids in the simpler systems such as KFASH or KMASH often seem more utilitarian than, say, grids in KFMASH, but for detailed analysis, the combination of the limiting systems with the full system is generally necessary.

For both KFASH and KMASH, we shall start with a model involving four degeneracies, and then remove them to attain a more realistic portrayal of the relations in the two systems. Thus the reader can readily adapt the less controversial, simpler, 'degenerate' grids according to his or her own opinions.

Thompson (1982, Figure 2) showed the non-degenerate relations between Qtz–Kfs–Mus–Als–Bio–L–V in KFASH (or KMASH) and Qtz–Kfs–Mus–Als–L–V in KASH and from this the present Figure 3.9 is taken. This should be compared with the analogous relations in Figures 3.10–3.13.

3.4.2 *KFASH*

First, let us consider quartz-bearing assemblages and chemographic relations between Kfs–Mus–Als–Bio–Cdt–Gar–Opx–L as shown in the 'AKF' diagram in Figure 3.10. (It should be remembered that orthopyroxene will be metastable with respect to quartz–fayalite in KFASH.) Immediately the question of liquid composition arises. Consideration of the low temperatures of melting of Qtz–Kfs–Cdt–V (in KMASH and KFMASH) relative to Qtz–Kfs–V (Seifert, 1976; Grant and Hoffer, 1977) and the data of Green (1976), Clemens and Wall (1981) and Vielzeuf (1983) favour a model liquid composition lying in Qtz–Kfs–Als–Bio–V, Qtz–Kfs–Cdt–Gar–V and Qtz–Kfs–Mus–Bio–V. This conforms to the choice of both Thompson (1982) and Grant (1985) but differs from that of Abbott and Clarke (1979), where the minimum melt was taken to lie in Qtz–Kfs–Bio–Gar–V, rather than Qtz–Kfs–Cdt–Gar–V. This is different, but not necessarily wrong, and is discussed further in Grant (1984*a*). However, the choice of Qtz–Kfs–Cdt–Gar–V is intermediate between the Al-poor field of Qtz–Kfs–Bio–Gar–V and the Al-rich field of Qtz–Kfs–Als–Cdt–V. Choice of this model liquid composition also favours the eutectic (or near-eutectic) melting of Qtz + Kfs + Cdt + V = L discussed previously. For the present, this is approximated by the collinearity Kfs–L–Cdt in the AKF diagram of Figure 3.10. Also in this diagram, note that L–Mus–Als and L–Bio–Opx are collinear. The former encompasses a Tschermak substitution ($Al_2Fm_{-1}\ Si_{-1}$) in muscovite (Tracy, 1978; Thompson, 1982), and the latter aluminous biotite (e.g. Grant, 1981; Thompson, 1982). These are improvements over the simple collinearities Kfs–Mus–Als and Kfs–Bio–Opx, which are often assumed in the bounding pseudo-binary systems. Finally, let us assume that cordierite is anhydrous.

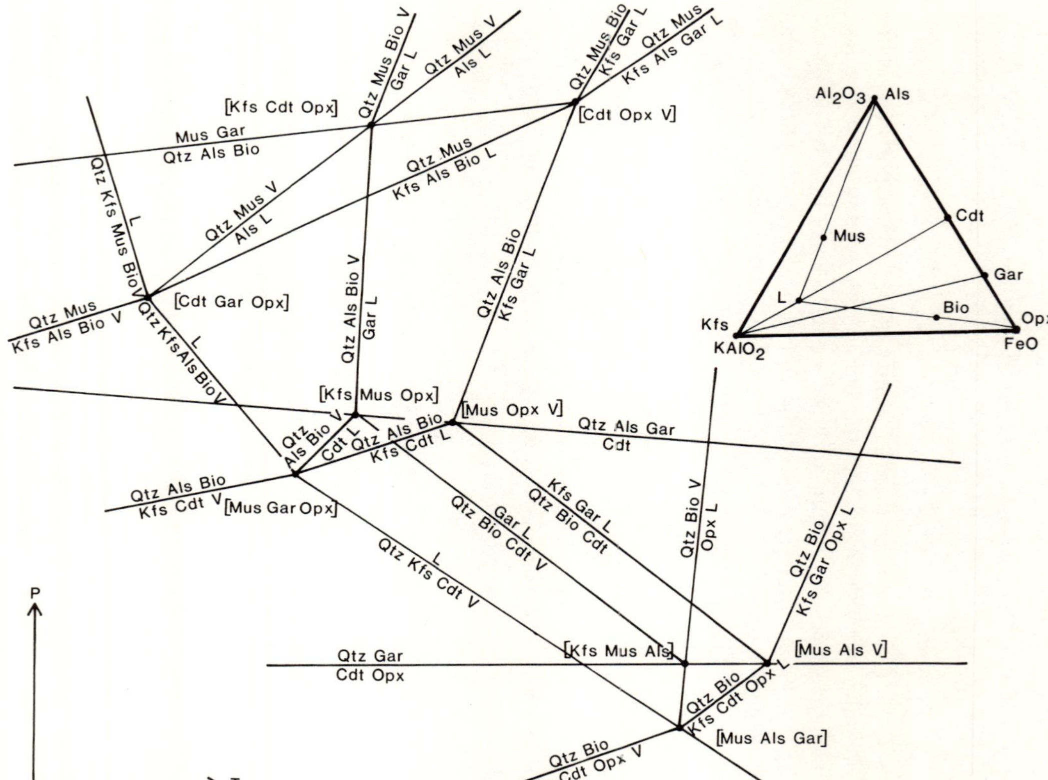

Figure 3.10 Simplified schematic pressure–temperature grid of invariant and univariant equilibria for quartz-bearing assemblages among Qtz–Kfs–Mus–Als–Bio–Cdt–Gar–Opx–L–V in KFASH. In the chemographic projection on to $KAlO_2$–Al_2O_3–FeO, L–Mus–Als, L–Bio–Opx and Kfs–L–Cdt are collinear. Cordierite is assumed to be anhydrous.

With these assumptions, there are four reactions involving micas that terminate in melting: Qtz + Bio = Kfs + Cdt + Opx + V, Qtz + Als + Bio = Kfs + Cdt + V, Qtz + Mus = Kfs + Als + Bio + V and the H_2O-conserving Mus + Gar = Qtz + Als + Bio. Equilibria at temperatures above Qtz + Mus = Kfs + Als + V have been considered recently by Holdaway and Lee (1977), Thompson (1982) and Grant (1983*a*, 1985), and relations on and below that isograd by Thompson (1982). The equilibria in the vicinity of Qtz + Bio = Kfs + Cdt + Opx + V where melting ensues require comment. If Qtz + Gar = Cdt + Opx intersects Qtz + Bio = Kfs + Cdt + Opx + V in the sub-solidus, the relations will be similar to those shown in Thompson (1982, Figure 4). However, this reaction, Qtz + Gar = Cdt + Opx, probably lies in the vicinity of 2.5 kbar with very shallow slope, as discussed in the previous section for the hydrous equivalent. Thus it seems more probable that this intersection will occur metastably and above the solidus. Moreover, with this alternative a topology similar to that of Thompson will be attained at $P_{H_2O} < P_{total}$, as shown in the next section after Grant (1985).

Because Qtz–Kfs–Gar–V is metastable in this model (at $P_{H_2O} = P_{total}$), the three vapour-present reactions at the beginning of melting are not sensitive to whether the liquid lies in Qtz–Kfs–Cdt–Gar–V or Qtz–Kfs–Bio–Gar–V. This grid of Figure 3.10 may be compared with that of Thompson (1982, Figure 4). At low pressure, the two grids differ because of the placement of Qtz + Gar = Cdt + Opx, the equilibria in the vicinity of the termination of Qtz + Als + Bio = Kfs + Cdt + V are the same (and in Grant, 1985, Figure 6) and at higher pressure, Thompson's reactions 35, 36 and 46 are replaced by Qtz + Mus + V = Als + L simply due to the degeneracy invoked in collinearity of L–Mus–Als in Figure 3.10, but the placement of [Cdt, Gar, Opx] should be as shown here according to the experimental data, both in Thompson (1982, Figure 5B) and in the present Figure 3.3.

Now let us proceed to remove the degeneracies. The assumption of hydrous cordierite does not affect the univariant equilibria shown by Thompson at pressures greater than Cdt = Qtz + Als + Gar + V, where Fe-cordierite is not stable. However, removal of this degeneracy does affect the equilibria at lower pressures, as shown in Figure 3.11, following Grant (1985, Figure 4). Removal of the assumption of collinear L–Mus–Als is not in contention (although the possibility of 'L' lying to the potassium feldspar side of the extended join muscovite–aluminosilicate is not ruled out), largely because the extent of Tschermak's substitution in muscovite near the ultimate breakdown of quartz–muscovite is likely to be small (Velde, 1965). However, removal of the analogous degeneracy L–Bio–Opx brings up a final question: the chemographic relations between these three phases. As noted above, the biotites encountered in this environment are normally significantly aluminous (e.g. Grant, 1983*b*, and 1981 and references therein; Thompson, 1982; Holdaway and Lee, 1977). As discussed by Grant (1985) and implied by the schematic relations in Thompson (1982, Figure 5B), the question is to which side of the

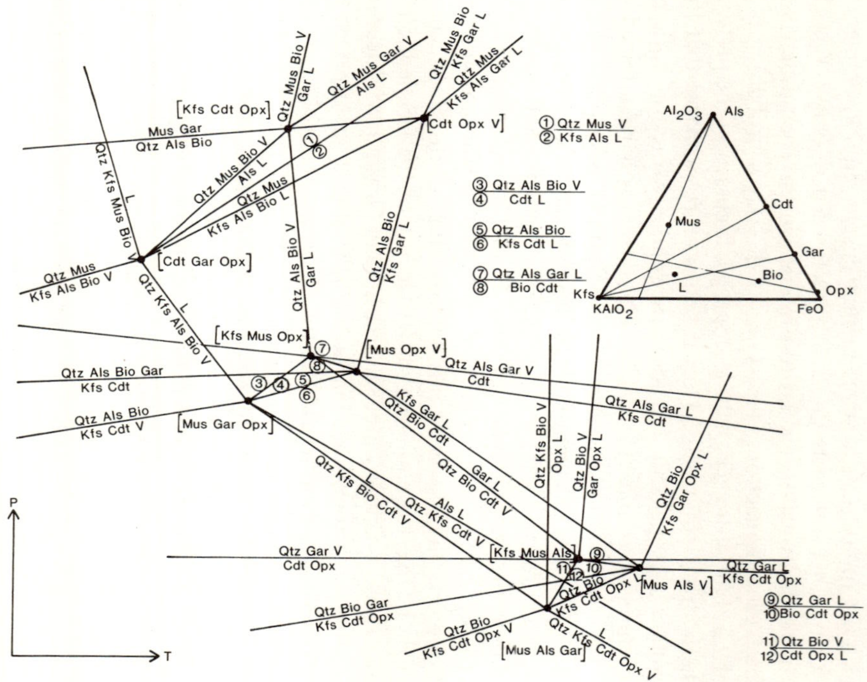

Figure 3.11 Schematic pressure–temperature grid of invariant and univariant equilibria for quartz-bearing assemblages among Qtz–Kfs–Mus–Als–Bio–Cdt–Gar–Opx–L–V in KFASH. In the chemographic projection onto $KAlO_2$–Al_2O_3–FeO, the liquid composition is chosen to lie in Qtz–Kfs–Mus–Bio–V, Qtz–Kfs–Als–Bio–V, Qtz–Kfs–Bio–Cdt–V and Qtz–Kfs–Cdt–Opx–V. The liquid is also chosen to lie on the iron-rich side of the extended join aluminosilicate-muscovite, and on the potassic side of the extended join orthopyroxene-biotite. Cordierite is assumed to be hydrous.

extended join biotite–orthopyroxene the liquid composition lies. Given the probability that low-temperature melts will lie relatively close to the potassium feldspar apex, this is the present choice, as it is in Grant (1985, Figure 5). Thompson (1982) chose the other case, with consequent changes in three reactions involving Bio–Opx–L.

With this choice of liquid composition, and removal of the other degeneracies, the schematic equilibria in the system KFASH are shown in Figure 3.11, in which muscovite-absent equilibria involving Qtz–Kfs–Als–Bio–Cdt–Gar–Opx–L–V are after Grant (1985, Figure 4) and muscovite-bearing equilibria involving Qtz–Kfs–Mus–Als–Bio–Gar–L–V conform to Thompson (1982, Figure 4).

3.4.3 *KMASH*

Earlier, it was concluded that the subsolidus relations in KMASH differ from those in KFASH, largely because of the restriction of quartz–pyrope to pressures greater than about 15 kbar. In this, the present analysis differs from that of Thompson (1982). It was also indicated that the equilibria in the vicinity of the termination of Qtz + Als + Bio = Kfs + Cdt + V in melting are questionable, but that probably this reaction has a small stable existence of less than 20 °C prior to melting, as suggested by Seifert (1976) and implied by the near coincidence of reactions about the invariant point in Thompson (1982, Figure 5A) and in the present Figure 3.3. With these caveats, the schematic phase relations in the system KMASH for the case of collinear L–Mus–Als, L–Bio–Opx, and Kfs–L–Cdt and anhydrous cordierite are as shown in Figure 3.12. If these assumptions are removed, as was done with KFASH, the equilibria are only slightly more complex, as shown schematically in Figure 3.13. One further comment is required here. As may be noted in Figure 3.3, for 'Bio' = phlogopite, Qtz + Bio + V = Opx + L intersects Qtz + Mus + V = Als + L in the vicinity of 750 °C and 8 kbar, that is, close to the high-pressure limit of anhydrous magnesian cordierite. Thus, for phlogopite, the invariant point equivalent to [Kfs, Mus, Gar] in Figure 3.12 may be metastable on the low-temperature side of Qtz + Mus + V = Als + L. In Figure 3.12, this invariant point is shown stable on the high-temperature side, taking into consideration the aluminous nature of biotite in this model. An analogous argument may be used for [Kfs, Mus, Gar] in Figure 3.13, but at higher pressure due to the involvement of hydrous cordierite. (One should not forget the back-bend in Qtz + Bio + V = Opx + L shown by Bohlen *et al.* (1983*b*, Figure 3): relations are far from certain here.)

3.4.4 *KFMASH*

The reactions in the system KFMASH involving Qtz–Kfs–Mus–Als–Bio–Cdt–Gar–Opx–L–V within the pressure-temperature region of interest and

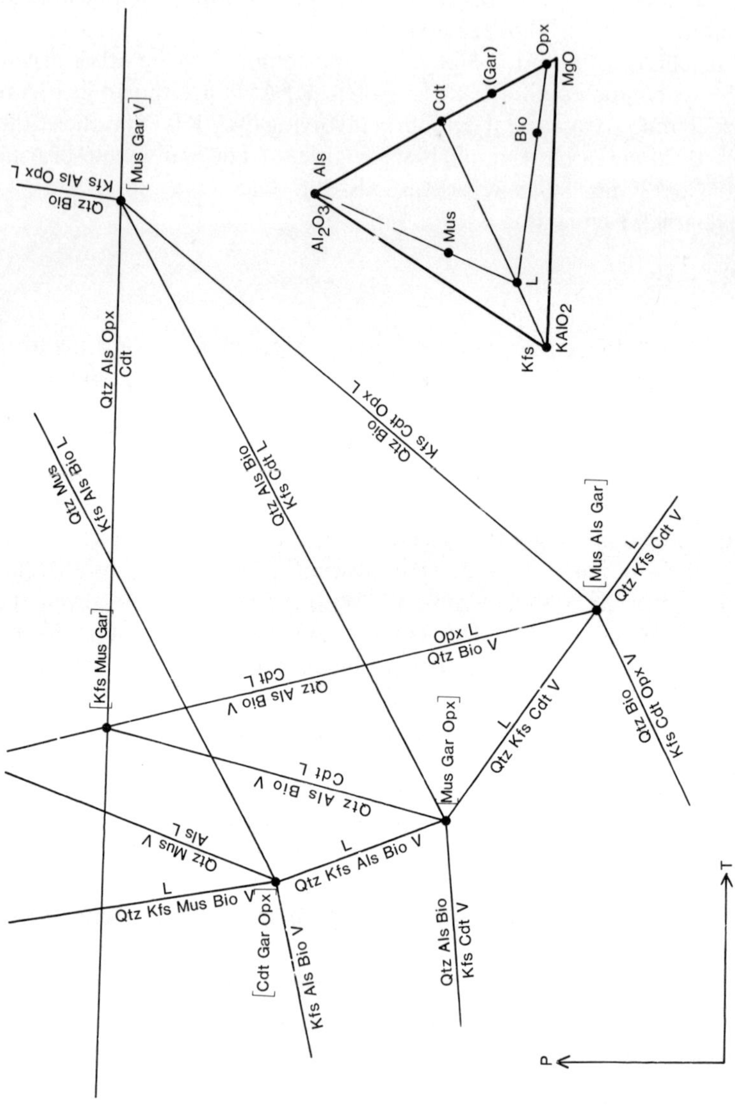

Figure 3.12 Simplified schematic pressure–temperature grid of invariant and univariant equilibria for quartz-bearing assemblages among Qtz–Kfs–Mus–Als–Bio–Cdt–Opx–L–V in KMASH. (Garnet is not stable in this system in this pressure range.) In the chemographic projection on to $KAlO_2$–Al_2O_3–MgO, L–Mus–Als, L–Bio–Opx and Kfs–L–Cdt are collinear. Cordierite is assumed to be anhydrous.

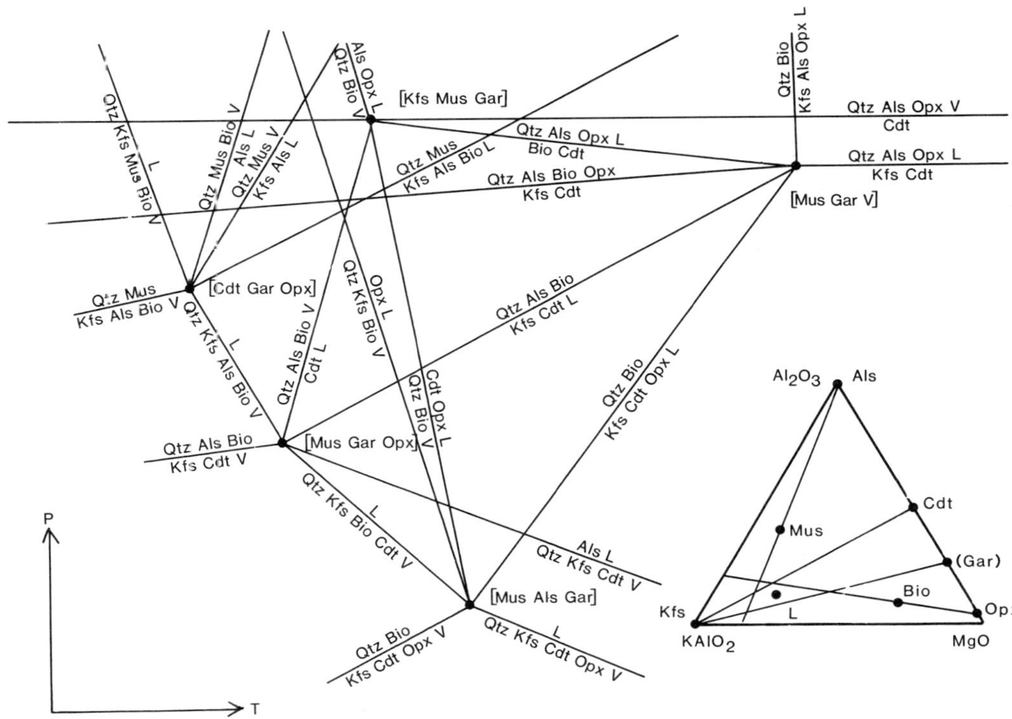

Figure 3.13 Schematic pressure–temperature grid of invariant and univariant equilibria for quartz-bearing assemblages among Qtz–Kfs–Mus–Als–Bio–Cdt–Opx–L–V in KMASH. (Garnet is not stable in this system in this pressure range.) The compositional assumptions are analogous to those for KFASH in Figure 3.11, and are shown in projection onto $KAlO_2$–Al_2O_3–MgO.

at $P_{H_2O} = P_{total}$ may be discussed in terms of the chemographic relations in the inset AFM projection from potassium feldspar in Figure 3.14. The relative chemistries of Bio–Cdt–Gar–Opx do not require much comment: they are as in Grant (1981). 'Mus' is depicted with an Mg/Fe ratio slightly greater than in coexisting 'Bio' (Tracy, 1978, p. 158), so that instead of the dehydration reaction Qtz + Mus = Kfs + Als + Bio + V of Figures 3.10–3.13, we have Qtz + Mus = Kfs + Als + Bio + Cdt + V. The liquid composition chosen lies in Qtz–Kfs–Als–Bio–Gar–V, Qtz–Kfs–Bio–Cdt–Gar–V, Qtz–Kfs–Cdt–Gar–Opx–V and Qtz–Kfs–Mus–Bio–Gar–V. This is based on the data discussed above in the context of KFASH, and implies that Fe/Mg in the melt is less than that in garnet. This is in accord with Clemens and Wall (1981) and (compare Thompson, 1982, p. 1587) *not* inconsistent with Green (1976, Table 1) who calculated '$K_{D\,\text{gar-liq}}^{\text{Fe}}$' in the range 1–4. Vielzeuf (1983, p. 304) reports garnet with Fe/Mg less than in adjacent glass in a xenolith, but it is surely questionable whether this situation represents an approach to equilibrium.

With these assumptions, the schematic grid is as shown in Figure 3.14, including hydrous cordierite. The only mutually stable invariant points are [Mus, Als], [Mus, Opx], [Gar, Opx], [Kfs, Mus] and [Mus, V], which compares well with the simpler version of Grant (1973, Figure 7), and the relations about [Mus, Opx] and [Mus, Als] are the same as in Grant (1985, Figure 7). To tie this into pressure–temperature space calls for considerable temerity, but nevertheless this is done in Figure 3.15, against a background of Figure 3.3. At most, two of the invariant points are stable, [Kfs, Mus] and [Mus, V]. The other three are metastable due to reactions in the systems KFASH and KMASH, so remarkably little of the grid is stable at $P_{H_2O} = P_{total}$. The few reactions stable at $P_{H_2O} = P_{total}$ are as follows. The shallow 'boiling' reaction, Qtz + Mus + L = Kfs + Als + Bio + V, runs from [Cdt, Gar, Opx] in KFASH (Figure 3.11) to the analogous point in KMASH (Figure 3.13), although, as was noted previously, the relations here are very 'busy' over a small pressure–temperature range, and a caution is necessary. Qtz + Kfs + Als + Bio + V = Cdt + L originates at [Mus, Gar, Opx] of Figure 3.11 and terminates at the analogous point in KMASH (Figure 3.13). That is, [Gar, Opx] and [Mus, Opx] of Figure 3.14 are occluded by reactions in the limiting subsystems. Qtz + Als + Bio + V = Cdt + Gar + L and Qtz + Bio + Gar + V = Cdt + Opx + L arise at [Kfs, Mus, Opx] and [Kfs, Mus, Als] respectively in KFASH (Figure 3.11), and terminate at stable [Kfs, Mus] on Cdt + Gar = Qtz + Als + Opx + V, which is very tentatively located as discussed in the section on experimental data. The vapour-absent reactions Qtz + Als + Bio = Kfs + Cdt + Gar + L and Qtz + Bio + Gar = Kfs + Cdt + Opx + L arise at [Mus, Opx, V] and [Mus, Als, V] respectively in Figure 3.11, and terminate at stable [Mus, V], presumably close to Cdt + Gar = Qtz + Als + Opx + V, as shown in Figure 3.14. This is used as an approximation in Figure 3.15, in view of the lack of

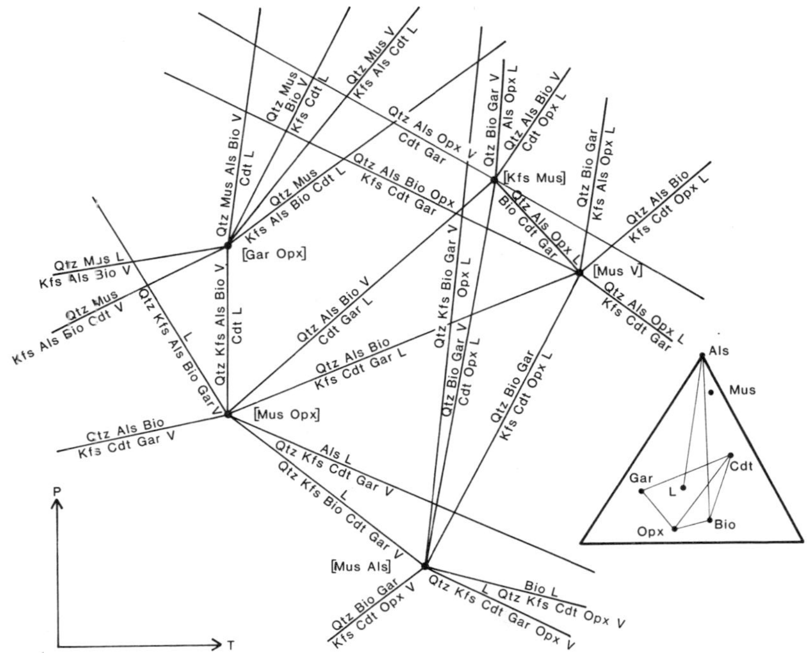

Figure 3.14 Schematic pressure–temperature grid of invariant and univariant equilibria for quartz-bearing assemblages among Qtz–Kfs–Mus–Als–Bio–Cdt–Gar–Opx–L–V in KFMASH. Chemographic relations on which this is based are shown in the AFM projection from potassium feldspar. Muscovite is assumed to be slightly more magnesian than coexisting biotite (Tracy, 1978), and cordierite is assumed to be hydrous.

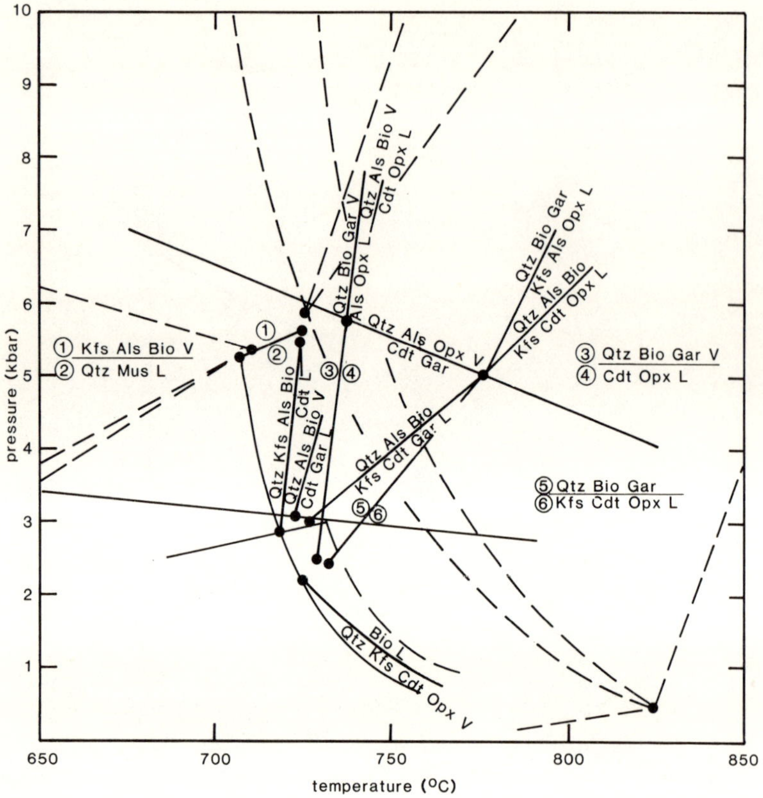

Figure 3.15 Grid of stable invariant and univariant equilibria in KFMASH, tentatively positioned in pressure–temperature space, and shown in heavy lines on a base from Figure 3.3.

constraint on even Cdt + Gar = Qtz + Als + Opx + V. Finally, Qtz + Kfs + Cdt + Opx + V = Bio + L starts at [Mus, Als, Gar] in KFASH and extends to the analogous point in KMASH, near the termination of Qtz + Bio = Kfs + Opx + V in KMASH.

3.4.5 *NKFMASH*

The addition of an albitic component to the system introduces the complexities of the alkali feldspars and the possible production of a more realistic granitic melt (Grant, 1973). The system NKASH has been treated in considerable detail by Thompson and Algor (1977), to which the reader is referred.

If it is assumed that, to a first approximation, this albitic component is confined to feldspar and liquid, then albitic plagioclase would be the only new phase introduced. Reactions similar to those discussed previously would occur

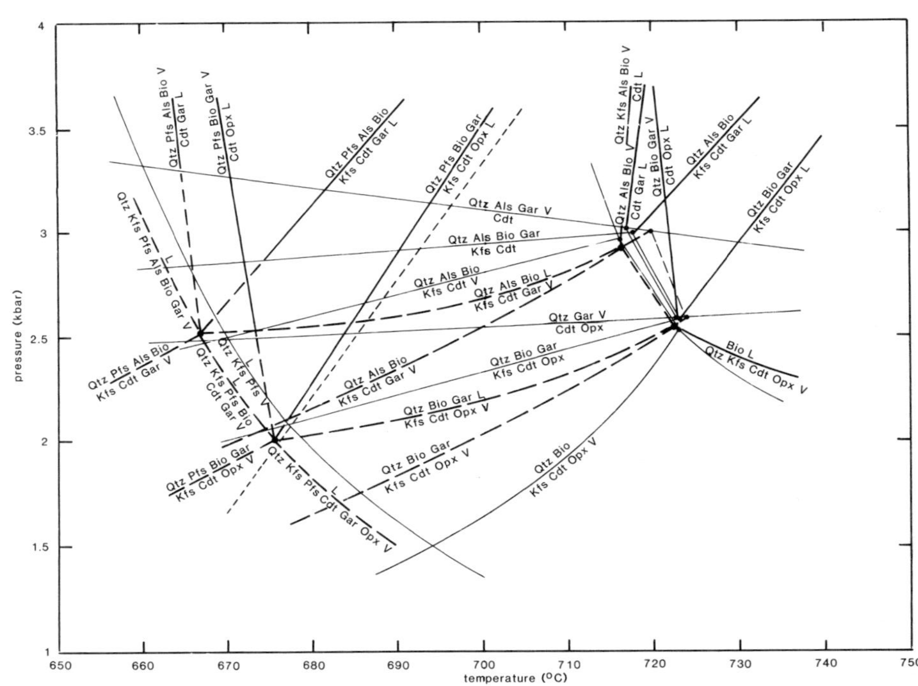

Figure 3.16 Pressure–temperature grid of invariant and univariant equilibria for quartz-bearing assemblages among Qtz–Kfs–Als–Bio–Cdt–Gar–Opx–L–V at $P_{H_2O} = P_{total}$. Light lines refer to the limiting system KFASH and to the haplogranite solidus 'Qtz + Kfs + Pfs + V = L'. The latter is after Tuttle and Bowen (1958) and Luth *et al.* (1964). The alkali feldspar critical curve is shown as a short-dashed line through the central part of the grid (Morse, 1970, p. 243). Heavy lines refer to the system KFMASH (reactions involving seven phases, generally on the high-temperature side of the grid) and to the system NKFMASH (reactions involving eight phases, generally on the low-temperature side of the grid). Dashed lines refer to metastable equilibria. Reproduced from Grant (1985, Figure 8). Note that on the far right of this grid, [Mus, Opx, L] and [Mus, Als, L] in KFASH are located 10°C lower than in Figure 3.3. On the left-hand side, the two metastable invariant points below the haplogranite solidus are after Hoffer and Grant (1980), which also constrains Qtz + Pfs + Als + Bio + V = Cdt + Gar + L and Qtz + Pfs + Bio + Gar + V = Cdt + Opx + L.

without change in variance, but generally at lower temperatures with albitic plagioclase on the low-temperature side. Such reactions are linked to those in KFMASH generally by reactions with shallow slopes, with liquid on the high-pressure side. These stem from invariant points involving two feldspars (below the alkali feldspar solvus) and terminate in KFMASH in tangency with analogous reactions in that system. This is illustrated for conditions above the breakdown of quartz–muscovite in Figure 3.16, from Grant (1983*a*, 1985) and relates the results of Hoffer and Grant (1980) on Qtz + Pfs + Als + Bio + V = Cdt + Gar + L and Qtz + Pfs + Bio + Gar + V = Cdt + Opx + L to the KFMASH system just discussed.

This may be extended to include an anorthitic component, again yielding a closer approximation to reality. In this case, generally no new phases are involved, and if so, the variance is increased by one, extending the stability fields of assemblages to higher temperatures. This is thoroughly discussed for the system CNKASH by Thompson and Tracy (1979) and, apropos granitic compositions, by Winkler *et al.* (1975).

While the addition of albitic and anorthitic components certainly makes the model more realistic in terms of temperatures, pressures and melt compositions, the system KFMASH remains most useful as far as understanding partial melting processes in pelitic rocks is concerned.

3.5 Melting at $P_{H_2O} < P_{total}$

Let us now turn to comparable equilibria in circumstances where there is a still a vapour present, but a vapour diluted by species which are not very soluble in a silicate liquid. Fluid inclusion studies suggest that CO_2 is the most common diluting species, but CO, CH_4 or N_2 are also reported (Touret, 1971; Hollister and Burruss, 1976; Kreulen, 1980; Kreulen and Schuiling, 1982; Yardley *et al.*, 1983).

If the diluting species occurs only in the liquid and vapour phases, dehydration reactions gain a degree of freedom, and generally are displaced to lower temperatures (e.g. Kerrick, 1972; Wyllie, 1977*b*). Vapour-present melting reactions also gain a degree of freedom and, insofar as the diluting species is dominantly in the vapour, are displaced to higher temperatures. (Reactions such as Bio + L = Kfs + Opx + V in Figure 3.7, with shallow slopes, liquid on the high-pressure side and vapour on the low-pressure side, would be displaced to higher pressures.) If the diluting species has negligible solubility in the melts, the vapour-absent reactions will show corresponding negligible change. If this solubility is not negligible, then even the vapour-absent melting reactions would gain a degree of freedom, and generally be displaced to lower temperatures (Wendlandt, 1981; Grant, 1985).

For the present, let us assume negligible solubility of the diluting species in the melts. The evolution of equilibria in KFMASH at low pressures under such circumstances has been discussed by Grant (1982, 1985), and will be but briefly considered here. As has been well shown by Holdaway and Lee (1977), Qtz + Als + Bio = Kfs + Cdt + Gar + V in KFMASH and Qtz + Bio + Cdt = Kfs + Gar + V in KFASH are metastable in the vicinity of 3 kbar at $P_{H_2O} = P_{total}$. In the model presented in the previous section this is also the case for Qtz + Bio + Gar = Kfs + Cdt + Opx + V, and Qtz + Bio + Cdt = Kfs + Gar + V is wholly metastable. Consider Figure 3.16. As P_{H_2O} decreases, Qtz + Bio + Cdt = Kfs + Gar + V (dashed line on the right of Figure 3.16) migrates to lower temperatures between the vapour- and melt-absent reactions Qtz + Als + Bio + Gar = Kfs + Cdt and Qtz + Bio + Gar = Kfs + Cdt + Opx. The two univariant dehydration reactions, Qtz + Als + Bio = Kfs + Cdt + Gar + V and Qtz + Bio + Gar = Kfs + Cdt + Opx + V become stable and expand their stability ranges between their low-temperature origins in KFASH and their high temperature terminations

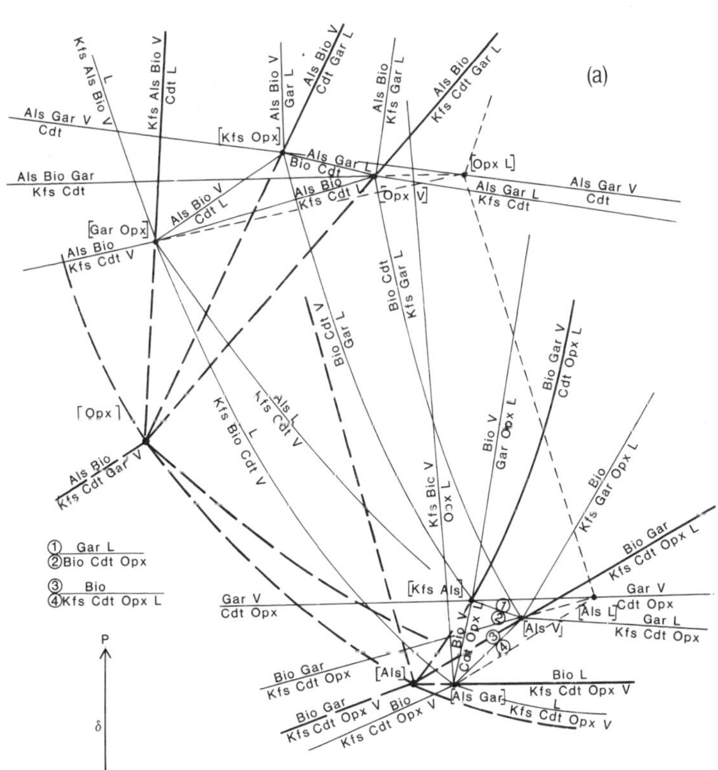

[Figure 3.17

on vapour-absent melting reactions in KFMASH (Qtz + Als + Bio = Kfs + Cdt + Gar + L and Qtz + Bio + Gar = Kfs + Cdt + Opx + L in Figure 3.16). In the present model, this results in the lowest-temperature melts evolving from KFASH into KFMASH, as P_{H_2O} decreases and Qtz–Kfs–Gar–V becomes stable, because of the assumption that Fe/Mg in garnet is greater than that in the liquids. This is illustrated in the schematic grids of Figure 3.17*a, b,* and *c*, at successively lower P_{H_2O}, from Grant (1985, Figure 9).

As also shown in that paper, it is not difficult to construct isobaric T–$x_{H_2O}^{vapour}$ grids with the aid of these diagrams, and such portrayals of equilibria are useful in the context of partial melting, as they are in normal metamorphism.

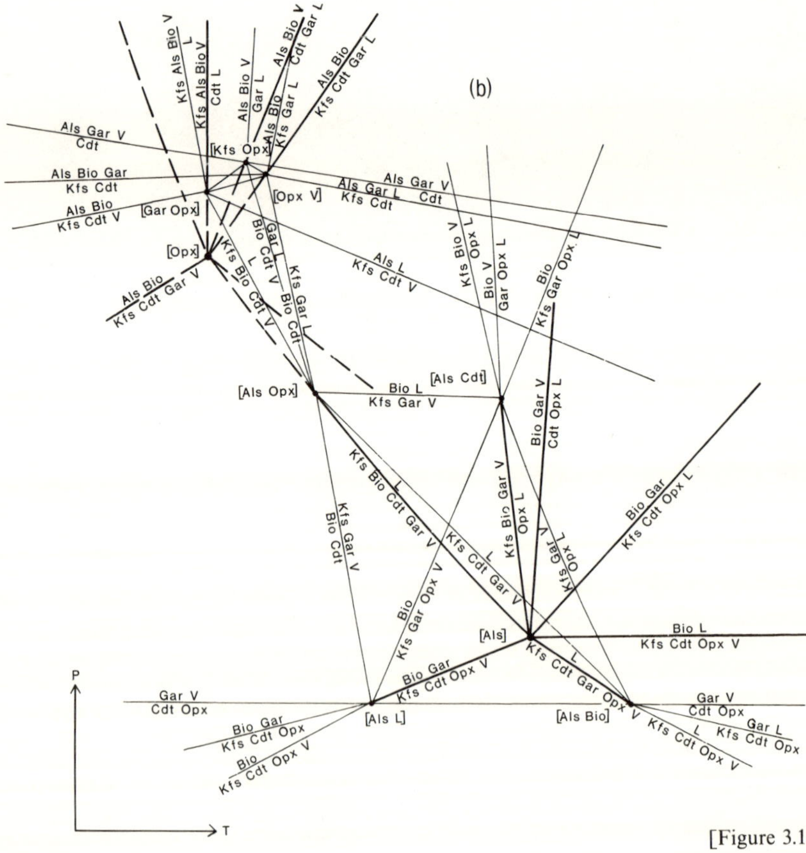

[Figure 3.17

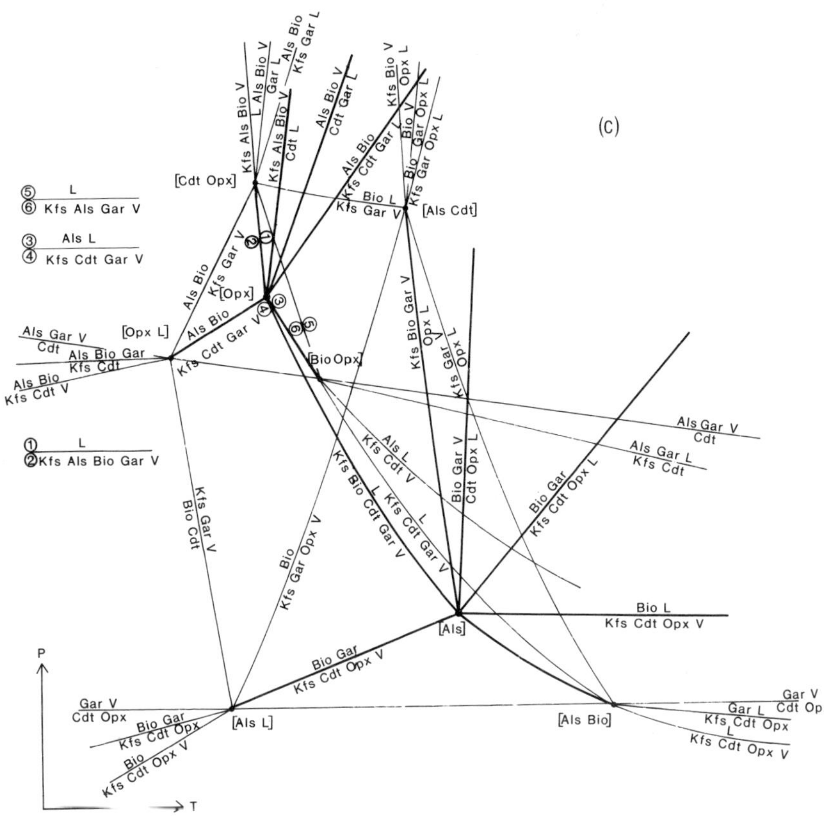

Figure 3.17 Schematic pressure-temperature grid of invariant and univariant equilibria for quartz-bearing assemblages among Qtz–Kfs–Als–Bio–Cdt–Gar–Opx–L–V in the systems KFASH (light lines) and KFMASH (heavy lines). Dashed lines represent selected metastable equilibria. *a*: $P_{H_2O} = P_{total}$, *b*: $P_{H_2O} < P_{total}$, *c*: $P_{H_2O} \ll P_{total}$. From Grant (1985, Figure 9A, B, and C). Part *a* shows the detailed relationships on the high-temperature side of Figure 3.16. The relationships in KFASH at low pressure in *b* are essentially the same as those in Thompson (1982, Figure 4), except for the difference in the chemographic relations between liquid, biotite and orthopyroxene assumed (see text for discussion).

Equilibria involving a binary vapour have been discussed in the classic papers of Greenwood (1962 and 1967) for normal metamorphism, and in the following paragraphs, this will be extended to encompass reactions involving melt. Following Greenwood, for a reaction written as $a\mathrm{A} = b\mathrm{B} + \nu_\mathrm{w}\mathrm{w} + \nu_\mathrm{d}\mathrm{d}$, in which a moles of A react to form b moles of B and ν_w and ν_d moles of components w (H_2O) and d (diluting species in an ideal vapour, with $\nu_\mathrm{w} + \nu_\mathrm{d}$ greater than or equal to zero, and $|\nu_\mathrm{w}| + |\nu_\mathrm{d}| = 1$, the slope of an equilibrium at constant pressure is given by:

$$\left(\frac{\partial T}{\partial x_\mathrm{d}}\right)_P = \frac{RT}{\Delta S_\mathrm{r}}\left(\frac{\nu_\mathrm{d}}{x_\mathrm{d}} - \frac{\nu_\mathrm{w}}{x_\mathrm{w}}\right)$$

and at constant temperature by:

$$\left(\frac{\partial P}{\partial x_\mathrm{d}}\right)_T = -\frac{RT}{\Delta V_\mathrm{r}}\left(\frac{\nu_\mathrm{d}}{x_\mathrm{d}} - \frac{\nu_\mathrm{w}}{x_\mathrm{w}}\right)$$

where x_w and x_d are the mole fractions of the two species in the ambient vapour, ΔV_r and ΔS_r are the volume and entropy changes for the reaction, and R is the gas constant. First, assume that the diluting species only occurs in vapour. Using the equilibria of Figure 3.7 for an example, the melting reactions have $\nu_\mathrm{d} = 0$ and $\nu_\mathrm{w} = 1$. For L = Qtz + Kfs + Bio + V, L = Qtz + Kfs + Opx + V, and Opx + L = Qtz + Bio + V, ΔS is negative, but for Bio + L = Kfs + Opx + V, which is essentially a dehydration reaction, ΔS is small and probably positive. Thus for the first three reactions, $\mathrm{d}T/\mathrm{d}x_\mathrm{d}$ will be positive and for the last, $\mathrm{d}T/\mathrm{d}x_\mathrm{d}$ will be negative (cf. Powell, 1983*b*, Figure 4, reaction mu + 1 = or + vapour). Thus a schematic isobaric T–x_{H_2O} diagram for equilibria at pressure greater than that at the invariant point will be of the form shown in Figure 3.18.

Although T–x_{H_2O} diagrams are familiar, it may well be that isothermal P–x_{H_2O} diagrams will be of considerable value in considering reactions encountered during relatively rapid uplift. Using the same example, for L = Qtz + Kfs + Bio + V, L = Qtz + Kfs + Opx + V, Opx + L = Qtz + Bio + V and Bio + L = Kfs + Opx + V, ΔV is positive. Thus all four reactions will have positive slopes as shown on the P–x_{H_2O} diagram of Figure 3.19, which applies to a temperature greater than that at the invariant point in Figure 3.7. Note that in Figure 3.18, the dehydration reaction Qtz + Bio = Kfs + Opx + V has a negative slope, whereas in Figure 3.19, the slope is positive.

Let us remove the restriction of negligible solubility of the diluting species in the liquids (e.g. Holloway, 1976) but retain the assumption of negligible solubility in the solid phases (e.g. no carbonates present, for the case of d being CO_2). In this case, the mole fraction of the diluting species in the vapour involved in all the liquid-present reactions will be non-zero and, of course, small if the solubility in the liquid is small. In this likely case, in both T–x and P–x space there would be reversals in slope close to the H_2O ordinate, where

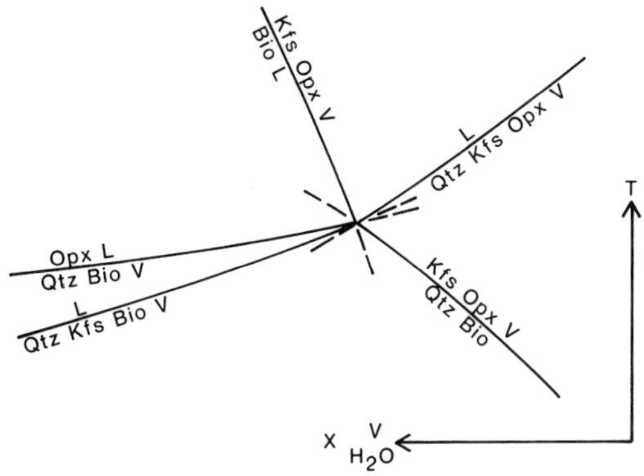

Figure 3.18 Schematic isobaric T–x_{H_2O} diagram for a section through the equilibria of Figure 3.7 at a pressure greater than at the invariant point, assuming a binary vapour phase.

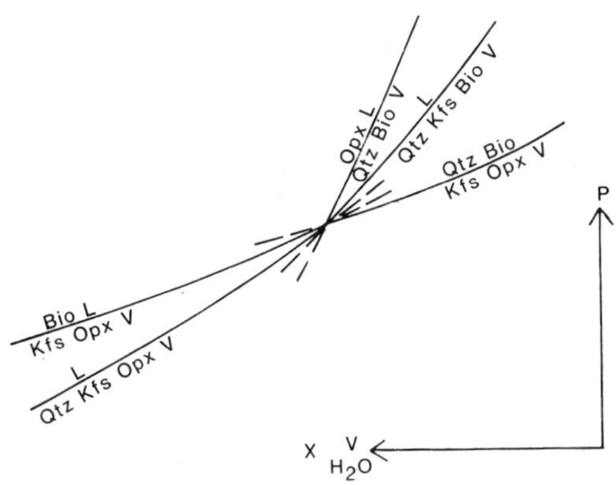

Figure 3.19 Schematic isothermal P–x_{H_2O} diagram for a section through the equilibria of Figure 3.7 at a temperature greater than at the invariant point, assuming a binary vapour phase.

the composition of vapour in the reaction equals that of the ambient vapour, maxima for those reactions with negative slopes in Figure 3.18 and 3.19, and minima for those with positive slopes.

Finally, if hydrates are present, as in this example, and if a liquid may evolve to low water contents, additional maxima or minima will appear where the vapour shifts from one side of the reaction to the other (Holloway, 1973).

As an example of the use of isobaric T–x_{H_2O} relations, let us consider briefly the present model for pelitic rocks of the aureole of the Laramie complex, Wyoming (Grant, 1985). In this aureole, a sequence of seven metamorphic facies can be interpreted from petrography and microprobe analyses. These are bounded by the several subsolidus reactions shown in Figure 3.20. In this area, although Qtz–Kfs–Bio–Cdt–Gar is common and Qtz–Als–Bio is found, no convincing coexistence of Als–Bio–Gar has been discovered. Further quartz-spinel and Qtz–Als–Bio are each in contact and exist in the same rock. The reaction Kfs + Als + Gar = Qtz + Bio + Spl and the breakdown of Qtz–Als–Bio occur before Kfs–Cdt–Gar becomes stable, and Kfs–Cdt–Opx appears in quartz-bearing assemblages while Kfs–Bio–Gar is still stable in quartz-absent assemblages. Thus the subsolidus relations in KFASH must be modified by extension of the quartz–spinel stability field to lower temperatures, presumably due to the presence of 5–10 mole % magnetite in the spinels in question. This is incorporated into the isobaric diagram of Figure 3.20, which is taken at a pressure between about 3–3.5 kbar, using the tightest constraints available from the cordierite compositions found.

It may be noted that this figure involves quartz-absent equilibria, in contrast to most of the equilibria presented here. The removal of the restriction to quartz-bearing assemblages is essential for many cases of partial melting of pelitic compositions, facilitating access to assemblages with corundum and spinel (as here), sapphirine or kornerupine.

The sequence of facies, of course, represents the results of the metamorphism and one has to interpret the evolution which led to these results. Several observations suggest that partial melting occurred in the aureole, particularly segregation of leucosomes rich in potassium feldspar and complementary silica-poor restites, and phase relations which suggest a succession of peritectic reactions on cooling. For clarity in Figure 3.20, only the reactions of interest are shown above the solidus. One possible path through T–x_{H_2O} space with time is shown, for an assemblage that would evolve to facies '6', in order to emphasize that what we now see is the result of a history of metamorphism, melting, segregation and re-crystallization to subsolidus conditions.

Powell (1983*a*,*b*) has used this approach effectively in considering internal and external buffering, of which more later. In the present context, it seems worth while to say that T–a_{H_2O} (or P–a_{H_2O}) relations can indeed be used in discussion of vapour-present and vapour-absent equilibria. As Powell notes, the activities of volatile species may be well-defined even in vapour-absent conditions. Further, as long as the standard chemical potential used is the same for H_2O in liquid and vapour, the activities of H_2O in these phases must be the same at equilibrium. However the relationship between activity and mole fraction or partial pressure can be drastically different for the two phases (e.g. Holloway, 1976). One encounters this general problem of relating thermodynamic and analytical concentrations even in Powell's discussion. The T–a_{H_2O} diagrams used are not noticeably different from T–x_{H_2O}. In Powell (1983*a*, Figure 3), several of the reactions will not balance in the absence of a vapour phase (*e.g.* hb + q = opx ± g, pl.): this really is a T–x_{H_2O} portrayal.

Powell (1983*a*, p. 137) also suggests that a water-undersaturated melt may be effective in dehydrating the rock body in which it forms or a rock body through which it passes *en route* to lower pressures. Whether dehydration would be effective in either case depends on the availability of H_2O, by

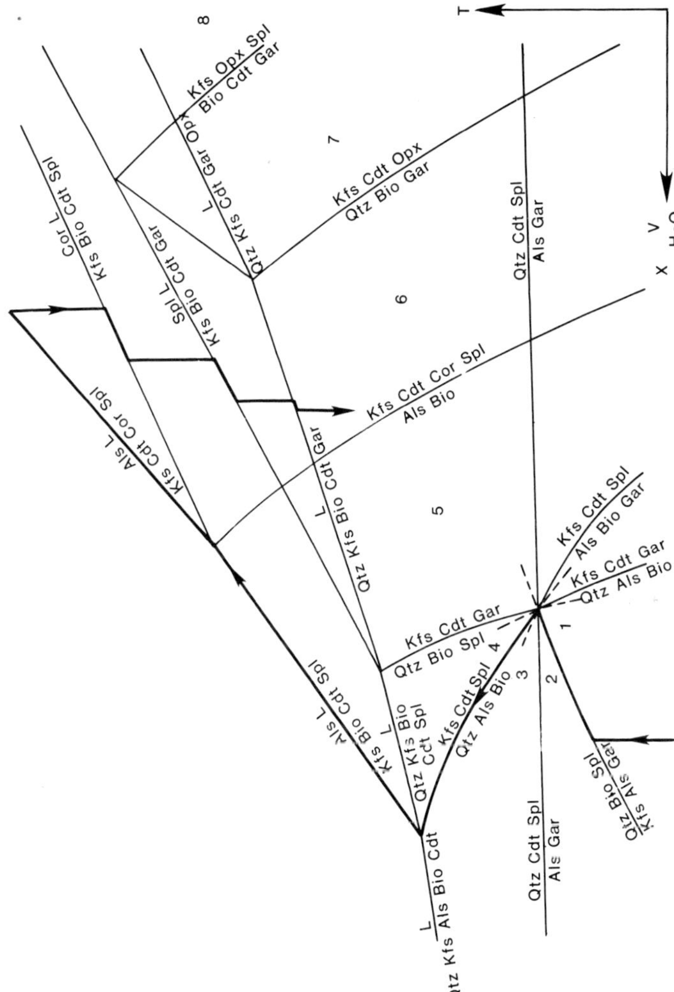

Figure 3.20 Schematic isobaric T–x_{H_2O} diagram for some of the equilibria in KFMASH pertinent to the development of the seven facies interpreted from pelitic rocks in the aureole of the Laramie anorthosite complex in Grant (1983*b*, 1985). The heavy line is but one path through time and T–x_{H_2O} space by which an assemblage now found in facies '6' may have evolved.

diffusion or fluid flow, and on the extent to which the melt is dispersed or localized (channelled in the case of a migrating melt). In '*in-situ*' dehydration, one would expect to find juxtaposed palaeosome and neosome differing chemically only in their H_2O contents. Although the possibility does indeed exist, examples of this are hard to find. More commonly, it seems that migmatites and 'ordinary' metamorphic rocks in close proximity are different in their major element compositions.

3.6 Vapour-absent melting

In the absence of aqueous vapour, melting in pelitic compositions depends upon (OH) from hydrates for an H_2O-component to yield low-temperature melts (e.g. Brown and Fyfe, 1970; Grant, 1973; Yoder, 1952; Robertson and Wyllie, 1971; Abbott and Clarke, 1979; Thompson, 1982).

This is what Thompson has called 'dehydration melting' (e.g. Thompson, 1982, p. 1586), so called because 'water' released from a hydrate is dissolved in melt without formation of a vapour phase. However, this is also true of reactions such as Qtz + Bio + V = Opx + L and Qtz + Kfs + Bio + V = L and therefore the phrase 'vapour-absent' melting will be used here.

The major characteristics of reactions such as Qtz + Bio = Kfs + Opx + L or Qtz + Mus = Kfs + Als + L are that (a) they do not require vapour, (b) they produce water-undersaturated liquids, and (c) these liquids may rise significantly through the crust before crystallizing (and may boil in reactions like Bio + L = Kfs + Opx + V). (See Thompson, 1982, and Wyllie, 1977*a*, *b*, for good discussions of this.) Moreover, compared with vapour-present equilibria, there are markedly fewer univariant reactions accessible to vapour-absent assemblages.

The vapour-absent equilibria in KFASH may be seen in Figures 3.10 and 3.11. The equilibria about [Mus, Opx, V] are the same as in Thompson (1982, Figure 6B) for the degenerate case and as in Abbott and Clarke (1979, Figure 4, 'e') for the non-degenerate case.

However in KMASH, because of the restriction of quartz-pyrope to high pressures, the invariant point involving Qtz–Kfs–Als–Bio–Cdt–Gar–L, which is equivalent to 'g' in Abbott and Clarke (1979, Figure 4) and to Thompson's (1982, Figures 6A and 7) invariant point at approximately 800–900 °C and 8–8.5 kbar, is considered to be metastable. The equilibria are, rather, as shown in Figures 3.12 and 3.13 here.

In this case and with the assumptions previously invoked, the relations between vapour-absent reactions in KFASH, KMASH and KFMASH for the 'degenerate' case are as shown schematically in Figure 3.21. The equilibria about the invariant point involving Qtz–Als–Bio–Cdt–Gar–Opx–L for the system KFMASH are the same as those shown for this point in Grant (1973, Figure 9). However the position in pressure–temperature space is more probably in the vicinity of 770 °C and 5 kbar, considering that the slope of

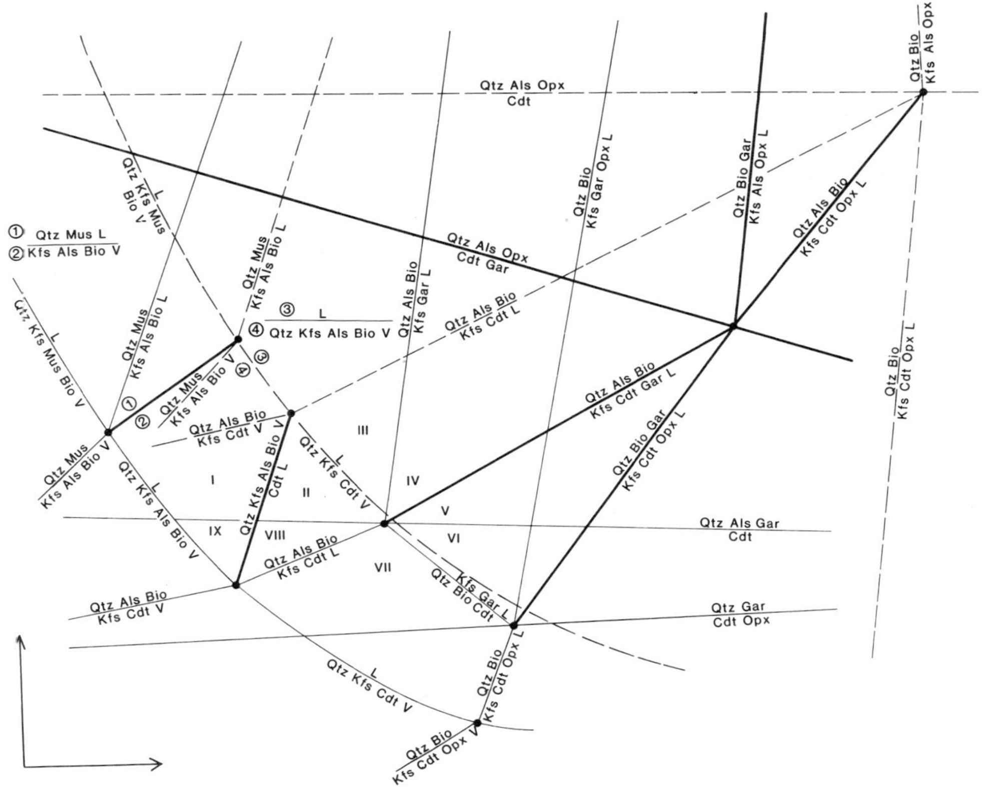

Figure 3.21 Schematic pressure–temperature grid of invariant and univariant vapour-absent equilibria among Qtz–Kfs–Als–Bio–Cdt–Gar–Opx–L in KMASH (dashed lines), KFASH (light lines) and KFMASH (heavy lines). Also shown are water-saturated solidi in KFASH and KMASH, the reactions Qtz + Mus + L = Kfs + Als + Bio + V and Qtz + Kfs + Als + Bio + V = Cdt + L in KFMASH, and the dehydration reactions Qtz + Mus = Kfs + Als + Bio + V Qtz + Als + Bio = Kfs + Cdt + V and Qtz + Bio = Kfs + Cdt + Opx + V in KFASH, Qtz + Mus = Kfs + Als + Bio + V and Qtz + Als + Bio = Kfs + Cdt + V in KMASH. Equilibria in KFASH and KMASH are from Figures 3.10 and 3.12 respectively, those in KFMASH are taken from Figure 3.14, with the assumption of anhydrous cordierite (which merely simplifies relations near the limit of the cordierite–garnet field).

Qtz + Als + Bio = Kfs + Cdt + Gar + L must be somewhat greater than the 25 bars/deg for Qtz + Als + Bio = Kfs + Cdt + Gar + V (Holdaway and Lee, 1977) and given the very approximate position of Cdt + Gar = Qtz + Als + Opx + V on Figure 3.15, which would limit the anhydrous equivalent to lower pressures and temperatures. (It is conceivable that Cdt + Gar = Qtz + Als + Opx will lie at sufficiently low pressures to alter the equilibria on the low-temperature side of Figure 3.2.1.)

Although there is no argument about the sets of divariant assemblages implied by Abbott and Clarke (1979, Figure 4) around 'e', their 'liquidus topologies' on that figure should be translated into schematic 'facies-types.' The remainder of their good discussion is not affected by this.

Thompson (1982, Figure 7) shows around his equivalent invariant point some topologies for assemblages bearing Qtz–Kfs–Als, which is certainly a useful method of portrayal and will be followed here. Although Thompson shows initial melting in KMASH (*op.cit.*, Figure 6A) in the vicinity of the point, and assumes that the H_2O-saturated solidi for KMASH and KFASH are coincident, he only shows Fe-rich liquids in his topologies. So it seems worthwhile to develop these more fully, to help give more insight into the relations between water-saturated and water-undersaturated equilibria.

This is shown schematically in Figure 3.22, for temperatures below the Qtz + Kfs + V = L solidus and for assemblages bearing Qtz–Kfs–Als. We note that, in Figure 3.21, the water-saturated solidus in KFASH lies below the invariant point involving Qtz–Kfs–Als–Bio–Cdt–Gar–L (Abbott and Clarke, 1979; Grant and Hoffer, 1977; Holdaway and Lee, 1977; Thompson, 1982). However it is moot whether the solidus Qtz + Kfs + Cdt + V = L in KMASH lies on one side or the other of this invariant point (in KFASH). Seifert (1976), Holdaway and Lee (1977), and Thompson (1982) show it slightly below, but if the point should be 10–20 °C lower than in Holdaway and Lee (1977), as suggested in Grant (1985), the solidus would lie above the invariant point, as shown on Figure 3.21. Since the question only arises within about 10 °C of the invariant point, the topologies shown are not affected by this minor uncertainty. Although, of course, we do not know the actual compositions of the liquids involved and all are shown more water-rich than the Qtz–Kfs–Als–Bio join, the topologies are at least internally consistent. The only 'other' major uncertainty is whether or not there is a thermal maximum in KFMASH involving Qtz + Kfs + Als + Cdt + V = L. This is not invoked in Figure 3.22, and the liquid field is assumed to extend continuously with increasing temperatures from KFASH to KMASH.

The sequence of facies types in Figure 3.22 may be compared with the discussion in Thompson (1982, p. 1589), noting that both there and in the present paper, Qtz + Als + Bio = Kfs + Gar + L in KFASH at least originates at a temperature *above* Qtz + Kfs + Als + Bio + V = Cdt + L if the vapour

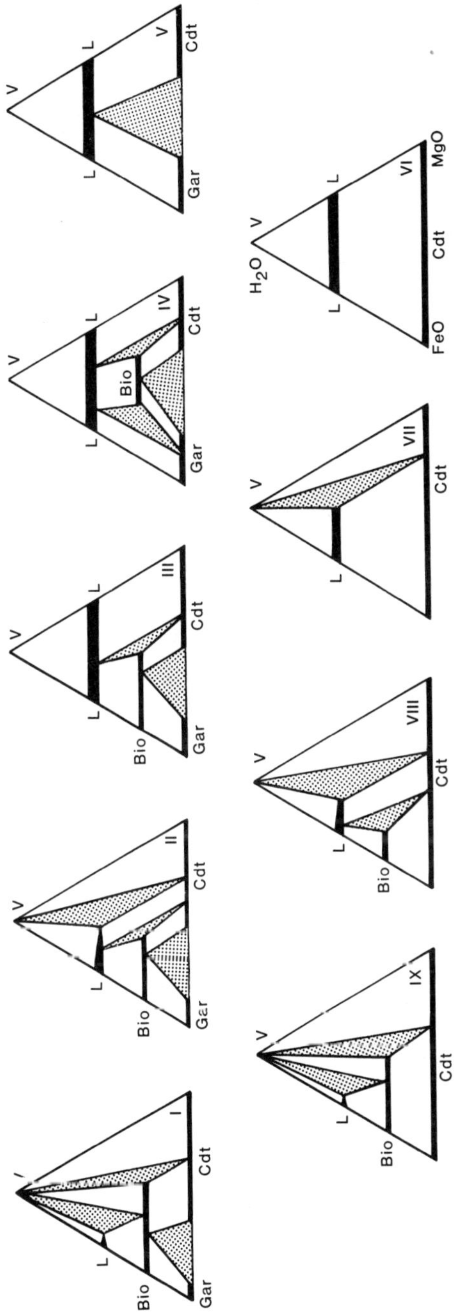

Figure 3.22 Schematic facies types for the divariant areas numbered on Figure 3.21. Only assemblages bearing quartz, potassium feldspar and aluminosilicate are shown, and projections are on to FeO–MgO–H_2O (cf. Thompson, 1982, Figure 7).

is essentially pure H_2O (although this will change with decreasing a_{H_2O} as shown in Figure 3.17).

3.7 Internal and external control of intensive variables

Buffering is reaction by a phase assemblage to inhibit variation in the value of some intensive variable. In the limit, this includes maintaining the variable at a constant value. This is critical to the evolution of hydrothermal fluids, discussed in particular by Helgeson (1967, 1968) and Helgeson *et al.* (1969). Greenwood (1975) discussed the buffering of metamorphic pore fluids by mineral equilibria, in one of his classic papers, using isobaric T–x relations (cf. Figure 3.18) for his illustrations. He pointed out that buffering of fluid compositions by a reaction is most effective where the initial fluid/solid ratio is low and where the composition of the ambient fluid is far removed from that of the fluid involved in the reaction (for in that case slight reaction has a drastic effect on the resultant fluid composition). In all cases, the composition of the fluid produced in the buffering reaction is approached, and the closer the composition of the ambient fluid is to this, the greater the extent of reaction required to change it.

Greenwood (1975) argued persuasively that for small fluid/rock ratios, the extent of isobaric univariant reactions will be small, and that major modal changes are to be expected at isobaric invariant points, i.e. where the composition of any fluid evolved equals that of the ambient fluid. (The same can be said for maxima on T–x_{H_2O} diagrams.) Greenwood emphasizes a distinction between 'univariant conditions' and 'univariant reactions'. Here, the point will be made that if the diluting species is only present in the vapour, the reaction at such an invariant point must be the same as the relevant vapour-absent reaction: none other can maintain the composition of the vapour isobarically invariant. Only to the extent that the two species in a binary fluid occur in more than that one phase, can there be a different net reaction at an isobaric invariant point, because only in this case can fluid of the required composition be *involved* in the reaction. Greenwood's conclusion that buffering of fluid compositions by mineral assemblages is likely to be widely effective seems well warranted, as shown in the excellent sequel by Rice and Ferry (1982), on which the next few paragraphs are based.

Buffering is one major end-member control on fluid composition. Another is infiltration, wherein copious amounts of fluid (usually considered to be of constant composition) from an external source pervade a rock body. In this case, the mineral assemblage must adjust to be compatible with the fluid composition. This situation corresponds to a vertical path on a T–x_{H_2O} diagram.

In the case of (internal) buffering, one would expect divariant (isobarically univariant) zones in which mineral assemblages buffer fluid compositions, separated by abrupt, discontinuous isograds corresponding to univariant (isobarically invariant) reactions, as earlier discussed by Greenwood (1975). Individual layers may develop different fluid compositions and chemical potential gradients across layer boundaries would be anticipated. In infiltration, the variance of assemblages would be expected to be greater (the number of condensed phases less) than in buffering, because at least one compositional variable is arbitrarily fixed by the fluid, and the spatial development of assemblages should reflect this, with zones of isobarically divariant assemblages separated by isograds corresponding to isobarically univariant reactions at the appropriate fluid composition: on these the major modal changes would occur (e.g. Carmichael, 1970). Evidence of local gradients in chemical potential would be lacking.

In nature, the complete spectrum of combined buffering and infiltration may be expected, and Rice and Ferry (1982) made a strong plea for the reporting of modal proportions, from which one can estimate the minimum amount of infiltration that may have occurred.

Finally, Rice and Ferry (1982) make the major point that internal and external control is not limited to compositional variables in a fluid, but, at least theoretically, is applicable to any intensive variable: mineral (or melt!) compositions, temperature and pressure are all valid candidates. (For this reason, in the present review, buffering by assemblages is contrasted with external control, be it due to copious fluid, heat transfer or pressure. For such as the latter, 'buffering' is chemically inappropriate, and 'infiltration' semantically discomforting.) Internal and external control of condensed phase compositional variables are not different in principle from control of fluid compositions. Control of temperature by either extreme process is unlikely to occur in metamorphism, rather an interplay of heat exchange with the surroundings and heat consumption or production by reactions is likely to be the rule. As for pressure, it is hard to envisage mineral assemblages buffering pressure, (and to envisage pressure 'infiltrating'), but it is not at all hard to conceive of pressure exerting control on mineral assemblages—as in the isobaric diagrams discussed above. In a sense, we are rediscovering the phase rule, but only in a sense, for the relative quantities of phases are indeed significant here, as they are in considering the extent of any heterogeneous reaction.

Of course, these considerations are just as applicable to the environment of partial melting as they are to 'normal'metamorphism (Powell, 1983*a,b*; Grant, 1983*b*, 1985). Powell's (1983*a*) Figures 5 and 6 illustrate fluid infiltration and buffering respectively. Where melt is involved, slopes of melting reactions in isobaric T–x diagrams like Figure 3.18, are commonly positive, and on such reactions buffering leads to lower x_{H_2O} in the associated vapour during melting. As before, for small fluid/condensed phase ratios, substantial reaction (melting) occurs at isobarically invariant points or at maxima. This is the basis for Powell's (1983*a,b*) emphasis on step-wise melting. If a dehydration reaction takes place with increasing temperature, x_{H_2O} will run to higher values and melting will ensue when the solidus is intersected at relatively high x_{H_2O} in the fluid, and with an abundance of the solid dehydration products as well as melt being produced (Powell, 1983*a*, p. 135). If additional components were added to the system, the variance of the assemblages could be increased (depending on whether additional phases formed), the stability fields of the assemblages expanded and the melting steps smoothed out, essentially due to solid (and/or liquid) solution (Powell, 1983*b*).

Powell (1983*b*, p. 633) suggests that rocks which are being partially melted are likely to have lower a_{H_2O} than those as yet unmelted, or melted to a lesser extent. However, as can be seen from Figure 3.18, this is not quite so. Consider an isothermal section at a temperature less than at the invariant point: an assemblage Qtz–Bio–V may be unmelted at higher x_{H_2O} while Qtz–Kfs–Bio–V may coexist with melt at intermediate x_{H_2O} and Qtz–Kfs–Bio–Opx–V may be involved in a dehydration reaction at low x_{H_2O}. (This yields a nice example of 'carbonic metamorphism' which will be discussed in the next section.)

As Powell (1983*b*) observes, restite and melt may follow different paths on segregation and subsequent retrogression. Again referring to Figure 3.18, consider what might happen if Qtz–Kfs–Opx were segregated quite thoroughly from melt at high temperature and low x_{H_2O}. For the restite, any small amount of liquid left might crystallize before the invariant point was regained, and the assemblage Qtz–Kfs–Opx–V cool unbuffered to the hydration reaction Kfs + Opx + V = Qtz + Bio. Thereafter, the isobarically univariant assemblage would be buffered to lower temperatures and lower x_{H_2O} until one of the reactants was consumed. For the liquid, if little of the solid phases was entrained, crystallization could continue until the invariant point was attained and here, the reaction Kfs + Opx + L = Qtz + Bio would consume either potassium feldspar or orthopyroxene. In the latter case, for example, crystallization would continue buffered along L = Qtz + Kfs + Bio + V to high x_{H_2O} and low temperature. To the extent that there was little ambient vapour, the final composition could approach $x_{H_2O} = 1$. (More vapour would result in cessation of crystallization before this.) If the melt contained a finite but small amount of CO_2, crystallization would cease at the minimum where the composition of the ambient vapour attained the composition of the vapour *released* by crystallization of the melt. Note that some of the H_2O from the melt would be used directly in formation of biotite in this example, thus inhibiting the increase of x_{H_2O} in the vapour.

These considerations really lead us to paths in partial melting, to which we shall turn our attention next.

3.8 Paths through time and space

The conventional interpretation of metamorphic mineral assemblages is that they are likely to represent approaches to equilibrium at or near the culmination of the event that produced them, and the same can be said for a metamorphic facies. Thus in a region of progressive metamorphism, a metamorphic facies series (Miyashiro, 1961) is often correlated with a pressure–temperature 'gradient' in the region (or better, on the surface now exposed) at the culmination of metamorphism.

Even granting that assemblages commonly represent a significant approach to equilibrium near the culmination of the metamorphic event, and that correlation of a facies series with a line through pressure–temperature space has its value, it must be true that such a 'gradient' is time-transgressive. This is so even in simple contact metamorphism (Jaeger, 1957, 1959), where maximum temperatures are attained at later times at farther distances from the contact. Recently, this has been considered in particular by England and Richardson (1977) and Figure 3.23 is based on their Figure 1 (see also Spear and Selverstone, 1983). In Figure 3.23, the heavy solid line represents the path through pressure–temperature space taken by a particular rock body; the heavy dashed lines are similar, but represent metamorphism at lower grades. The light line represents the locus of culminations at different grades, and is the 'pressure-temperature gradient' referred to above. This is termed the 'metamorphic geotherm' by England and Richardson (1977). A close association of high-grade rocks now found at the surface might have undergone an evolution like that represented by the heavy solid line. In this scenario for pelites, they would form under surficial conditions, be buried and rise in temperature along a 'normal' geothermal gradient, be subjected to a metamorphic event involving increase in temperature, rise to lower pressures, and cool to surficial temperatures. For short-lived contact metamorphism, cooling may occur

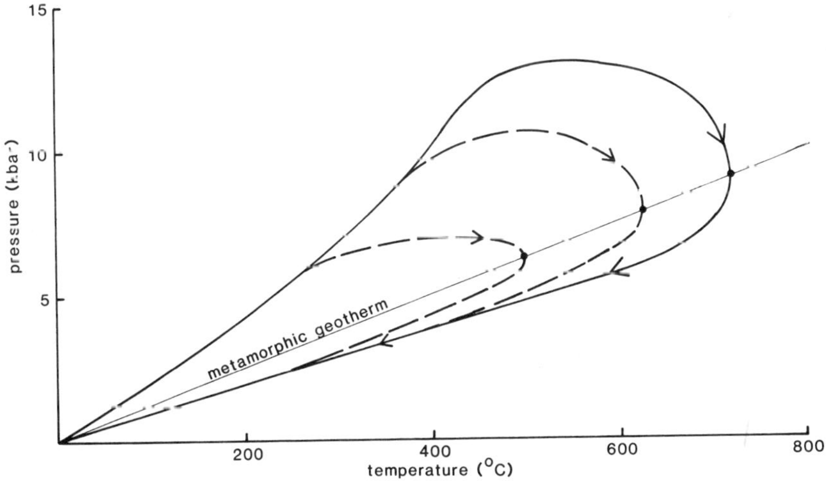

Figure 3.23 Hypothetical pressure–temperature–time paths for three regionally metamorphosed rock bodies, shown in heavy lines. Locus of the culminations of metamorphism at different grades is indicated by the light line, which is the 'metamorphic geotherm' of England and Richardson (1977), and the figure is based on their Figure 1.

before significant decrease in pressure. In regional metamorphism, especially where melting is involved, uprise may start before substantial cooling, and even before the maximum temperature is attained, as in Figure 3.23. In the case considered by England and Richardson, uprise is due to erosion, but where melt is present, it may also be aided by gravitational instability. Further, in their example, which involves crustal thickening due to overthrusting, higher maximum temperatures are attained at later times (and at greater depths) whereas in some other situations the opposite may be true (as in the case of contact metamorphism noted above).

For the present, the point is that what we see in high-grade rocks is the result of an eventful history: this is particularly true for the environment of partial melting (e.g. Grant, 1985), and on this we shall focus our attention.

Although we have been discussing partial *melting* of pelitic rocks, more of what we now see in a pelitic migmatite is probably attributable to *re-crystallization*. To quote Fyfe, Turner and Verhoogen (1958, p. 57) '... it may be anticipated that a reaction with a positive entropy change will, in general, have a positive entropy of activation, and will be faster, if all other factors remain the same, than the opposite reaction involving a decrease in entropy. This is a matter of common experience:... crystallization is slower than melting.' A rock that is to become a migmatite deserves more than to be simply taken to high temperature, partially melted, and left there: in reality, it is likely to be the relatively slow re-crystallization at *hypersolidus* conditions that is more effective in influencing the final character of the migmatite.

This is implied in the path shown in Figure 3.20 (Grant, 1985), on which a rock undergoing high-grade metamorphism in the presence of a mixed vapour is buffered up to the solidus (and relatively high x_{H_2O}) where melting goes on until one phase is exhausted (in this case, quartz). Melting then continues along reactions with positive slopes, to high temperature and low x_{H_2O}. Thereafter, the migma re-crystallizes through a series of peritectic reactions, in this case thought to be terminated by mantling of the reactants, which would explain the zoned character of the restites found here, and could result in isobarically invariant points being bypassed. At last, the solidus is reached, and we have a rock that belongs to facies '6', and from which we have to try to interpret a history like that just proposed. As discussed in Grant (1985), the sequence of events proposed here is not distant from what one could anticipate from considering the *liquidus* relationships, such as in Levin *et al.* (1966) for the systems FeO– or MgO–Al_2O_3–SiO_2. Indeed, for partial melting situations it is most helpful to 're-invent' liquidus diagrams (e.g. Figures 3.8 and 3.26).

Time is of the essence in discussing the evolution of metamorphic rocks, including migmatites. This may be implied by consideration of isobaric evolution in T–x space (e.g. Figure 3.18) during increasing and subsequent decreasing temperature, as in Greenwood (1967, p. 565), Powell (1983*b*, p. 633) and Grant (1985). However, this could also be done in terms of isothermal

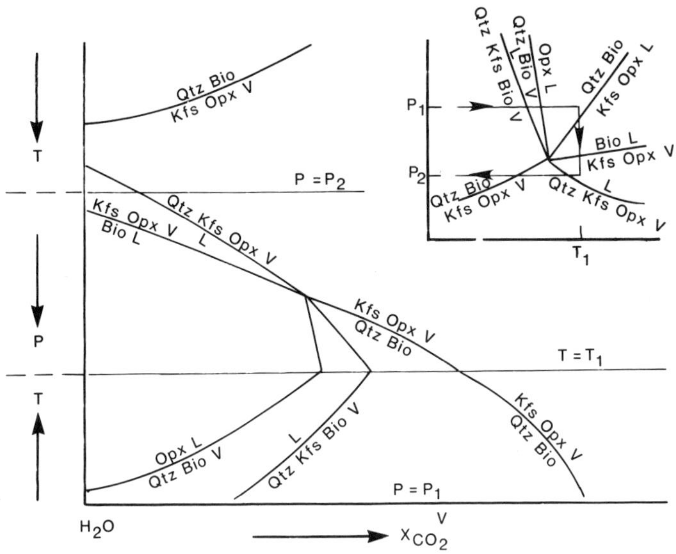

Figure 3.24 The inset repeats the schematic pressure–temperature grid of univariant reactions about the invariant point involving Qtz–Kfs–Bio–Opx–L–V in the system $KAlSi_3O_8$–MgO–SiO_2–H_2O from Figure 3.7, with part of a very hypothetical pressure–temperature–time path superimposed. The main portion of the figure illustrates sequential 'stacking' of the *T–x, P–x* and *T–x* equilibria at P_1, T_1 and P_2 respectively, following Figures 3.18 and 3.19, for the same phases, Qtz–Kfs–Bio–Opx–L–V, with CO_2 added to the system.

evolution in *P–x* space, (e.g. Figure 3.19), and, indeed, the two can be combined, as follows.

Figure 3.24 shows, in the inset, a version in Figure 3.7. Consider the very hypothetical square path shown in the inset: isobaric heating at pressure P_1 to temperature T_1, followed by isothermal decrease in pressure to P_2, and then isobaric cooling at P_2. If we allow a binary vapour to be present, say (H_2O–CO_2), we can apply the relations shown in Figures 3.18 and 3.19 to the relevant segments of this evolutionary path. Let us assume that CO_2 occurs in the vapour phase only. First, from Figure 3.18, we have *T–x* relations as temperature rises to just short of the vapour-absent reaction Qtz + Bio = Kfs + Opx + L, and thus, just short of the invariant point in Figure 3.18. At this time, the system is at P_1 and T_1, and the pressure is greater than that at the invariant point of Figure 3.19. As pressure *decreases,* the invariant point is attained and passed. Since the relationships at P_1, T_1 must be the same in both, and since time is continuous, we can 'stack' the relations in the isothermal *P–x* segment on 'top' of those in the isobaric *T–x* segment. Thereafter, at pressure P_2, less than at the invariant point of Figure 3.19, we revert to the isobaric relations of Figure 3.18, and add the final segment of the *T–x* equilibria, as temperature *decreases* from just above the solidus to below the hydration reaction, Kfs + Opx + V = Qtz + Bio.

If we turn from this square path to one more of the form shown in Figure

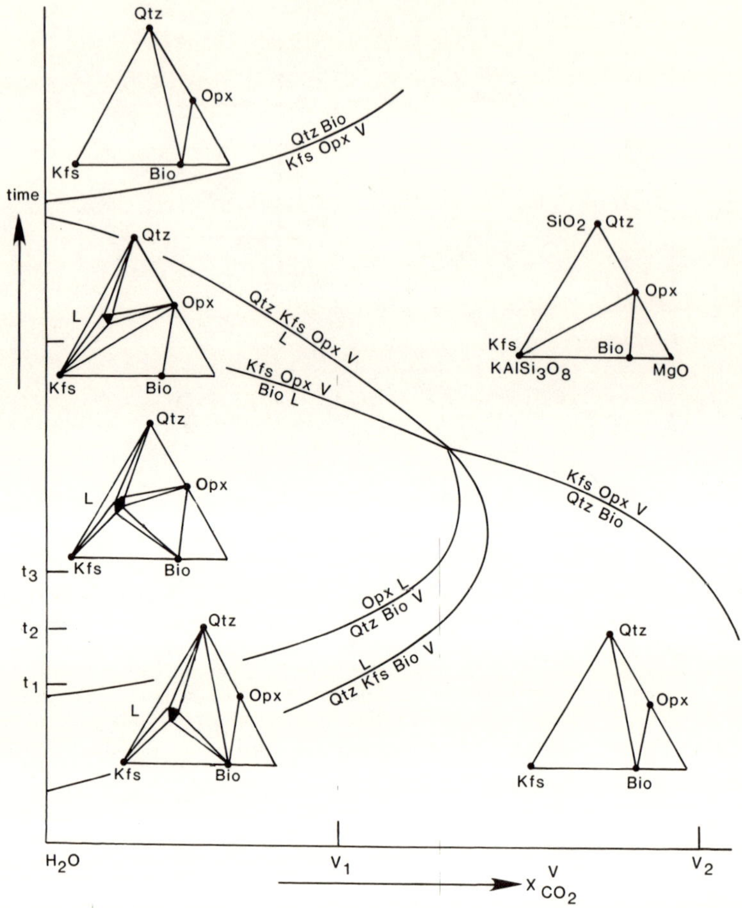

Figure 3.25 'Time–x' diagram, corresponding to a more realistic version of Figure 3.24, with the metamorphic path now continuous, as on the high-temperature side of Figure 3.23. Sets of divariant assemblages are shown in projection from H_2O onto the plane $KAlSi_3O_8$–MgO–SiO_2. (Note that this projection is not the same as in the related Figure 3.7.)

3.23, the result will be qualitatively as shown in Figure 3.25, a 'time–x' evolutionary diagram. Such a diagram is just as valid as either T–x or P–x representations of the equilibria, although the time element is much more difficult for us to quantify. The points already made with regard to T–x and P–x relations, and internal and external control of variables also apply here, and will not be repeated. However, there are some features of interest:

A horizontal line in a time–x diagram represents a point on a pressure–temperature path such as in Figure 3.23, and thus this horizontal line is an isobaric isothermal isochron. The suite of isobaric isothermal assemblages on this isochron comprises a facies in which variation in phase assemblages is correlated with variation in bulk composition, including the ambient vapours.

Of course, the term isochron is used somewhat loosely here, for a system may remain at a given pressure, temperature and vapour composition for more than an instant in time.

For reactions such as Qtz + Kfs + Bio + V = L or Qtz + Bio + V = Opx + L, with negative dP/dT, such a path of increasing temperature and decreasing pressure will lead to a 'back-bend' where the most CO_2-rich vapour in equilibrium with liquid is attained, and where the reaction changes from melting to crystallization, with enhancement of the vapour in H_2O. In so far as only the vapour contains CO_2, the reaction at the invariant point is Qtz + Bio = Kfs + Opx + L, for otherwise the composition of the vapour at this invariant point cannot remain constant.

Consider a few examples. An assemblage Qtz–Kfs–Bio–V, with only a little vapour of composition V_1, will start to melt at time t_1 on Qtz + Kfs + Bio + V = L, pass through the turning point at maximum x_{CO_2}, and crystallize liberating H_2O to the vapour, buffered all the way to the invariant point, presuming that none of the condensed phases are so scarce that they are eliminated by the small extent of reaction expected *en route*. At the invariant point the stability limit of quartz–biotite is reached, and quartz–biotite will melt to form not only liquid, but the 'dehydration products' potassium feldspar and orthopyroxene as well. Once quartz or biotite is exhausted, further evolution is along [Qtz] or [Bio]. If this were along [Qtz], peritectic reaction of biotite and liquid would continue until one or the other was used up, and Kfs–Bio–Opx–V or Kfs–Opx–L–V would continue unbuffered, either past L = Qtz + Kfs + Opx + V or on to L = Qtz + Kfs + Opx + V respectively. Eventually the two possible four-phase assemblages under subsolidus conditions, Kfs–Bio–Opx–V and Qtz–Kfs–Opx–V, attain the hydration reaction Kfs + Opx + V = Qtz + Bio. This buffer reaction proceeds, to lower temperature and higher x_{CO_2}, until all the orthopyroxene is consumed and Qtz–Kfs–Bio–V is again produced.

If we started instead with Qtz–Bio–V under the same conditions, melting would not start until time t_2 on Qtz + Bio + V = Opx + L. The buffered assemblage would evolve to the invariant point and thereafter in a manner similar to that just discussed. However, if at time t_3 there were a sudden influx of abundant CO_2-rich vapour (V_2), we would have the kind of 'carbonic metamorphism' envisaged by Newton *et al.* (1980): the assemblage Qtz–Bio–Opx–L–V would undergo peritectic crystallization to quartz–biotite, the assemblage Qtz–Kfs–Bio–L–V would crystallize to Qtz–Kfs–Bio, both releasing some H_2O to the vapour. Eventually quartz–biotite in both assemblages would dehydrate on Qtz + Bio = Kfs + Opx + V until either quartz or biotite was eliminated and finally there would be Qtz–Kfs–Opx–V or Kfs–Bio–Opx–V in equilibrium with the vapour V_2. This, of course, is also an example of the control to be expected in infiltration, discussed in the previous section. Note that crystallization would precede any possible dehydration under any such circumstances in this model.

If the initial vapour were again low in quantity, but high in CO_2 (e.g. V_2), an assemblage bearing quartz–biotite would simply dehydrate to the invariant point, where a melting 'step' would occur, immediately followed by crystallization along [Qtz] or [Bio]: the invariant point would be the only condition for melting along this evolutionary path.

Finally, *any* assemblage bearing quartz and biotite, with or without potassium feldspar, and a very wide (but not complete) range of vapour compositions and little vapour, would be funnelled to the invariant point, where melting would be most productive and would terminate in this model.

3.9 Separation of solid, liquid and vapour

The timing and efficiency of the segregation of solid, liquid and vapour relative to possible reactions are patently important in the development of migmatites, but little is really known about these processes (see Hargraves, 1980; Fyfe *et al.*, 1978; Walther and Orville, 1982; Arzi, 1978; Thompson, 1983). With regard to the solids, it should be noted at the outset that segregation is dominantly a physical phenomenon, and is not concerned with whether a solid was unmelted or the product of previous crystallization: restites left from segregation of melt may well contain both.

Consider first solid and liquid. If separation is complete, the restite may undergo little change, except for that due to continuous reactions. The availability of discontinuous solid–solid reactions is less than for liquid- or vapour-present reactions. Meanwhile, the liquid will tend to move upward, with greater or lesser success, eventually to crystallize. Even if restite and liquid remain in the same environment, the restite may appear to be of higher grade than the crystallized liquid, but this may only mean that the restite had nothing (vapour or trapped liquid) to react with. The results of incomplete separation of liquid and crystals depend upon the extent to which some solids are dispersed through the liquid (or vice versa) or whether the two are cleanly segregated but juxtaposed. In either case, limited reaction between solid and liquid is possible. In the first case, the products would be dispersed (and unmelted grains may be mantled with material crystallized from the melt), in the second the products would be localized near the contact, producing a selvage compatible with both solid and liquid. Continuous separation of liquid from solid seems rather unlikely at low rates of melting. Apparently a threshold of some 20% melt is required before significant mobility is possible (Arzi, 1978). Thus, if melting on vapour-absent reactions (even in the passive *presence* of vapour—see the previous section) is likely to be the most efficient means of producing liquid, other than in the presence of copious quantities of vapour, segregation is likely to *follow* such reactions.

Separation of vapour from solid is, of course, the rule rather than the exception in 'normal metamorphism', probably in steps or spurts (Thompson, 1983). For low rates of devolatilization, the volatile species may depart by diffusion along grain boundaries, without saturation (Walther and Orville, 1982). At higher rates, the fluid phase may leave via microcracks (Fyfe *et al.*, 1978).

But what of separation of solid and liquid and vapour? Surely vapour must escape more readily than viscous silicate melt, and surely melt 'charged' with bubbles of vapour is possible. One may talk of liquid being segregated in terms of, say, a Time–x diagram like Figure 3.25, but would solid *and vapour* remain? This seems implausible for more than locally trapped vapour. Simply let us note the common situation in hydrothermal melting, where, on opening a capsule the presence of vapour is readily confirmed by its escape, leaving glass round restitic solid grains. On this basis, it is expected that vapour will be most readily lost to lower pressures, and if there is no source of fresh vapour, only solids and liquid will be left. That is, vapour-present melting reactions cannot be very effective if permeability is such that vapour is lost, but if permeability is effective, vapour may be 'topped up' from an external source. If permeability is poor, vapour will tend to be retained and infiltration insignificant. This implies that vapour-absent reactions will be effective where local vapour has been incorporated into melt or could escape and no alternative source is available, or at invariant points where some component only occurs in vapour. Vapour-present melting reactions may occur where local vapour is retained or where infiltration of vapour takes place.

3.10 Summary liquidus relations

One means of visualizing the relations between crystals, liquids and vapours under water-saturated, vapour-saturated and vapour-absent conditions involves a schematic quaternary model for quartz-bearing assemblages among Qtz–Kfs–Bio–Opx–L–V in the system KMASH–CO_2, as in Grant (1984). A model for approximately 2 kbar is shown in the tetrahedron of Figure 3.26. The rear face of the tetrahedron is equivalent to Figure 3.8. In Figure 3.26, this has been extended with the addition of CO_2 as a possible component in vapour, and to a much lesser extent in liquids. Although liquid compositions for the H_2O-undersaturated region are not well known, the temperatures of several of the isobaric invariant points are constrained by the data of Shaw (1963), Wones and Dodge (1977), Wendlandt (1981) and Bohlen *et al.* (1983*b*). In this figure, a central primary field for biotite is flanked above by one for orthopyroxene and below by one for potassium feldspar. The surface for Qtz–Opx–L–V is omitted for clarity, and there is a large volume in which vapours can coexist with liquids in the front portion of the diagram. (The comments made with regard to the field for vapour in Figure 3.8 may be extended to this figure.) Ornamental tie lines between liquids and vapours are shown only on the base, and these are qualitatively compatible with the results of Holloway (1976) for albite –H_2O–CO_2. In this figure, a maximum is shown on the Qtz–Bio–Opx–L surface, following the discussion of Figure 3.8. In a

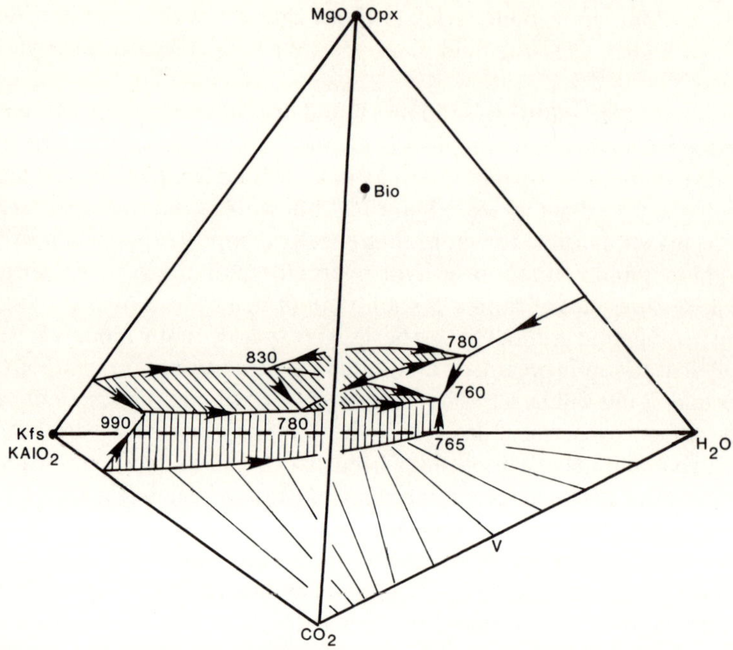

Figure 3.26 Schematic 'liquidus' diagram for quartz-bearing assemblages among Qtz–Kfs–Bio–Opx–L–V in the system KMASH–CO_2, analogous to Figure 3.8, with CO_2 added to the system. MgO and CO_2 contents of liquids are exaggerated for clarity, but otherwise, where known or estimated, compositions are plotted in mole percent. Temperatures at the isobaric invariant points are after the data of Shaw (1963), Wones and Dodge (1977), Wendlandt (1981) and Bohlen *et al.* (1983*b*) for approximately 2kbar. (After Grant, 1984).

similar vein, we note that the quaternary reaction reported by Wendlandt (1981) is Qtz + Kfs + Bio + V = Opx + L, as it is here, and that the reaction Qtz + Bio = Kfs + Opx + L shown on his figure (op. cit., Figure 4) is from Luth (1967) rather than from Wendlandt's own experiments. However, this reaction has been recently determined by Bohlen *et al.* (1983*b*). Compositions of liquids in equilibrium with solids and a binary vapour are on the surfaces extending from the H_2O-saturated solidus for KMASH, via the vapour-saturated quaternary isobaric invariant point labelled 780 °C, to the isobaric invariant point in KMAS—CO_2, labelled 990 °C. There are three such surfaces Qtz–Kfs–L–V (vertical ornament), Qtz–Bio–L–V (blank) and Qtz–Opx–L–V (not shown).

Unless carbonates are encountered, vapour-present and CO_2-absent relations are probably topologically similar, except in the range of pressure, temperature and vapour compositions in which the isobaric invariant melting reactions involving Qtz–Kfs–Bio–Opx–L, with or without vapour, change from one side of the extended join orthopyroxene–biotite to the other. The situation shown here is that probable at low pressures. With increasing pressure, solubility of H_2O in the melts increases, and one would expect that the CO_2-absent reaction Qtz + Kfs + Bio = Opx + L would change to Qtz + Bio = Kfs + Opx + L before the CO_2-present Qtz + Kfs + Bio + V = Opx + L changes to Qtz + Bio + V = Kfs + Opx + L (cf. Wendlandt, 1981, Figure 4). Major development of melts in common protoliths, in the absence of copious quantities of water-rich fluid, are unlikely until the isobaric univariant curve Qtz–Kfs–Bio–Opx–L, or the invariant point Qtz–Kfs–Bio–Opx–L–V are attained. Segregation of melt from Qtz–Kfs–Opx–L, with or without vapour, leaves a charnockitic assemblage, quartz–potassium feldspar–orthopyroxene, which can be stable in the presence of a high-x_{CO_2} vapour, whereas subsequent crystallization of the mobile H_2O-undersaturated melt may be accompanied by generation of high-x_{H_2O} vapour.

3.11 Conclusions

1. In the subsolidus equilibria among quartz, potassium feldspar, muscovite, aluminosilicate, biotite, cordierite, garnet, orthopyroxene and vapour, in the systems KMASH and KFASH, orthopyroxene-absent equilibria are topologically similar but muscovite-absent equilibria are different. This is largely due to the restriction of quartz–pyrope to high pressures compared with quartz-almandine.
2. At $P_{H_2O} = P_{total}$, melting in assemblages bearing quartz–potassium feldspar–cordierite–vapour in KMASH, KFASH and KFMASH occurs at temperatures significantly lower than for melting in quartz–potassium feldspar–vapour, at least at low pressure.
3. At $P_{H_2O} = P_{total}$, many of the possible equilibria in KFMASH are occluded by reactions in the sub-systems KFASH and KMASH. This changes significantly with decreasing P_{H_2O}.
4. Although melting may indeed start in the system KFASH at $P_{H_2O} = P_{total}$, the liquid field extends into KFMASH over a small temperature interval, and at $P_{H_2O} < P_{total}$, the beginning of melting may be in KFMASH, due to the emergence of garnet as a phase on the solidus. (This, of course, depends on the partition of Fe/Mg between garnet and melt.)
5. Unless aqueous vapour is readily available, which probably implies infiltration, significant melting is likely to occur on vapour-absent reactions. However, such reactions can occur in the presence of a mixed vapour, if its composition is the same as at the isothermally or isobarically invariant point corresponding to the reaction.
6. Internal and external control apply to more than just vapour composition: they may apply to any intensive variable.
7. A metamorphic rock is the product of an eventful history, of which the culmination of metamorphism is but an important event. Considering the changes in pressure, temperature and phase compositions that occur especially near culmination, the use of *P*-*x* diagrams and even of 'Time–*x*' diagrams is advocated.

8. Because of their relative timing and different rates, melting may be less influential than high-temperature re-crystallization on the final products of the partial melting process.

Acknowledgement

Support by grants EAR 76-10986 and 82-07429 from the National Science Foundation is gratefully acknowledged.

References

Abbott, R.N., Jr. and Clarke, D.B. (1979) Hypothetical liquidus relationships in the subsystem Al_2O_3–FeO–MgO projected from quartz, alkali feldspar and plagioclase for a $(H_2O) \leqslant 1$. *Can. Miner.* **17**, 549–560.

Ackermand, D., Seifert, F. and Schreyer, W. (1975) Instability of sapphirine at high pressures. *Contrib. Miner. Petrol.* **50**, 79–92.

Althaus, E., Karotke, E., Nitsch, K.H. and Winkler, H.G.F. (1970) An experimental re-examination of the upper stability limit of muscovite plus quartz. *Neues Jb. Miner. Mh.* 325–336.

Arzi, A.A. (1978) Critical phenomena in the rheology of partially melted rocks. *Tectonophysics* **44**, 173–184.

Ashworth, J.R. (1975) Staurolite at anomalously high grade. *Contrib. Miner. Petrol.* **53**, 281–291.

Barrow, G. (1893) On an intrusion of muscovite-biotite gneiss in the east Highlands of Scotland, and its accompanying metamorphism. *Q.J. Geol. Soc. Lond.* **49**, 330–358.

Berg, J.H. (1977) Dry granulite mineral assemblages in the contact aureoles of the Nain Complex, Labrador. *Contrib. Miner. Petrol.* **64**, 33–52.

Berg, J.H. and Wheeler, E.P. (1976) Osumilite of deep-seated origin in the contact aureole of the anorthositic Nain Complex, Labrador. *Am. Miner.* **61**, 29–37.

Bird, G.W. and Fawcett, J.J. (1973) Stability relations of Mg-chlorite, muscovite and quartz between 5 and 10 kb water pressure. *J. Petrol.* **14**, 415–428.

Bohlen, S.R. and Dollase, W.A. (1983) Calibration of hercynite-quartz stability. *Geol. Soc. Am. Abstr. Programs* **15**, 529.

Bohlen, S.R., Wall, V.J. and Boettcher, A.L. (1983*a*) Geobarometry in granulites. In *Kinetics and Equilibrium in Mineral Reactions*, ed. Saxena, S.K., Springer-Verlag, New York, 141–171.

Bohlen, S.R., Boettcher, A.L., Wall, V.J. and Clemens, J.D. (1983*b*) Stability of phlogopite-quartz and sanidine-quartz: a model for melting in the lower crust. *Contrib. Miner. Petrol.* **83**, 270–277.

Boyd, F.R. and England, J.L. (1959) Pyrope. *Yb. Carnegie Instn Wash.* **58**, 83–87.

Brown, G.C. and Fyfe, W.S. (1970) The production of granite melts during ultra-metamorphism. *Contrib. Miner. Petrol.* **28**, 310–318.

Burnham, C.W., Holloway, J.R., and Davis, N.F. (1969) Thermodynamic properties of water to 1000 °C and 10,000 bars. *Geol. Soc. Am. Spec. Pap.* **132**.

Carmichael, D.M. (1970) Intersecting isograds in the Whetstone Lake area, Ontario. *J. Petrol.* **11**, 147–181.

Chappell, B.W. and White, A.J.R. (1974) Two contrasting granite types. *Pacific Geol.* **8**, 173–174.

Chatterjee, N.D. and Johannes, W. (1974) Thermal stability and standard thermodynamic properties of synthetic $2M_1$–muscovite, $KAl_2[AlSi_3O_{10}(OH)_2]$. *Contrib. Miner. Petrol.* **48**, 89–114.

Clarke, D.B. (ed.) (1981) Peraluminous granites. *Can. Miner.* **19**, 1–216.

Clemens, J.D. and Wall, V.J. (1981) Origin and crystallization of some peraluminous (S-type) granitic magmas. *Can. Miner.* **19**, 111–131.

Day, H.W. (1971) A Theoretical and Experimental Study of Some Equilibria in the System K–Fe–Al–Si–O–H. Unpubl. Ph.D. Thesis, Brown University.

Day, H.W. (1973) The high temperature stability of muscovite plus quartz. *Am. Miner.* **58**, 255–262.

Eggler, D.H. (1973) Principles of melting of hydrous phases in silicate melt. *Yb. Carnegie Instn Washington* **72**, 491–495.

Ellis, D.J., Sheraton, J.W., England, R.N., and Dallwitz, W.B. (1980) Osumilite-sapphirine-quartz granulites from Enderby Land, Antarctica—mineral assemblages and reactions. *Contrib. Miner. Petrol.* **72**, 123–143.

England, P.C., and Richardson, S.W. (1977) The influence of erosion upon the mineral facies of rocks from different metamorphic environments. *J. Geol. Soc. Lond.* **134**, 201–213.

Essene, E.J. (1982) Geologic thermometry and barometry. *Rev. Miner.* (*Min. Soc. Amer*) **10**, 153–206.

Eugster, H.P. and Wones, D.R. (1962) Stability relations of the ferruginous biotite, annite. *J. Petrol.* **3**, 82–125.

Evans, B.W. (1965) Application of a reaction-rate method to the breakdown equilibria of muscovite and muscovite plus quartz. *Am. J. Sci.* **263**, 647–667.

Fyfe, W.S., Turner, F.J., and Verhoogen, J. (1958) Metamorphic reactions and metamorphic facies. *Geol. Soc. Am. Mem.* **73**.

Fyfe, W.S., Price, N.J., and Thompson, A.B. (1978) *Fluids in the Earth's Crust.* Elsevier, Amsterdam.

Ganguly, J. (1972) Staurolite stability and related parageneses: theory, experiments and applications: *J. Petrol.* **13**, 335–365.

Gibbs, G.V. (1966) The polymorphism of cordierite: I. The crystal structure of low cordierite. *Am. Miner.* **51**, 1068–1087.

Grant, J.A. (1968) Partial melting of common rocks as a possible source of cordierite anthophyllite assemblages. *Am. J. Sci.* **266**, 908–931.

Grant, J.A. (1973) Phase equilibria in high-grade metamorphism and partial melting of pelitic rocks, *Am. J. Sci.* **273**, 289–317.

Grant, J.A. (1981) Orthoamphibole and orthopyroxene relations in high-grade metamorphism of pelitic rocks. *Am. J. Sci.* **281**, 1127–1143.

Grant, J.A. (1982) Low pressure equilibria in anatexis of pelitic rocks. *Geol. Soc. Am. Abstr. Programs* **14**, 501.

Grant, J.A. (1983*a*) Equilibria in low-pressure melting of pelitic rocks (Abstract). In *Migmatites, Melting and Metamorphism*, eds. Atherton, M.P. and Gribble, C.D., Shiva, Nantwich, 292–293.

Grant, J.A. (1983*b*) Quartz-present and quartz-absent equilibria in partial melting of pelitic rocks. *Geol. Soc. Am. Abstr. Programs* **15**, 584.

Grant, J.A. (1984) Relationships between vapor-present and vapor-absent melting in the system KMASCH. *Geol. Soc. Am. Abstr. Programs* **16**, 523.

Grant, J.A. (1985) Phase equilibria in low-pressure partial melting of pelitic rocks. (*Am. J. Sci.*, in press.)

Grant, J.A. and Hoffer, E. (1977) Reactions in pelitic migmatites in the light of hydrothermal melting experiments involving Mg–Fe cordierite. *Geol. Soc. Am. Abstr. Programs* **9**, 994.

Green, T.H. (1976) Experimental generation of cordierite- or garnet-bearing granitic liquids from apelitic composition. *Geology* **4**, 85–88.

Greenwood, H.J. (1962) Metamorphic reactions involving two volatile components. *Yb. Carnegie Instn. Wash.* **61**, 82–85.

Greenwood, H.J. (1967) Mineral equilibria in the system $MgO–SiO_2–H_2O–CO_2$. In *Researches in Geochemistry*, vol. 2, ed. Abelson, P.H., Wiley, New York, 542–567.

Greenwood, H.J. (1975) The buffering of pore fluids by metamorphic reactions. *Am. J. Sci.* **275**, 573–593.

Grew, E.S. (1981) Granulite-facies metamorphism at Molodezhnaya Station, East Antarctica. *J. Petrol.* **22**, 297–336.

Haggerty, S.E. (1983) Oxide-silicate reactions in lower crustal granulites from Liberia, West Africa. *Geol. Soc. Am. Abstr. Programs* **15**, 589.

Hargraves, R.B. (ed.) (1980) *Physics of Magmatic Processes.* Princeton Univ. Press, Princeton, N.J.

Harris, N. (1981) The application of spinel-bearing metapelites to P/T determinations: an example from South India. *Contrib. Miner. Petrol.* **76**, 229–233.

Harris, N.B.W., Holt, R.W. and Drury, S.A. (1982) Geobarometry, geothermometry, and Late Archean geotherms from the granulite facies terrain of South India. *J. Geol.* **90**, 509–527.

Helgeson, H.C. (1967) Solution chemistry and metamorphism. In *Researches in Geochemistry*, vol. 2, ed. Abelson, P.H., Wiley, New York, 362–404.

Helgeson, H.C. (1968) Evaluation of irreversible reactions in geochemical processes involving minerals and aqueous solutions–I. Thermodynamic relations. *Geochim. Cosmochim. Acta* **32**, 853–877.

Helgeson, H.C., Garrels, R.M. and Mackenzie, F.T. (1969) Evaluation of irreversible reactions in geochemical processes involving minerals and aqueous solutions–II. Applications. *Geochim. Cosmochim. Acta* **33**, 455–481.

Helgeson, H.C., Delany, J.M., Nesbitt, H.W. and Bird, D.K. (1978) Summary and critique of the thermodynamic properties of rock-forming minerals. *Am. J. Sci.* 278A.

Hensen, B.J. (1971) Theoretical phase relations involving cordierite and garnet in the system MgO–FeO–Al_2O_3–SiO_2. *Contrib. Miner. Petrol.* **33**, 191–214.

Hensen, B.J. (1977) The stability of osumilite in high grade metamorphic rocks. *Contrib. Miner. Petrol.* **64**, 197–204.

Hensen, B.J., and Essene, E.J. (1971) Stability of pyrope-quartz in the system MgO–Al_2O_3–SiO_2. *Contrib. Miner. Petrol.* **30**, 72–83.

Hoffer, E. and Grant, J.A. (1980) Experimental investigation of the formation of cordierite-orthopyroxene parageneses in pelitic rocks. *Contrib. Miner. Petrol.* **73**, 15–22.

Holdaway, M.J. and Lee, S.M. (1977) Fe–Mg cordierite stability in high-grade pelitic rocks based on experimental, theoretical, and natural observations. *Contrib. Miner. Petrol.* **63**, 175–198.

Hollister, L.S. and Burruss, R.C. (1976) Phase equilibria in fluid inclusions from the Khtada Lake metamorphic complex. *Geochim. Cosmochim. Acta* **40**, 163–175.

Holloway, J.R. (1973) The system pargasite–H_2O–CO_2: a model for melting of a hydrous mineral with a mixed-volatile fluid–I. Experimental results to 8 kbar. *Geochim. Cosmochim. Acta* **37**, 651–666.

Holloway, J.R. (1976) Fluids in the evolution of granitic magmas: consequences of finite CO_2 solubility. *Geol. Soc. Am. Bull.* **87**, 1513–1518.

Huang, W.L. and Wyllie, P.J. (1974) Melting relations of muscovite with quartz and sanidine in the K_2O–Al_2O_3–SiO_2–H_2O system to 30 kilobars and an outline of paragonite melting relations. *Am. J. Sci.* **274**, 378–395.

Jaeger, J.C. (1957) The temperature in the neighborhood of a cooling intrusive sheet. *Am. J. Sci.* **255**, 306–318.

Jaeger, J.C. (1959) Temperatures outside a cooling intrusive sheet. *Am. J. Sci.* **257**, 44–54.

Johannes, W. (1983) On the origin of layered migmatites. In *Migmatites, Melting and Metamorphism*, eds. Atherton, M.P. and Gribble, C.D., Shiva, Nantwich, 234–248.

Kars, H., Jansen, J.B.H., Tobi, A.C. and Poorter, R.P.E. (1980) The metapelitic rocks of the polymetamorphic Precambrian of Rogaland, SW Norway, Part II. *Contrib. Miner. Petrol.* **74**, 235–244.

Keesmann, I., Matthes, S., Schreyer, W. and Seifert, F. (1971) Stability of almandine in the system FeO–(Fe_2O_3)–Al_2O_3–SiO_2–(H_2O) at elevated pressures. *Contrib. Miner. Petrol.* **31**, 132–144.

Kerrick, D.M. (1972) Experimental determination of muscovite + quartz stability with $P_{H_2O} < P_{total}$. *Am. J. Sci.* **272**, 946—958.

Kreulen, R. (1980) CO_2-rich fluids during regional metamorphism on Naxos (Greece): carbon isotopes and fluid inclusions. *Am. J. Sci.* **280**, 745–771.

Kreulen, R. and Schuiling, R.D. (1982) N_2–CH_4–CO_2 fluids during formation of the Dôme de l'Agout, France. *Geochim. Cosmochim. Acta* **46**, 193–203.

Lal, R.K., Ackermand, D., Seifert, F. and Haldar, S.K. (1978) Chemographic relationships in sapphirine-bearing rocks from Sonapahar, Assam, India. *Contrib. Miner. Petrol.* **67**, 169–187.

Lambert, I.B., Robertson, J.K. and Wyllie, P.J. (1969) Melting reactions in the system $KAlSi_3O_8$–SiO_2–H_2O to 18.5 Kbars. *Am. J. Sci.* **267**, 609–626.

Levin, E.M., Robbins, C.R. and McMurdie, H.F. (1966) *Phase Diagrams for Ceramists*. American Ceramic Society, Columbus, Ohio.

Lonker, S.W. (1981) The P-T-X relations of the cordierite-garnet-sillimanite-quartz equilibrium *Am. J. Sci.* **281**, 1056–1090.

Luth, W.C. (1967) Studies in the system $KAlSiO_4$–Mg_2SiO_4–SiO_2–H_2O: I, inferred phase relations and petrologic applications. *J. Petrol.* **8**, 372–416.

Luth, W.C., Jahns, R.H. and Tuttle, O.F. (1964) The granite system at pressures of 4 to 10 kilobars. *J. Geophys. Res.* **69**, 759–773.

Martignole, J. and Sisi, J.-C. (1981) Cordierite-garnet–H_2O equilibrium: a geological thermometer, barometer and water fugacity indicator. *Contrib. Miner. Petrol.*, **77**, 38–46.

Miyashiro, A. (1961) Evolution of metamorphic belts. *J. Petrol.* **2**, 277–311.

Morse, S.A. (1970) Alkali feldspars with water at 5 kb pressure. *J. Petrol.* **11**, 221–251.

Naney, M.T. (1983) Phase equilibria of rock-forming ferromagnesian silicates in granitic systems. *Am. J. Sci.* **283**, 993–1033.

Newton, R.C. (1972) An experimental determination of the high-pressure stability limits of magnesian cordierite under wet and dry conditions. *J. Geol.* **80**, 398–420.

Newton, R.C. (1983) Geobarometry of high-grade metamorphic rocks. *Am. J. Sci.* **283-A**, 1–28.

Newton, R.C., and Wood, B.J. (1979) Thermodynamics of water in cordierite and some petrologic consequences of cordierite as a hydrous phase. *Contrib. Miner. Petrol.* **68**, 391–405.

Newton, R.C., Smith, J.V. and Windley, B.F. (1980) Carbonic metamorphism, granulites and crustal growth. *Nature* **288**, 45–50.

Olsen, S.N. (1983) A quantitative approach to local mass balance in migmatites. In *Migmatites, Melting and Metamorphism*, eds. Atherton, M.P. and Gribble, C.D., Shiva, Nantwich, 201–233.

Powell, R. (1983*a*) Processes in granulite-facies metamorphism. In *Migmatites, Melting and Metamorphism*, eds. Atherton, M.P. and Gribble, C.D., Shiva, Nantwich, 127–139.

Powell, R. (1983*b*) Fluids and melting under upper amphibolite facies conditions. *J. Geol. Soc. Lond.* **140**, 629–633.

Rao, B.B. and Johannes, W. (1979) Further data on the stability of staurolite + quartz and related assemblages. *Neues Jb. Miner. Mh.* 437–447.

Rice, J.M. and Ferry, J.M. (1982) Buffering, infiltration, and the control of intensive variables during metamorphism. *Rev. Mineralogy* (*Min. Soc. Amer.*) **10**, 263–326.

Richardson, S.W. (1968) Staurolite stability in a part of the system Fe–Al–Si–O–H. *J. Petrol.* **9**, 467–488.

Robertson, J.K. and Wyllie, P.J. (1971) Rock-water systems, with special reference to the water-deficient region. *Am. J. Sci.* **271**, 252–277.

Robie, R.A., Bethke, P.M. and Beardsley, K.M. (1967) Selected X-ray crystallographic data, molar volumes, and densities of minerals and related substances. *U.S. Geol. Surv. Bull.* 1248.

Robie, R.A., Hemingway, B.S. and Fisher, J.R. (1978) Thermodynamic properties of minerals and related substances at 298.15 K and 1 bar (10^5 pascals) pressure and at higher temperatures. *U.S. Geol. Surv Bull.* 1452.

Robinson, P. and Tracy, R.J. (1983) Ilmenite-rutile-ferromagnesian silicate equilibria in medium- to high-grade graphitic schists. *Geol. Soc. Am. Abstr. Programs* **15**, 672.

Rutherford, M.J. (1969) An experimental determination of iron biotite–alkali feldspar equilibria. *J. Petrol.* **10**, 381–408.

Schairer, J.F. (1954) The system K_2O–MgO–Al_2O_3–SiO_2. *J. Am. Ceram. Soc.* **37**, 501–533.

Schreyer, W. (1968) A reconnaissance study of the system MgO–Al_2O_3–SiO_2–H_2O at pressures between 10 and 25 kb. *Yb. Carnegie Instn. Wash.* **66**, 380–392.

Schreyer, W. (1976) Experimental metamorphic petrology at low pressures and high temperatures. In *The Evolution of the Crystalline Rocks*, eds. Bailey, D.K. and Macdonald, R. Academic Press, London, 261–331.

Schreyer, W. and Seifert, F. (1969) Compatibility relations of the aluminum silicates in the systems MgO–Al_2O_3–SiO_2–H_2O and K_2O–MgO–Al_2O_3–SiO_2–H_2O at high pressures. *Am. J. Sci.* **267**, 371–388.

Seifert, F. (1970) Low-temperature compatibility relations of cordierite in haplopelites of the system K_2O–MgO–Al_2O_3–SiO_2–H_2O. *J. Petrol.* **11**, 73–99.

Seifert, F. (1974) Stability of sapphirine: a study of the aluminous part of the system MgO–Al_2O_3–SiO_2–H_2O. *J. Geol.* **82**, 173–204.

Seifert, F. (1975) Boron-free kornerupine: a high pressure phase. *Am. J. Sci.* **275**, 57–87

Seifert, F. (1976) Stability of the assemblage cordierite + K feldspar + quartz. *Contrib. Miner. Petrol.* **57**, 179–185.

Shaw, H.R. (1963) The four-phase curve sanidine-quartz-liquid-gas between 500 and 4000 bars. *Am. Miner.* **48**, 883–896.

Spear, F.S. and Selverstone, J. (1983) Quantitative P-T paths from zoned minerals: theory and tectonic applications. *Contrib. Miner. Petrol.* **83**, 348–357.

Spry, P.G. (1982) An unusual gahnite-forming reaction, Geco base-metal deposit, Manitouwadge, Ontario. *Can. Miner.* **20**, 549–553.

Storre, B. and Karotke, E. (1971) An experimental determination of the upper stability limit of muscovite + quartz in the range 7–20 kb water pressure. *Neues Jb. Miner. Mh.* 237–240.

Storre, B. and Karotke, E. (1972) Experimental data on melting reactions of muscovite + quartz in the system K_2O–Al_2O_3–SiO_2–H_2O to 20 kb water pressure. *Contrib. Miner. Petrol.* **36**, 343–345.

Thompson, A.B. (1976) Mineral reactions in pelitic rocks: II. Calculation of some P–T–X (Fe–Mg) phase relations. *Am. J. Sci.* **276**, 425–454.

Thompson, A.B. (1982) Dehydration melting of pelitic rocks and the generation of H_2O-undersaturated granitic liquids. *Am. J. Sci.* **282**, 1567–1595.

Thompson, A.B. (1983) Fluid-absent metamorphism. *J. Geol. Soc. Lond.* **140**, 533–547.

Thompson, A.B. and Algor, J.R. (1977) Model systems for anatexis of pelitic rocks. I. Theory of melting reactions in the system $KAlO_2$–$NaAlO_2$–SiO_2–H_2O. *Contrib. Miner. Petrol.* **63**, 247–269.

Thompson, A.B. and Tracy, R.J. (1979) Model systems for anatexis of pelitic rocks. II. Facies series melting and reactions in the system CaO–$KAlO_2$–$NaAlO_2$–Al_2O_3–SiO_2–H_2O. *Contrib. Miner. Petrol.* **70**, 429–438.

Touret, J. (1971) Le facies granulite en Norvège méridionale: II. Les inclusions fluides. *Lithos* **4**, 423–436.

Tracy, R.J. (1978) High grade metamorphic reactions and partial melting in pelitic schist, west-central Massachusetts. *Am. J. Sci.* **278**, 150–178.

Tuttle, O.F. and Bowen, N.L. (1958) Origin of granite in the light of experimental studies in the system $NaAlSi_3O_8$–$KAlSi_3O_8$–SiO_2–H_2O. *Geol. Soc. Am. Mem.* **74**.

Velde, B. (1965) Phengitic micas: synthesis, stability, and natural occurrence. *Am. J. Sci.* **263**, 886–913.

Vielzeuf, D. (1983) The spinel and quartz associations in high grade xenoliths from Tallante (S.E. Spain) and their potential use in geothermometry and barometry. *Contrib. Miner. Petrol.* **82**, 301–311.

Vielzeuf, D. and Boivin, P. (1984) An algorithm for the construction of petrogenetic grids: application to some equilibria in granulitic paragneisses. *Am. J. Sci.* **284**, 760–791.

Walther, J.V. and Orville, P.M. (1982) Volatile production and transport in regional metamorphism. *Contrib. Miner. Petrol.* **79**, 252–257.

Wendlandt, R.F. (1981) Influence of CO_2 on melting of model granulite facies assemblages: a model for the genesis of charnockite. *Am. Miner.* **66**, 1164–1174.

Winkler, H.G.F. (1979) *Petrogenesis of Metamorphic Rocks.* 5th edn., Springer-Verlag, New York.

Winkler, H.G.F., Bocsc, M. and Marcopoulos, T. (1975) Low temperature granitic melts. *Neues Jb. Miner. Mh.* 245–268.

Wones, D.R. and Dodge, F.C.W. (1966) On the stability of phlogopite. *Geol. Soc. America Ann. Mtg.* 1966, *Abstr.* 243.

Wones, D.R. and Dodge, F.C.W. (1977) The stability of phlogopite in the presence of quartz. In *Thermodynamics in Geology*, ed. Fraser, D.G., Reidel, Boston, 229–247.

Wyllie, P.J. (1977*a*) Crustal anatexis: An experimental review. *Tectonophysics* **43**, 41–71.

Wyllie, P.J. (1977*b*) Effects of H_2O and CO_2 on magma generation in the crust and mantle. *J. Geol. Soc. Lond.* **134**, 215–234.

Wyllie, P.J. and Tuttle, O.F. (1960) The system CaO–CO_2–H_2O and the origin of carbonatites. *J. Petrol.* **1**, 1–46.

Yardley, B.W.D. Shepherd, T.J. and Barber, J.P. (1983) Fluid inclusion studies of high-grade rocks from Connemara, Ireland. In *Migmatites, Melting and Metamorphism*, eds. Atherton, M.P. and Gribble, C.D., Shiva, Nantwich, 110–126.

Yoder, H.S., Jr. (1952) The MgO–Al_2O_3–SiO_2–H_2O system and the related metamorphic facies. *Am. J. Sci.* **250a**, 569–627.

Yoder, H.S., Jr. and Kushiro, I. (1969) Melting of a hydrous phase: phlogopite. *Am. J. Sci.* **267-A**, 558–582.

Zen, E-an (1967) Time and space relationships of the Taconic allochthon and autochthon. *Geol. Soc. Am. Spec. Pap.* **97**.

4 Mass balance in migmatites

SAKIKO N. OLSEN*

4.1 Introduction

The migmatite problem has always been interrelated with the granite problem. The long and sometimes heated debates of the granite controversy between 'magmatists' and 'transformationists' were ended, for the time being, by Tuttle and Bowen (1958). Their study also convinced many geologists that migmatites were the product of simple partial melting and that the leucosomes in migmatites represented squeezed-out minimum melts. However, it quickly became clear that this assumption had many difficulties when applied to regional migmatites. The problems most often cited were: (1) the leucosomes do not have the minimum melt compositions (for example, Misch, 1968); (2) certain migmatitic structures, such as relict mineral orientations (for example, Hopson, 1964), metamorphic texture of and replacement textures in the leucosome (for example, King, 1965), cannot be reconciled with the idea of the leucosome as a melt. Winkler and Breitbart's melting experiment study (1978) illustrates clearly the type of dilemma that crops up in migmatite research. They found 'with surprise' that at least six out of nine leucosomes of '*in-situ* migmatites' probably were not completely liquid at the temperature reasonable for a high-grade metamorphism even at $P_{H_2O} = P_{total} - 5$ kbar. It is very unlikely that such crystal mush could be filter-pressed out of the wall rocks, especially when the melt must contain plagioclase and quartz but leave mafic minerals in the selvages. They also found that there are different types of *in-situ* leucosomes and more than one type present in one area. The migmatite problem, it seems, is far from solved.

One important use of mass balance in migmatites is as a quantitative tool in determining the mechanism of migmatization for each migmatite. The mechanisms of migmatization postulated in the works of Sederholm (1967), Mehnert (1968), Misch (1968) and many others, fall into four categories: (1) partial melting and segregation of the initial melt; (2) metamorphic differenti-

---

*Department of Earth and Planetary Sciences, The Johns Hopkins University, Baltimore, MD 21218, USA.

ation (subsolidus) by mechanical and/or chemical processes; (3) lit-par-lit injection of a granitic magma along the foliation planes; (4) metasomatism, with introduction of potassium, or less commonly, sodium. If a rock is shown to have been closed to all (except possibly volatile) components during migmatization, that is if the migmatization was isochemical, partial melting or metamorphic differentiation must have been the cause of its migmatization. Conversely, if a rock is shown to have been open to external components, metasomatism or igneous injection was most likely the sole or dominant mechanism for the migmatization.

The first great debate on migmatites was between Sederholm (1967, p. 418) and Holmquist (1921) on the migmatite as 'arterite' *v.* 'venite'. That debate, in a sense, was on the question of mass balance in migmatites. Sederholm, observing the field relations of migmatites, or veined gneisses, and granites in Finnish Precambrian basement concluded that all granitic veins in the migmatites were somehow connected with a deep-seated 'eruptive' magma. He likened the relationship to that of the human arteries to the heart and called the rock 'arterite'. Whatever the material was, a magma or an 'ichor', at least some of the vein material was thought to have been introduced from outside source(s), or the migmatite was an open system. In 'venite' of Holmquist, on the other hand, the vein material was thought to have formed by 'exudation' from immediately adjacent rocks, or the migmatite was a closed system.

Most of the early attempts at mass-balance calculations on migmatites were on a regional scale. The first step in such a study would be to correlate a migmatite with its premigmatization equivalent. The difference in their bulk compositions then is assumed to be the mass change due to migmatization: for example, Cheng (1943); Engel and Engel (1958); Brown (1967); Kays (1976); MacRae and Nesbitt (1980). In many cases, however, it is difficult or impossible to find the unmigmatized, lower-grade equivalent and an assumed protolith composition must be used. Even when one can be traced to the other, there is a possibility of a facies change or other original bulk chemical change over a large distance. Such mass-balance calculations on regional scale therefore are semiquantitative at best.

More quantitative and smaller-scale mass-balance calculations have been attempted by some workers. If a leucosome and its mafic selvages in a migmatite formed from the adjacent palaeosome in a locally closed system, they formed by simple segregation of the components without any gain or loss except possibly for volatiles. The net (leucosome + selvage) composition in such migmatites should be identical to the palaeosome compositions. The homogeneous-appearing gneiss between the leucosome-selvage sets is generally assumed to be the palaeosome or the original rock whose composition remained unaffected by the migmatization. It should be kept in mind however, that it is also possible that the 'palaeosome' (called 'mesosome' by Johannes and Gupta, 1982) remained unmigmatized because of its unsuitable composition and that the leucosome and selvages formed from a layer with a more suitable composition. Various methods are used to compare the net and palaeosome compositions as discussed below. Mehnert (1951) in his study of the Black Forest migmatites containing pegmatitic veins or leucosomes with mafic-rich selvages or 'extraction zones', calculated the sum of average compositions for the vein and extraction zones combined at the ratio of 3 to 1. The ratio roughly matched the volume ratio of the two zones on polished rock surface. Comparison of the net composition to the average palaeosome (host rock) composition convinced him that nothing was lost or added during the migmatization. The distances of migration of the matter within the migmatite were a few to a few tens of centimetres. The leucosome material was assumed to be a silicate melt.

In a similar study, Ljunggren (1957) used petrographic and modal data taken along a 230 mm profile containing 7 leucosomes, 5 to 20 mm thick, to assess mass balance in a banded gneiss or layered migmatite at Gothenburg, Sweden. He found that enrichment of mafic components in the proximity of most leucosomes can be reasonably accounted for by depletion in the leucosome of such components except in one case. His data showed, however, that although some of depletion of felsic components in the mafic selvages can be accounted for by the mass in the leucosome, the leucosome contained a surplus in microcline and a deficiency in plagioclase. Relying heavily on textural evidence such as 'replacement' textures, he concluded that these leucosomes formed by K-metasomatism along certain layers which resulted in replacement of plagioclase by K-feldspar and simultaneous metamorphic differentiation of mafic from felsic components.

Kretz (1966) in his now classic study of Precambrian migmatites from Einasleigh, N. Queensland, arrived at conclusions similar to Ljunggren's. His data indicated that in a biotite–quartz–feldspar migmatite, nearly all materials in the vein are derived from the host rock up to 2 to 3 cm from the leucosome boundary or within the 'extraction' zone enriched in mafic minerals. In an amphibolite and a quartz–plagioclase–biotite gneiss, a considerable amount, but not all of the vein-forming matter was locally derived. He also found that some migmatites showed constant mineral compositions in all zones but others exhibited slight differences in mineral composition. He rejected partial melting as the cause of segregation because there was no (or not enough) difference in plagioclase composition between the vein and palaeosome. The mechanism proposed was metamorphic differentiation by exchange of minerals by a diffusive process, but the cause and kinetics of the process were left for further study. It should be noted that in the studies of Ljunggren and Kretz, the differentiation process is assumed to occur at constant volume. The validity of such an assumption is discussed in detail below. Similar studies of migmatites also presented the results indicating that the migmatite forms in both closed and open systems (for example, White, 1966; Hughes, 1970; Ashworth, 1976; Mehnert and Büsch, 1982; Kenah and Hollister, 1983). Results of the studies of mass balance in migmatites on hand specimen scale have been presented recently in series of papers by Olsen (1977, 1982, 1984) and by Dougan (1979, 1981, 1983) which are discussed in detail below.

4.2 Baltimore Gneiss and Front Range migmatites

The Precambrian migmatites in the Baltimore Gneiss Domes in Maryland, and in the Colorado Front Range have been studied by Olsen using modal and mineral data and least squares analysis. The geological settings of the Baltimore Gneiss and Front Range migmatites are similar in several aspects: they are both Precambrian in age, from regionally migmatized terrains and their *P-T* conditions estimated are high enough for at least some partial melting if $P_{H_2O} \cong P_{total}$. The differences are: (1) the Baltimore Gneiss occurs above the kyanite isograd while the Front Range region contains sillimanite and no kyanite; (2) the Baltimore Gneiss suffered a regional metamorphic event during the Palaeozoic (Hopson, 1964) but the Front Range rocks were affected only by local thermal events after the main migmatization episode (Peterman *et al.*, 1968).

There are several general textural and mineralogical characteristics in these migmatites which are worth noting. First, all zones have xenoblastic texture in thin section and no relict igneous texture has been found. Mineral compositions generally stay constant in each sample within the limits of

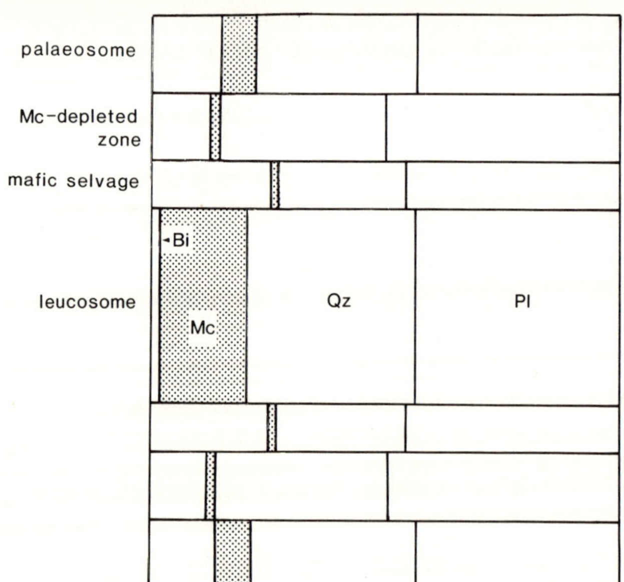

Figure 4.1 A typical modal variation of migmatitic zones. Microcline-depleted zone is present in some of the Front Range rocks but not in the Baltimore Gneiss. See text.

accuracy in probe analysis. Zoning in plagioclase is rare and diffuse when present, the core being 2 to 3% more An-rich than the rim. The palaeosome in these rocks is basically medium-grained, equigranular, and lepidoblastic, containing mainly quartz and feldspar with biotite flakes in subparallel orientation. The grain size, especially that of biotite, in the selvage is in general coarser than in the palaeosome. However, biotite also occurs as trains of fine-grained flakes. Biotite foliations wrap around the nose of any minor fold which may be present. Many of the leucosomes tend to be bimodal in grain size containing large plagioclase and microcline as elongated irregular grains with many embayments and interstitial fine quartz and feldspar. Minerals in the leucosome are also more altered than those in the palaeosome or selvage in general. Some of the Front Range rocks have a microcline-depleted zone, up to a few centimetres thick, between the mafic selvage and the palaeosome, similar in composition and texture to the palaeosome except that it contains more plagioclase and less microcline.

The zones in these migmatites seem to be in mutual equilibrium mineralogically (generally constant mineral composition, compatible mineral assemblage) but not texturally. Because it is not likely that a retrograde process would affect only the leucosome, the textural differences between the zones are probably related to the migmatization process. All samples discussed contain basically biotite-quartz-plagioclase-microcline-magnetite assemblages. A typical modal variation in a sample is shown in Figure 4.1.

4.3 Mass-balance calculations: methods and assumptions

The samples in the studies range in size from about 150 to 10 000 cm^3. A slab has been cut from each sample roughly perpendicular to the main foliation and leucosomes, and normal to the lineation (generally in the form of the axes of minor folds). For the larger samples several smaller slabs were cut perpendicular to the main slab. Thin sections were cut covering almost entire surfaces of the slabs. The sampling ensured that: (1) any significant modal variation away from the main slab would be detected (none was); (2) the palaeosome, selvage, and leucosome whose compositions were used in mass-balance calculations were always in direct contact. Modal compositions obtained by point-counting and mineral compositions by electron microprobe were used to calculate bulk compositions (see Olsen, 1977). The estimated uncertainties in the bulk composition are on the order of 1 wt % for a major oxide (Olsen, 1982). The method has an advantage over a bulk chemical analysis in obtaining the composition of zones with very irregular boundaries.

The net (leucosome + selvage) and palaeosome compositions are compared by the use of least-squares analysis code by Bryan *et al.* (1969; also Olsen, 1977). A least-squares calculation gives the estimated proportion of the leucosome to selvage i^{est}/j^{est} in the equation

$$j^{est}(\text{leucosome}) + j^{est}(\text{selvage}) = \text{palaeosome} + R \pm \text{volatiles} \qquad (4.1)$$

which yields the smallest sum of squares of residuals, R, summed over nine major components. A residual is the difference between the amounts of each component in the estimated-net and palaeosome compositions. Zone compositions in cation-mole % were used in all calculations in order to minimize the effects of components present in large amount, especially SiO_2. If a rock was migmatized in a closed system except for the volatile components, the estimated proportions of the leucosome and selvage should match the observed ones measured by planimetric analyses of slabs and thin sections and R must be zero. Because of the errors inherent in data and of minor inhomogeneities in the rock, it is necessary to determine the level of significance in the data. Less than 2 vol. % difference between estimated and measured proportions of the zones, about the same order of magnitude as the error in modal analysis, was considered not significant. R was considered significantly different from zero if it was larger than the sum of squares of the uncertainties in the palaeosome composition (Olsen, 1982, Table A1). The value of a significant R depends on the bulk composition and is about 2.00 on the average. The leucosomes which occur between two palaeosomes with different compositions probably formed from both palaeosomes (Figure 4.2): $L_1 + S_1$ and $L_2 + S_2$ formed from some of P_1 and P_2 respectively. It was assumed that the volume proportion of P_1 and P_2 converted is equal to that of S_1 to S_2. The leucosomes are homogeneous as a rule and the division into L_1 and L_2 in Figure 4.2 has no physical basis.

In calculating mass transport in the metamorphic rocks, it is generally assumed that the transport occurred either at constant volume or conserving at least one component, most commonly alumina (for example, Ferry, 1982) or oxygen (for example, Barth, 1948) in the rock (see also Babcock, 1973). In general, neither of these 'reference frames' can be adopted in the case of migmatites without reservations: neither of the assumptions above would be valid if the transport of a silicate melt occurred. Unless some reference frame is adopted, however, it is impossible to calculate any mass transport. Approaches tried in these studies are illustrated in Figure 4.3. The relationship is similar to what has been presented by Olsen (1984, Figure 5) which was a slightly simplified representation of the actual mass balance. The relationship in Figure 4.3 includes the possibility of excess mass present in the selvage as

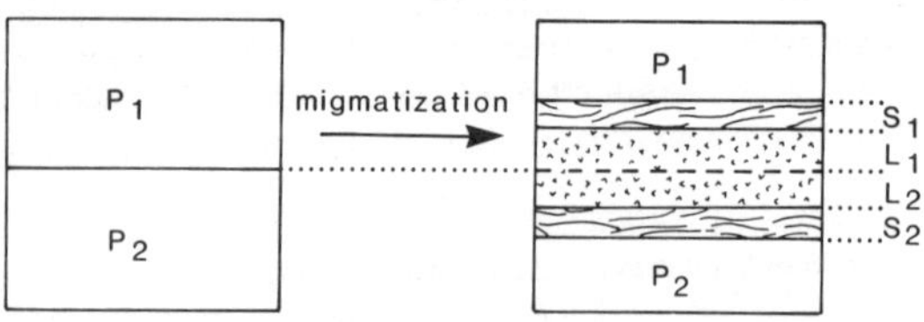

Figure 4.2 Postulated formation of a leucosome from two dissimilar palaeosomes.

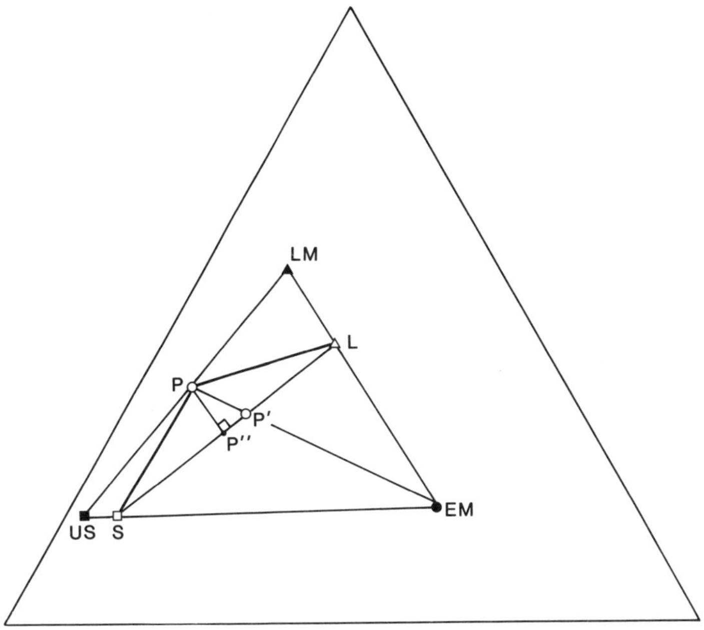

Figure 4.3 Graphical representation of a method used in calculating excess mass (EM) and locally derived mass (LM) in the leucosome (L) and that in the selvage (US). P, palaeosome; S, selvage; P′, net L + S. See text for explanations.

well as in the leucosome. The unadulterated original selvage (US in Figure 4.3) then is the residue after subtraction of LM from P and observed selvage (S) is US plus some excess mass (EM). A least squares analysis test for a closed system checks to see how close P is to P′. In treating mass balance between a migmatite and outside source(s) it may be considered that there are 5 variables (not all independent), P, L, S, P′ and EM. Of these five, four are known or measurable, P, L, S and P′, the last calculated using the measured ratio of L to S. Only one equation or constraint is needed to define the system and to find EM. The constraint, in this case, can be provided by the assumption of Al conservation in the rock,

$$Al_{EM} = Al_{P'} - Al_{P}$$

in which Al_{EM}, $Al_{P'}$, Al_{P} are Al contents of EM, P′, and P respectively. The constraint may be conservation of oxygen or any other component or constant volume in other cases: conservation of Al, oxygen, or volume can be a reasonable constraint if the mass transport in and out of the rocks is indicated to be by a fluid or diffusion.

When mass balance within a migmatite itself is considered, two more variables, LM and US, and another constraint, LM + US = P, are added. There remains the need for one more constraint. Because of the possibility of the transport of the partial melt from the selvage to leucosome (see below), conservation of any component or volume is no longer a reasonable constraint. The constraint used was the assumption that excess mass (EM) was confined in the leucosome or S = US (see Olsen, 1984, Figure 5). The composition of LM is calculated as

$$\mathrm{LM_c} = \mathrm{P_c} - [n]_{\mathrm{Al}}\, j^{\mathrm{obs}}\mathrm{S_c}$$

in which LM_c, P_c, S_c are the numbers of cations in LM, palaeosome, and selvage respectively; j^{obs} is the measured proportion of the selvage in cation mole %; $[n]_{Al}$ is the ratio of the number of aluminium ions in the palaeosome to that in the net leucosome + selvage. In Figure 4.3, the distance between LM and P is proportional to $[n]_{Al}\, j^{obs}$, and EM = L − LM.

In the case of a possible melt injection from an outside source, the assumption of Al conservation used for the migmatite as a whole is no longer valid. The constraint necessary was provided instead by (1) assumption of excess mass confined in the leucosome (or S = US); (2) minimization of the difference between L and LM or minimization of excess mass. Because P and S are fixed by assumption (1), the least-squares analysis will minimize the difference between L and LM. R, the residual in the calculation may be considered to be proportional to the distance between P and P″. The proportion of LM to P(= distance S–P) is nearly equal to S–P″ if R is small. S–P″ is used to calculate LM and EM. It should be noted that the distance S–P is not measurable in the rock because it is the proportion of LM to P. Numerically, EM can be calculated as

$$c_x = 1/100\left(\frac{\mathrm{S_{est}}}{\mathrm{S_{obs}}}\cdot \mathrm{L_{obs}} - \mathrm{L_{est}}\right)c_{\mathrm{L}} + \mathrm{R_c}$$

in which S_{est} and L_{est} are relative amounts of the selvage and leucosome estimated by a least squares calculation, S_{obs} and L_{obs} are those obtained by planimetric analysis, c_L is the amount of c in the leucosome and R_c is the residual for c in the least-squares calculation. A negative value for c_x represents a loss of material. It can be inferred from the geometry in Figure 4.3 that the method is a fairly good approximation if the line L–S is close to P, or the residual in the least-squares calculation is fairly small. In addition, S must be close to US or the selvage had to be nearly closed, which also seem reasonable assumptions in most cases. The assumption of the confinement of gains and losses to the leucosomes is consistent with the facts that (1) the selvage composition is best explained as the residues after the extraction of some felsic material similar to the leucosome from the palaeosome; (2) two mafic selvages bordering each leucosome have almost identical compositions, the symmetry suggesting the introduction of material along the leucosome.

In addition to the assumptions of palaeosomes as the unaltered rocks, mass-balance calculations used in these studies were based on two other assumptions: (1) present leucosomes are still in contact with the original selvages, which in turn are in contact with the original palaeosomes: (2) the subsequent deformation, if any, did not alter the relative volume proportion of the leucosome to selvage. Although the validity of these assumptions cannot be demonstrated conclusively, it is supported by some evidence, at least for the study samples: (a) the sharp boundary between the palaeosome and migmatized zone, and chemical and textural homogeneity of the palaeosome, suggest little interaction between the two zones; (b) the systematic relationships in the results of the calculations below, especially the fact that all of the observed leucosome proportions are larger than the estimated ones, are difficult to reconcile with large displacements of the leucosomes from the original selvages; (c) in some cases, the composition of leucosomes varies slightly but systematically with minor variation in the composition of immediately adjacent palaeosome; (d) the texture of selvages with delicate wisps and cuspate boundaries is difficult to reconcile with a much greater attenuation of the selvage than the leucosome as may be expected in the case of an intense deformation and displacement. In addition, both leucosome and selvages stay fairly constant in composition along the strike at least on the order of a metre. Mass-balance calculations, therefore, would be valid even if there was some displacement of the leucosome from the selvage provided that such displacement was small.

4.4 Mass-balance calculations: results

4.4.1 *Closed-system migmatites*

The net compositions, leucosome + selvages, of six leucosomes (two samples of the Baltimore Gneiss and four samples of the Front Range migmatites) closely approximate the respective palaeosome compositions (Figure 4.4). The results for four other small samples containing one leucosome each and two large samples containing six and three leucosomes each, all from the Front Range, are quite different. They show that none of the thirteen leucosome + selvage sets could have formed from the adjacent palaeosome(s) isochemically: either the estimated leucosome/selvage proportions differ significantly from the measured ones, or *R* was significantly large, or both (Figure 4.5). The results for these 'open-system' migmatites indicate that some mass transport occurred in these rocks.

Closed-system leucosomes are 3 to 30 mm thick and bordered by mafic-rich selvages between 0.5 and 6 mm thick. They must have formed by either metamorphic differentiation or partial melting and not by injection or metasomatism. These migmatites contain evidence for both subsolidus metamorphic differentiation and partial melting. Metamorphic differentiation may occur by either mechanical or chemical differentiation. Mechanical differentiation seems unlikely for these migmatites because the development of the leucosomes involved not only the segregation of felsic and mafic minerals but also of

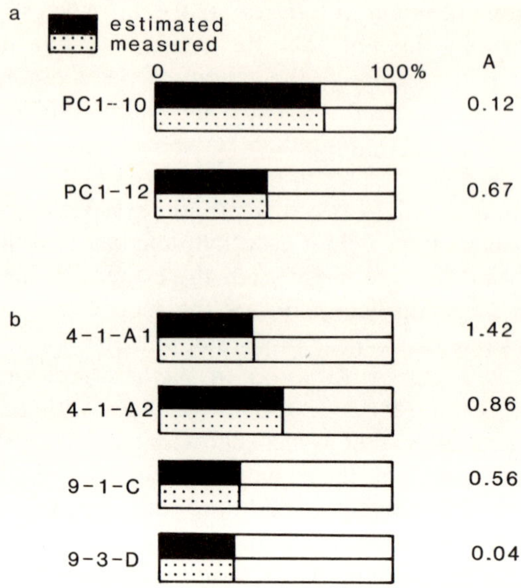

Figure 4.4 Results of mass-balance calculations for two closed-system migmatites in (*a*) the Baltimore Gneiss, (*b*) the Front Range migmatites: estimated (solid) and measured (shaded) volume proportions of the leucosome. A = the sum of squares of residuals in the least-squares calculation.

microcline from plagioclase. Because the two feldspars are structurally similar, it is unlikely that their mechanical behaviour would differ enough to permit this separation. The evidence indicates chemical differentiation by diffusion of constituents as a mechanism, probably as ions or associated species through an intergranular phase (e.g., Carmichael, 1969).

The main differences for the migmatitic zones in these rocks are in their SiO_2, FeO and MgO contents, which correspond to the differences in the amounts of biotite, microcline, and quartz. A subsolidus ionic reaction written in simplified form as

$$\text{biotite} + 6H^+ = \text{microcline} + 3(Fe^{2+}, Mg^{2+}) + 4H_2O \qquad (4.2)$$

was assumed to have been responsible for the formation of biotite in the selvage and microcline in the leucosome. Reaction (4.2) is written conserving alumina in the solid because alumina stays fairly constant in the three zones of these migmatites (Olsen, 1977, Figure 4).

Decreasing f_{H_2O} at constant a_{H^+} or increasing a_{H^+} at constant f_{H_2O} will drive the reaction to right and increase activities of Fe and Mg. If some independent factor could lower f_{H_2O} or raise a_{H^+} selectively in one part of the rock, the biotite there would break down to microcline releasing Fe and Mg ions. The resultant increase in activities of Fe and Mg forces both ions to diffuse outward.

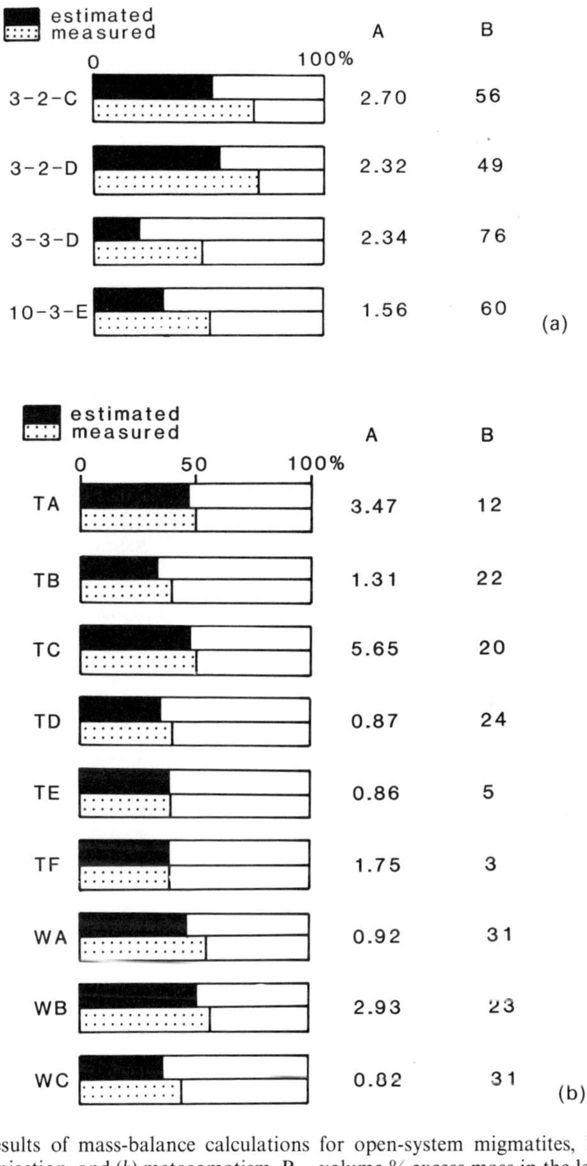

Figure 4.5 Results of mass-balance calculations for open-system migmatites, Front Range: formed by (*a*) injection, and (*b*) metasomatism. B = volume % excess mass in the leucosome. See the caption to Figure 4.4 and text for other explanations.

This zone of low f_{H_2O} or high a_{H^+} becomes enriched in microcline and depleted in biotite to form a leucosome. Increased activities of Fe and Mg in the adjacent palaeosome would cause biotite to replace microcline by reaction 4.2, forming a mafic selvage along the leucosome.

The modal amounts of biotite and microcline in the zones of a migmatite

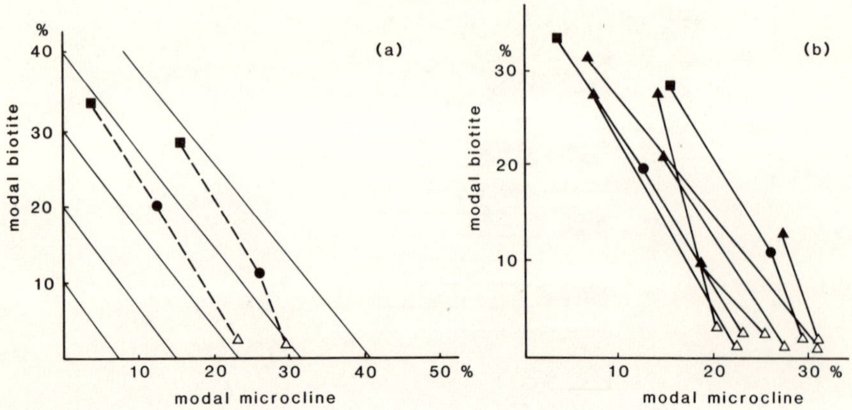

Figure 4.6*a* The modal biotite-microcline plot for the biotitic layered migmatites in the Baltimore Gneiss. In the migmatites formed by metamorphic differentiation, via the reactions biotite ⇄ microcline in the leucosomes and microcline ⇄ biotite in the selvage, the palaeosome (solid circle), leucosome (open triangle) and selvage (solid square) must lie on the same predicted tie line (solid). The tie lines have been constructed using reaction 4.3 (see text). The dashes are the observed tie lines for the two closed-system migmatites at Piney Creek.
b Observed biotite-microcline tie lines between the leucosome and its mafic host rock (solid triangle), which may be either the selvage or palaeosome, in the layered and the veined migmatites at Piney Creek containing biotite as the only major dark phase. Other symbols as in (*a*). Note the general parallelism of the tie lines.

formed in this manner will result in parallel tie lines between the leucosome and selvage through different palaeosome compositions in a modal biotite-microcline plot. Figure 4.6*a* shows the tie lines predicted by the model using reaction 4.3 below and the observed tie lines for the Baltimore Gneiss migmatites at Piney Creek, Maryland.

$$1\% \text{ biotite} + 0.024\% \text{ quartz} = 1.02\% \text{ microcline } (Or_{93}Ab_7) \qquad (4.3)$$

Reaction 4.3 is written as a constant-volume reaction with Al conserved assuming SiO_2 migration from the selvage to the leucosome to compensate for the volume decrease of about 20% in the conversion of biotite to microcline.

There is only a fair agreement between the slopes of observed and predicted tie lines in Figure 4.6*a* but the slopes are influenced significantly by minor variation in the reaction. (For the variations of reaction 4.3, see Olsen, 1977). More important, however, there is a general parallelism of observed tie lines in Figure 4.6*a* and also in Figure 4.6*b* which shows available modal data including other migmatites from the same locality. Mass-balance calculations for the latter were inconclusive mainly because these rocks lack either well-developed selvages or host rocks which can be identified with confidence as the unaltered palaeosome. The data strongly suggest that the segregation of biotite and microcline occurred as outlined above: by the coupled biotite–microcline reactions in the leucosome and selvage with diffusion of compo-

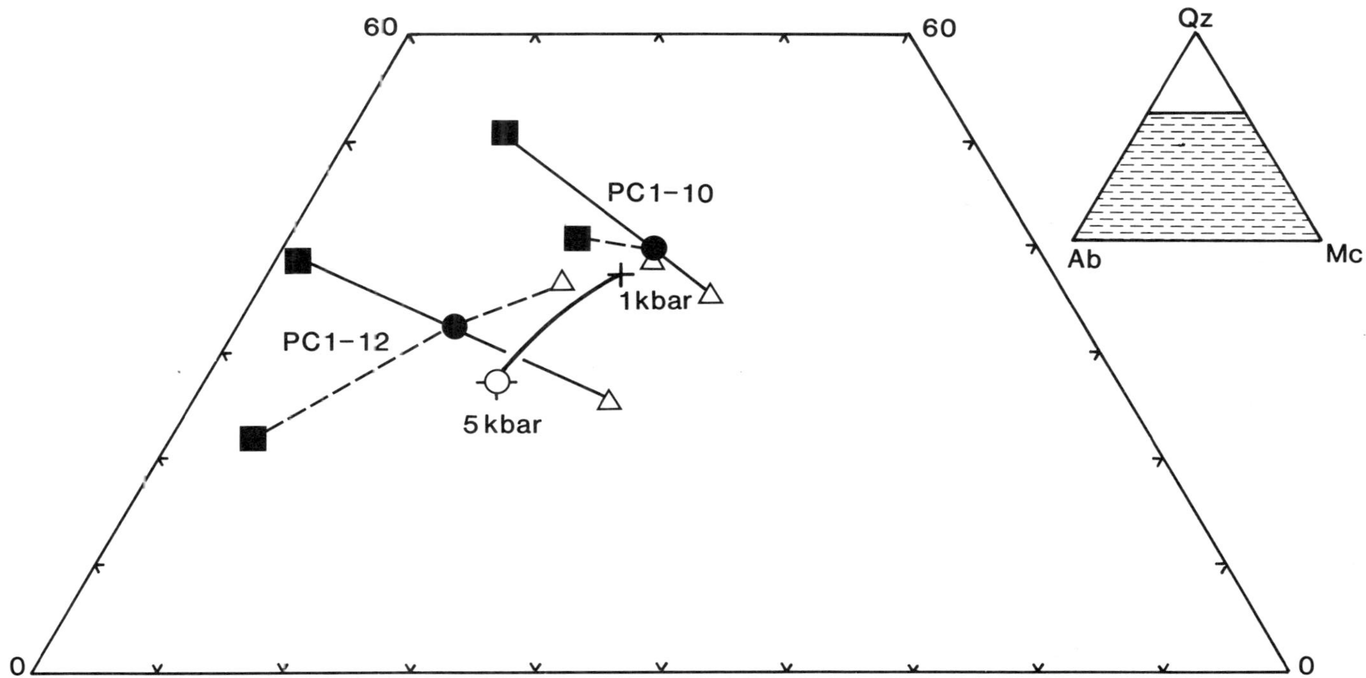

Figure 4.7 Tie lines in the modal granite ternary diagram for the two closed-system migmatites in the Baltimore Gneiss. Dashed, observed; solid, calculated tie lines. All data in the granite ternary plots in this chapter are in the modal percent, although the granite minimum is in weight percent. This is adequate because the maximum difference between the volume and weight percents in these plots, about 1 percent, should hardly be noticeable.

nents between the zones. However, the biotite–microcline reactions alone will not result in the observed leucosome and selvage compositions.

The predicted felsic compositions of the leucosomes and selvages calculated according to the model and the observed compositions do not agree (Figure 4.7). They would agree quite well, however, if there had been a migration of quartz from the selvage to leucosome in addition to that assumed for the reaction. However, the granite-minimum compositions of the leucosomes seem to be more than a coincidence. Furthermore, the discrepancy between the predicted and observed compositions in Figure 4.7 may also be explained by a mechanical segregation of a quartz-rich, sodic partial melt from the selvage rather than by the migration of quartz. This evidence suggests that partial melting may have been also involved in the formation of these leucosomes. Another feature suggests that partial melting did occur in the two

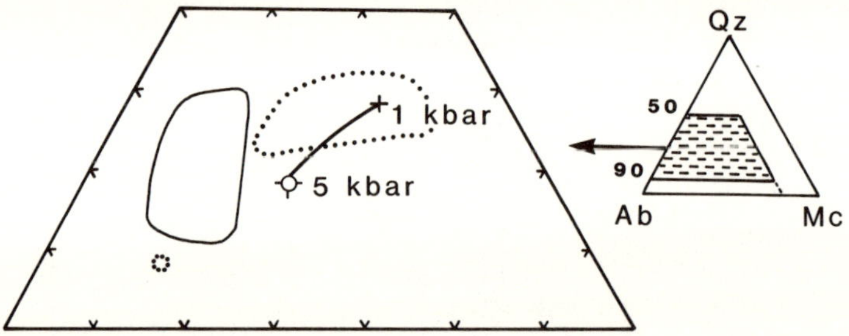

Figure 4.8 Relationship between the palaeosome composition and the degree of migmatization in the Baltimore Gneiss. Note that the palaeosomes of the more extensively migmatized, layered migmatites (bounded by the dotted line) are closer to the granite minimum (the heavy solid line, from Luth *et al.*, 1964) than the palaeosomes of the less extensively migmatized veined gneisses (bounded by the light solid line).

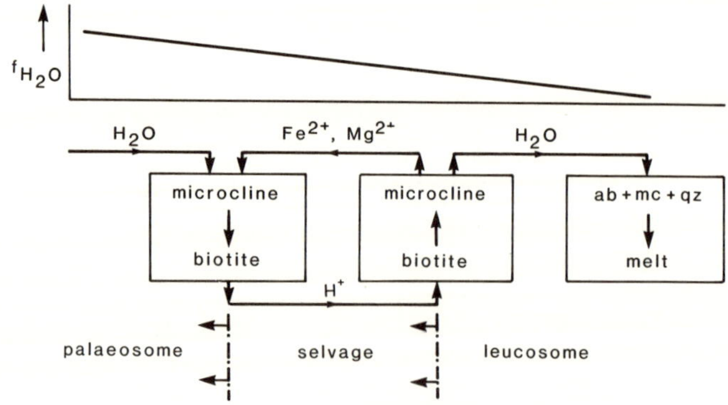

Figure 4.9 Schematic representation of the mechanism of migmatization by metamorphic differentiation induced by partial melting. See text for explanation. It is not implied here that the f_{H_2O} gradient is linear.

closed-system migmatites and most or all of the other migmatites at the same locality. That is a correlation between the palaeosome composition and the degree of migmatization: the palaeosomes in the layered migmatites at Piney Creek plot closer to the granite minimum than those in the veined gneisses (Figure 4.8). In other words, the closer the palaeosome compositions are to the granite minimum, the more voluminous the leucosomes become, or the rock is more thoroughly migmatized.

The evidence for both subsolidus metamorphic differentiation and partial melting in the two closed-system migmatites suggests that these migmatites formed by a combination of the two processes. A model of migmatization by metamorphic differentiation induced by partial melting is shown in Figure 4.9. As the initial melt forms, it must absorb all of the available water in the intergranular phase in its vicinity because of a large solubility of H_2O in such melt. The intergranular fluid probably contained at least two components: CO_2 inclusions which appear to be related to the migmatization process have been found in Front Range migmatites (Touret and Olsen, this volume). Assuming that the initial melts form at scattered sites in a rock, this causes activity of water in the immediate vicinity of the melt to fall below that elsewhere. The decrease in f_{H_2O} initiates the biotite breakdown reaction forming a leucosome. The water released by the reaction combines with quartz, albite, and microcline to form still more melt. The Fe and Mg released by the same reaction diffuse into the adjacent palaeosome, where biotite replaces microcline and a selvage is formed. The water needed for the selvage reaction is brought in from farther out in the palaeosome. As long as melting continues in the leucosome, some gradient in f_{H_2O} would be maintained between the leucosome and the palaeosome, and this gradient continues driving the metamorphic differentiation. This mechanism should create the parallel tie lines of Figure 4.6*a,b*. At the same time, it should contribute to the scatter of leucosome compositions to the microcline-rich side of the granite minimum as observed in these rocks (Olsen, 1977, Figure 9): if the leucosome contains a minimum melt, microcline forming from biotite would modify its

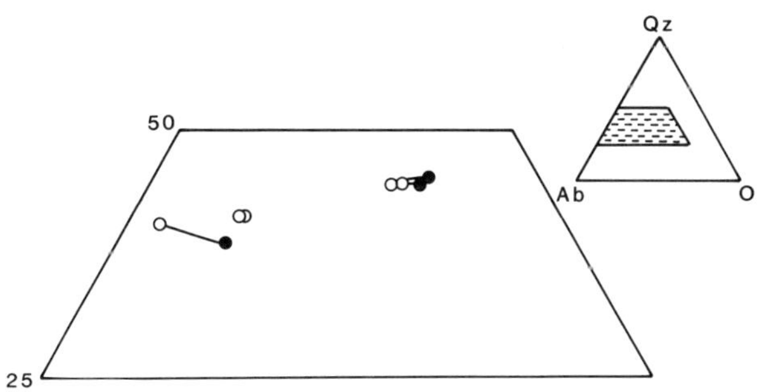

Figure 4.10 Possible K-enrichment of the closed-system migmatites, Front Range. Open circle, observed palaeosome; solid circle, net, leucosome plus selvage, composition.

minimum composition to that intermediate between the minimum and the microcline apex. The closed-system migmatites from the Front Range may have formed as postulated for the Baltimore Gneiss above (Olsen, 1982). But, it is also possible that some or all of these migmatites were slightly open to the external components. Each leucosome plus selvage composition is more K-rich than that of the palaeosome (Figure 4.10), suggesting a minor metasomatism by either an introduction of K-spar components or K-Na exchange.

4.4.2 *Open-system migmatites*

Four of the small samples of biotite–quartz–feldspar migmatites from the Clear Creek area with granitic leucosomes 8 to 30 mm and mafic selvages 3 to 15 mm thick show relatively large residuals for the least-squares calculations and large difference between the measured and estimated zone proportions (Figure 4.5*a*). An addition of some components to each leucosome is indicated because the measured leucosome proportion is considerably larger than the estimated one in every case. The excess masses calculated for the four leucosomes are granitic to tonalitic in composition and compose 49 to 76% of the respective total leucosome (Figure 4.11). Their low-temperature melt compositions (compare Winkler *et al.*, 1975) and fairly large volumes strongly suggest that the excess masses were externally-derived silicate melts and that these migmatites formed by the injection of magma.

The difference between the conventional model of lit-par-lit injection, or a simple mechanical injection of magma along the foliation planes (Misch, 1949,

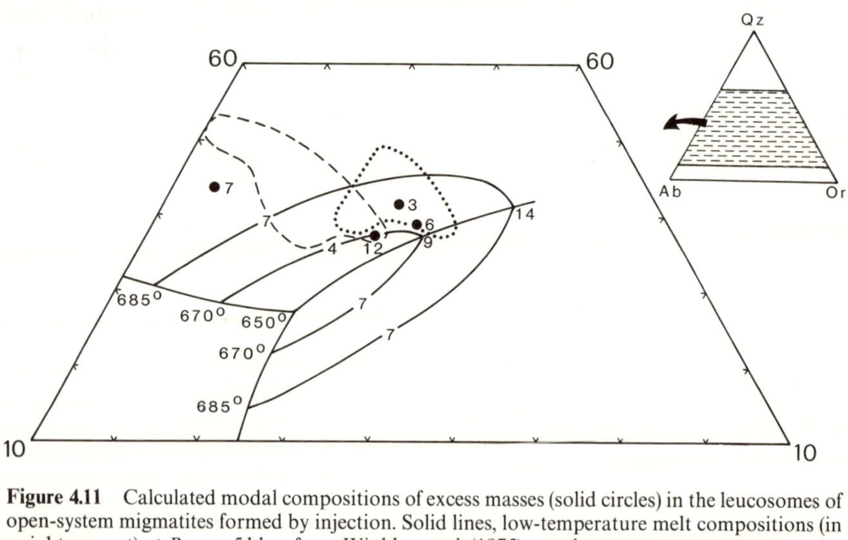

Figure 4.11 Calculated modal compositions of excess masses (solid circles) in the leucosomes of open-system migmatites formed by injection. Solid lines, low-temperature melt compositions (in weight percent) at $P_{H_2O} = 5$ kbar from Winkler *et al.* (1975); numbers on contours represent An content in mole percent. Dashed line, the ranges of normative compositions of trondhjemitic rocks, and dotted, of red leucogranites in the area, respectively, both from Hedge (1969).

p. 230), and the model postulated here is that a leucosome in these rocks does not represent the original melt injected. The conventional model mostly neglects the possible interaction between the melt and the wall rocks, but the mass balance in these rocks requires that some felsic components were segregated from the palaeosome and combined with the injected melt to form present leucosome (Figure 4.12). Interaction between the melt and the wall rocks may have been partial melting of the palaeosome by the influx of H_2O from a hydrous melt, or a diffusional process that modified the injected melt toward the composition of the melt in equilibrium with the palaeosome. The selvages must be the residues of the segregation process. The calculated melt compositions are similar to those of small plutons and dykes present in the area (see Hedge, 1969) indicating that these intrusives or their sources are likely sources of the injected melts in the migmatites.

Of the two large samples of open-system migmatites, one (10-3-T) contains

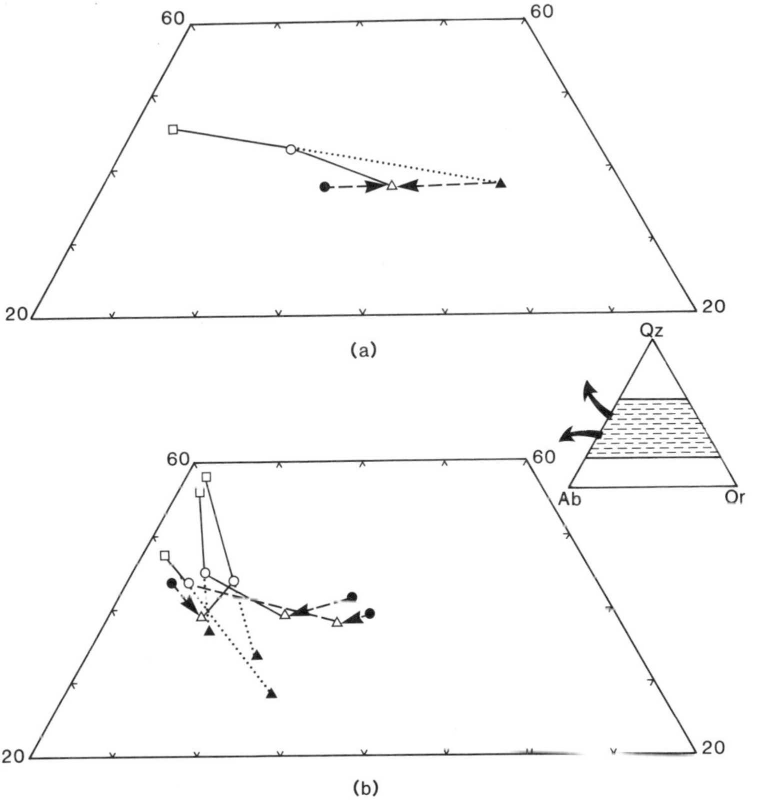

Figure 4.12 Mass balance in open-system migmatites formed by injection: (*a*) sample 10-3-E; (*b*) samples 3-2-C, 3-2-D, 3-3-D. Symbols: open circle, palaeosome; open triangle, leucosome; open square, total selvage; solid triangle, calculated locally derived mass in the leucosome; solid circle, calculated excess mass in the leucosome; all in modal percent.

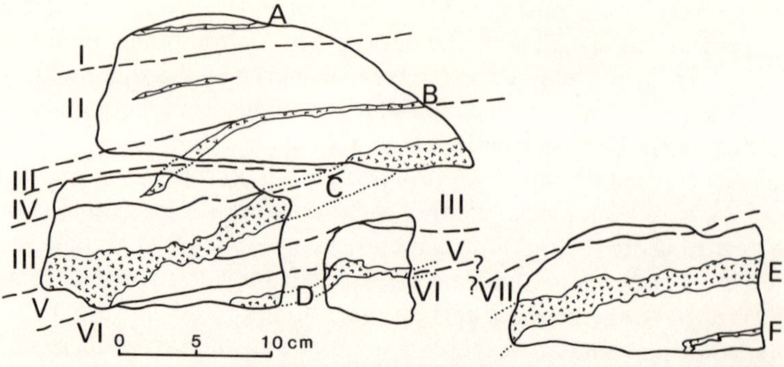

Figure 4.13 Sample 10-3-T. Roman numerals, palaeosome; solid lines, boundaries between different palaeosomes, dashed where inferred. Hatched pattern, leucosome.

six (Figure 4.13) and the other (10-3-W), three leucosomes with average thickness from 2 to 20 mm rimmed on both sides by mafic-rich selvages 1 to 10 mm thick. The palaeosome in 10-3-T, although homogeneous in appearance, can be divided into seven layers based on the difference in microcline and hornblende contents. The contacts between the layers are quite sharp and modal composition within a layer is constant. The leucosomes are generally parallel to the layering but at least three are partly cross-cutting. The three leucosomes, or at least parts of them, occur at the contact between two layers. The layering is interpreted as due to original compositional layering and not to migmatization.

All measured leucosome proportions for the nine leucosomes are larger than those estimated by the least-squares analyses (Figure 4.5*b*). For two leucosomes (10-3-T-E and T-F), the measured and estimated proportions agree within the uncertainties in data but the sum of squares of residuals is fairly large. None of the leucosome + selvage sets therefore is likely to have formed from the adjacent palaeosome(s) isochemically. The results can be interpreted in either of two ways: (1) these leucosomes formed in an open system by exchanging masses with external source (s); (2) each of the present leucosome + selvage sets formed from a rock whose composition was different from that of the palaeosome. The second possibility is considered first.

Johannes and Gupta (1982) concluded that the leucosome + selvages in migmatites of Arvika, Sweden, formed by selective partial melting of the paragneiss layers with more suitable (= more granitic) compositions than the unmigmatized mesosomes or host rocks. If the same was true for these rocks, the leucosome must be a low-temperature melt of a rock whose composition was equal to the net leucosome + selvage composition assuming an isochemical migmatization. The possibility was considered unlikely for several reasons: (1) some of the leucosomes seem to cut across compositional layering (Figure 4.13), which is probably primary; (2) the ranges of net, leucosome + selvage, and palaeosome compositions overlap and at least one

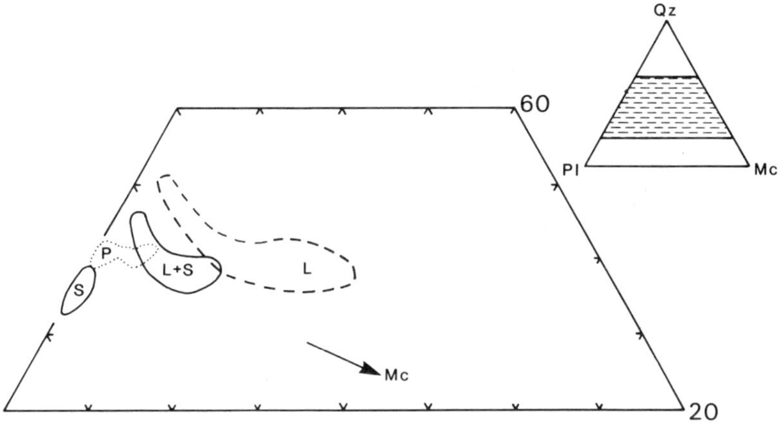

Figure 4.14 Ranges of total compositions of the leucosome (L), selvage (S), palaeosome (P) and net leucosome + selvage (L + S) in 10-3-T and 10-3-W.

of the palaeosomes is more granitic than some of the net, leucosome + selvage, compositions (Figure 4.14); (3) leucosome compositions are much more varied than those of the net, leucosome + selvage (Figure 4.4), not a likely relationship between the partial melts and their host rocks. The spread in leucosome composition may indicate a temperature range for the leucosome formation (see Winkler *et al.*, 1975). Such melting interval, however, is not consistent with the model that the leucosome + selvage zones were layers with compositions more favourable for melting. Feldspar compositions stay constant within each migmatite. If the local system is completely closed, even to volatiles, all leucosome + selvage zones and mesosome (or palaeosome in this paper) must begin to melt at the same temperature, although the amounts of melts produced would differ (e.g. Henkes and Johannes, 1981). Neither is it likely that the scatter of the leucosome composition has been caused by the formation of microcline in the leucosome by breakdown of biotite as postulated above for the closed-system migmatites: the leucosome compositions cannot be reproduced by any combination of the compositions predicted to form by such a mechanism and a reasonable composition for partial melt (Olsen, 1984, Figure 8). It is most likely that these leucosomes formed in an open system and the rocks contain introduced masses from external source(s). The net excess mass (gains – losses), calculated assuming that they were contained in the leucosome and minimizing the difference between the measured and estimated leucosome compositions as discussed above, range in volume from 3 to 30% of the leucosome (Figure 4.15). The larger a leucosome is, the more material seems to have been introduced into the leucosome: in general, the net excess mass increases with the average thickness of the leucosome. Figure 4.16*a* indicates that, in general, the leucosome + selvage zone gained Si, Al, K and Fe, and lost Mg relative to

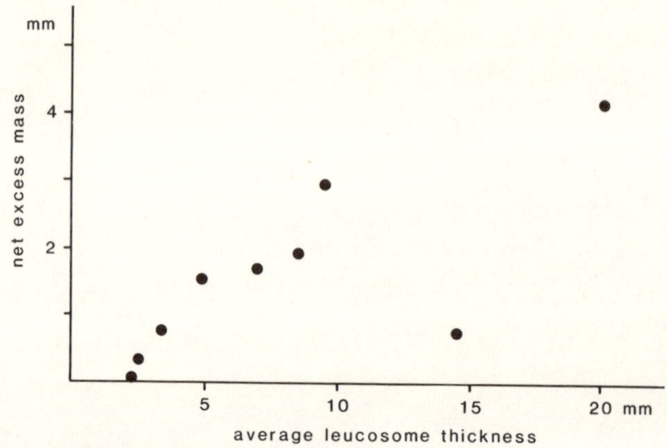

Figure 4.15 Relationship between the amount of excess mass in the leucosome and leucosome thickness, assuming the mass to be contained and present as a layer in the leucosome.

the palaeosome. The calculated gains and losses are not consistent with large-scale introductions of silicate melts (compare above) but reasonably consistent with metasomatic introductions, probably by a fluid. The possibility of a melt injection was rejected because it requires a wide range in melt composition (Olsen, 1984, Figure 12). The melt injection was rejected also on the basis of field evidence: some 'melts' must be less than 5 mm thick and be introduced over a distance of at least 20 m or more because no obvious source of the melts is found in the outcrop. It seems physically improbable that such injection can occur without a rapid quenching of the melt. For the same reason, it is not likely that the mass transport occurred by mass exchanges between the rock and melt flowing through the rock.

If the transport medium for the components was a fluid, Al gains and losses were probably very small: fluid inclusion studies (e.g. Roedder, 1972) indicate that Al concentrations in most metamorphic fluids should be low, and Al seems to be relatively immobile in many metamorphic environments (e.g. Carmichael, 1969). Accordingly, the gains and losses were recalculated on constant Al basis. The results in Figure 4.16*b* are interpreted to show that some components, especially K, were introduced into the rock by a fluid along what is now a leucosome and were fixed in the rock by a reaction with the minerals. Other components, especially Na and Mg, and probably Ca, were liberated by the same reaction and removed from the rock.

The metasomatic reaction which could account for fixing and liberating the gained and lost components was calculated from the modal and mineral data. Figure 4.17 shows volumes of minerals formed (+) or consumed (−) per 1000 cm^3 of the palaeosome in the process of its transformation to a leucosome and selvages. The main effect of the metasomatic reaction is clearly an increase in microcline. In addition, plagioclase and biotite decrease while magnetite increases in most. The reactions seem to have been replacement of plagioclase

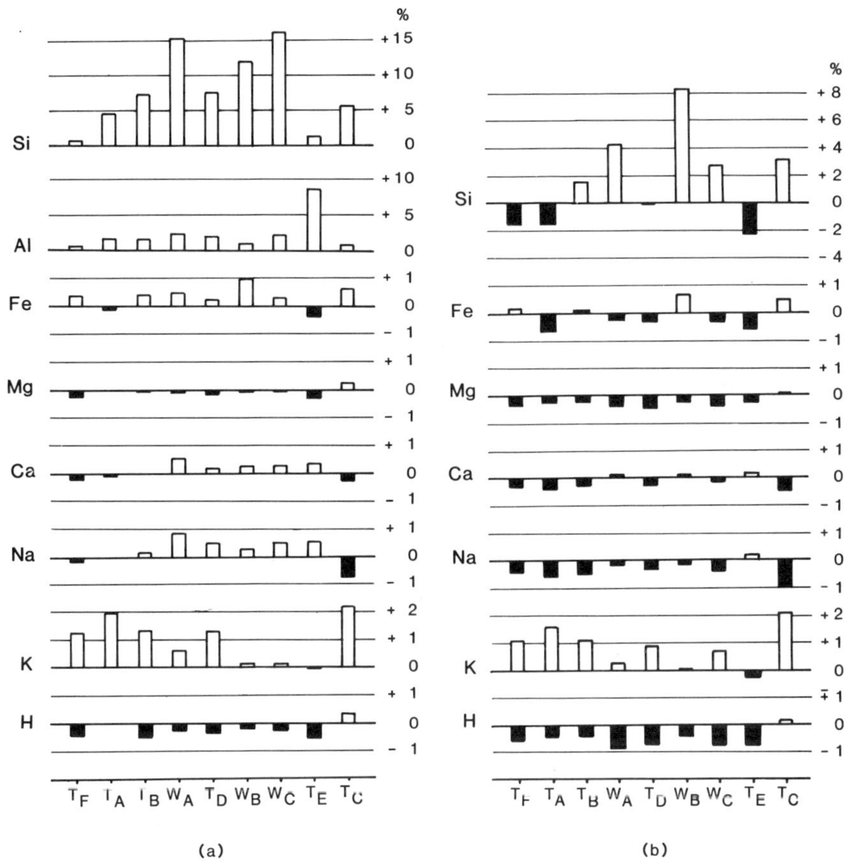

Figure 4.16*a* Gains and losses of components in samples 10-3-T and 10-3-W from Front Range under the assumption that they were contained in the leucosome (see text). In cation-mole % per 1 mole palaeosome.
b Gains and losses of components on constant-Al basis. In cation-mole % per 1 mole palaeosome.

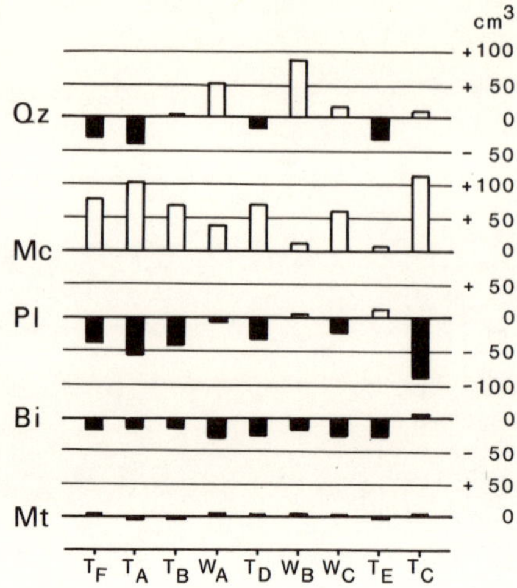

Figure 4.17 Modal gains and losses of minerals in the net leucosome + selvage, in cm^3 per 1000 cm^3 palaeosome, Al conserved.

by microcline and breakdown of biotite, such as

$$\text{plagioclase} + \text{quartz} + KO_{1/2} \rightarrow \text{microcline} + NaO_{1/2} + CaO \qquad (4.4)$$

$$\text{biotite} + \text{quartz} \rightarrow \text{microcline} + \text{magnetite} + MgO + H_2O \qquad (4.5)$$

Because of uncertainties in the mass-balance calculations, exactly balanced reactions are probably not very meaningful. However, the net reaction below calculated for the smallest leucosome (T-F, Figure 4.13) is representative of these reactions.

$$\begin{aligned} &0.11\,\text{Bio} + 0.37\,\text{Pla} + 1.36\,\text{Qz} + 0.04\,\text{Hb} + 0.53\,KO_{1/2} \\ &\quad + 0.11\,FeO + 0.002\,TiO_2 \rightarrow \\ &0.70\,\text{Mc} + 0.11\,\text{Mt} + 0.02\,\text{Sph} + 0.67\,SiO_2\,(\text{sol.}) + 0.15\,MgO \\ &\quad + 0.16\,CaO + 0.19\,NaO_{1/2} + 0.17\,HO_{1/2} \end{aligned} \qquad (4.6)$$

It seems likely that incipient migmatization occurred by such an exchange reaction. The reactions are nearly constant volume reactions ($\Delta V = -2$ to $+2\%$) except in the case of the leucosomes in 10-3-W ($\Delta V = +3$ to $+11\%$) for which a later introduction of Si is suspected, and that of 10-3-TC ($\Delta V = 5\%$) for which an igneous injection cannot be ruled out. All except one are also dehydration reactions (Figure 4.16*b*).

The larger part of the leucosome must represent the mass which was already present in the rock. Such locally derived mass in the leucosome most likely was extracted from the palaeosome forming the mafic selvages as residues. The composition of the mass (LM, Figure 4.3) was calculated assuming that Al transport occurred within the leucosome + selvage zone but not in or out of

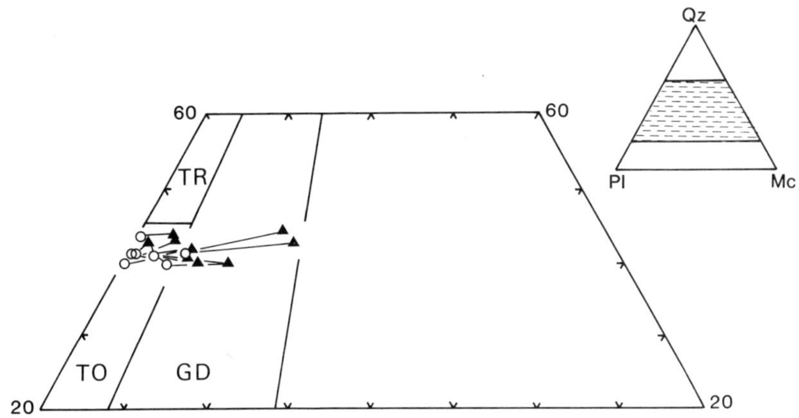

Figure 4.18 Calculated modal compositions of locally derived masses (solid triangles) and palaeosomes (open circles) (see also text). Immediately adjacent palaeosome and leucosome are connected by a tie line. TR, trondhjemite; TO, tonalite; GD, granodiorite fields.

the rock. The first assumption is necessary because partial melting and transport of the partial melt within these rocks are likely. The excess mass must be assumed to be in the leucosome as the necessary constraint for the mass-balance calculation (see Figure 4.3). The positive numbers in the calculated compositions (Olsen, 1984, Table 3) represent the phases lost by the palaeosome to the leucosome. The positive biotite and negative magnetite values in most of the calculated compositions, however, include the amounts of biotite consumed and magnetite produced in the migmatization reactions such as reaction 4.5 (compare Figure 4.17): it is not necessary to assume physical transport for entire amounts of these minerals. The felsic compositions of calculated locally derived masses (Figure 4.18) which include the microcline formed from biotite by reaction 4.5, vary fairly systematically with those of adjacent palaeosome except perhaps for one. Their tonalitic to granodioritic compositions and the correlation between their compositions and those of the palaeosomes suggest strongly that these locally-derived masses were partial melts formed in the palaeosomes.

When all the available modal compositions of the palaeosome-mafic selvage pairs in the Front Range rocks are combined, the evidence for partial melting is even more convincing (Figure 4.19). Each arrow in Figure 4.19*a* starts at the selvage composition and ends at the palaeosome composition at the point. The palaeosome-mafic selvage pairs are divided roughly by line A–A′ into two groups: above the line, the selvage is more quartz-rich than the palaeosome, and below the line, the selvage is more plagioclase-rich (compare Ashworth, 1976, Figures 4 and 8). A few which fall on the wrong side of the line will be discussed below. The relationship is to be expected if the selvages formed from the palaeosome by extraction of the partial melts similar to those on the cotectic surface plagioclase-quartz in the system Ab–An–Or–Qz–H_2O

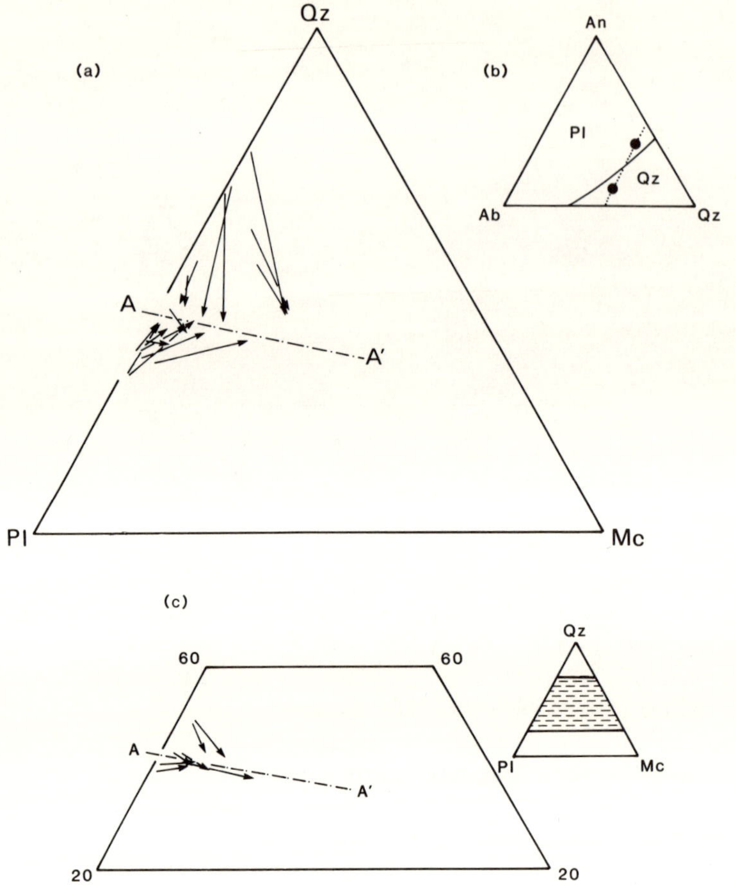

Figure 4.19*a* Modal compositions of palaeosome (at the point of arrow) and mafic selvage (at the end of arrow) in the Front Range migmatites. Some data below line A–A′ have been omitted for clarity.
b System Qz–An–Ab–H_2O after Yoder (1968): solid line, qz–plag cotectic; solid circles, bulk compositions. All compositions on dotted line have same qz–plag ratio.
c Modal compositions of palaeosome (at the point of arrow) and microcline-depleted zone (at the end of arrow).

at a moderate P_{H_2O} (e.g. Winkler *et al.*, 1975). The overlap of the data points in Figure 4.21*a* can be explained by differences in An content of the palaeosomes. Those which fall below line A–A′ but have quartz-rich selvages have low An content. The trace of the plagioclase–quartz surface becomes more quartz-rich with increasing An content in the An–Ab–Qz–H_2O system (Yoder, 1968). Of the two rocks with the same quartz-plagioclase ratio shown in Figure 4.19*b*, one, with a low An content, is in the quartz volume and therefore should produce quartz-rich selvages while the other, with a higher An content, is in plagioclase volume and should produce plagioclase-rich selvages.

The microcline-depleted zone present between the mafic-rich selvage and palaeosome in some of the rocks (Figure 4.1) is similar in composition to the palaeosome but contains less microcline. It could have formed by an extraction of a melt, although depletion in K-spar and enrichment in plagioclase by Na-K exchange cannot be ruled out. The palaeosome-microcline-depleted zone pairs (Figure 4.19*c*) show a relationship similar to that between the palaeosome-mafic selvage pairs in Figure 4.19*a* except that the melt extracted must have been similar to a melt on the cotectic curve quartz-two-feldspars. However, the palaeosome and microcline-depleted zone in the current study samples have compositions too similar to show a clear trend. The calculated compositions of locally derived masses in the leucosomes in the Front Range rocks plot in the general area of line A–A′ as expected but with the spread reasonable for the cotectic melts with different An contents (Figure 4.19*a*). Although the observed leucosome compositions in these migmatites overlap those of the locally derived masses, their maxima are shifted toward a more microcline-rich and quartz-poor composition from those of the calculated locally derived masses (Olsen, 1983, Figure 27). The shift indicates that the Front Range migmatites as a whole may not have been a closed system.

The evidence for both metasomatism and partial melting in these rocks suggests that the introduction of a fluid caused a metasomatic reaction (Figure 4.20*a*, Stage 1) and partial melting in the wall rocks (Figure 4.20*a*, Stage 2). The partial melt segregated from the wall rocks combined with the introduced material to form the leucosome (Figure 4.20*a*, Stage 3). As more material was introduced (= more fluid), more partial melt was produced and segregated

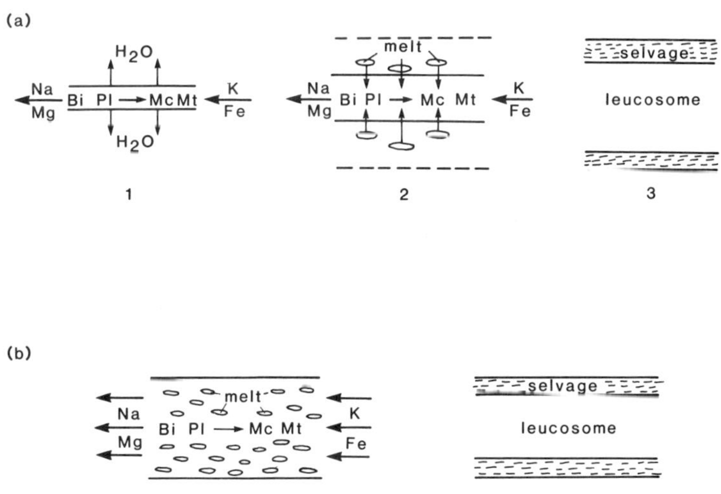

Figure 4.20 Possible mechanisms of migmatization by metasomatism and partial melting.

into the leucosome, and the leucosome grew. The transfer of the partial melt from the selvage to the leucosome may have been mechanical (filter-pressing) or diffusional. Stages 1 and 2 above need not be separate steps but could have occurred simultaneously (Figure 4.20*b*). The effect of such a process is to create a layer with a composition more suitable for melting, similar to the selectively melted paragneiss layer as described by Johannes and Gupta (1982). The spread in leucosome composition (Figure 4.14) can be explained as a result of modification in varying degrees of the partial melts by metasomatism. Alternatively, the leucosomes may have been equilibrium partial melts formed over a temperature range (see above).

The approach of Eugster and Gunter (1981) and Eugster (1981) has been used to estimate the composition of the fluid in equilibrium with those migmatites using available experimental data (see Olsen, 1984, Appendix 1). The fluid is assumed to be an OH-Cl solution and all species in the solution to be associated (e.g. Frantz and Popp, 1979). The pressure effect on the equilibria has been neglected except for quartz solubility and K/Na ratio. The calculations were made at 0.2, 1.0, and 2.0 m total Cl at 650°, 700°, and 750°C at 3.5 kbar (Figure 4.21). The general orders of species abundance in Figure 4.21 are similar to those known from the calculated values for lower temperatures and for some natural occurrences (Eugster, 1982).

If the reactions and mass transfers in these rocks were caused by the infiltration of a fluid, they must be accomplished by reasonable changes in fluid composition and reasonable fluid fluxes. The observed $K \leftrightarrows Na$ (+Ca) exchange can be accounted for, at least qualitatively, by a temperature gradient but not other mass transfers. The reaction for breakdown of biotite to microcline and magnetite (reaction 4.5) is similar to the muscovite-K feldspar reaction (Gunter and Eugster, 1980) and could be caused by a change in Cl concentration. It can also involve a redox reaction (Wones and Eugster, 1965). Obviously, there is no simple answer to the problem of which gradient in fluid property accounts for the observed mass transport without knowing the source of the fluid and the mineral assemblage with which the fluid was initially in equilibrium: such information cannot be obtained from the rocks. Temperature, Cl and f_{O_2} gradients are simply some of many possibilities. The minimum fluid needed for the transport observed in 10-3-TF at 700 °C ranges from 300 cm^3 of 2 m Cl solution (for Na) to 4 000 000 cm^3 of 0.2 m Cl solution (for Mg) per 1000 cm^3 of the palaeosome with a median value of roughly 5000 cm^3 of 1 m Cl solution (Olsen, 1984, Table 4). In comparison, the amount of water needed from the outside source(s) to saturate the melt in 10-3-TF is about 105 cm^3 per 1000 cm^3 of the rock (see, e.g., Burnham, 1979). These model calculations and speculations indicate that the metamorphic solutions with the compositional range as calculated above could have caused the metasomatic reactions and mass transfers observed in these rocks, but many more experimental and field data are needed for a definite conclusion.

The two samples discussed in this section are intermediate between the

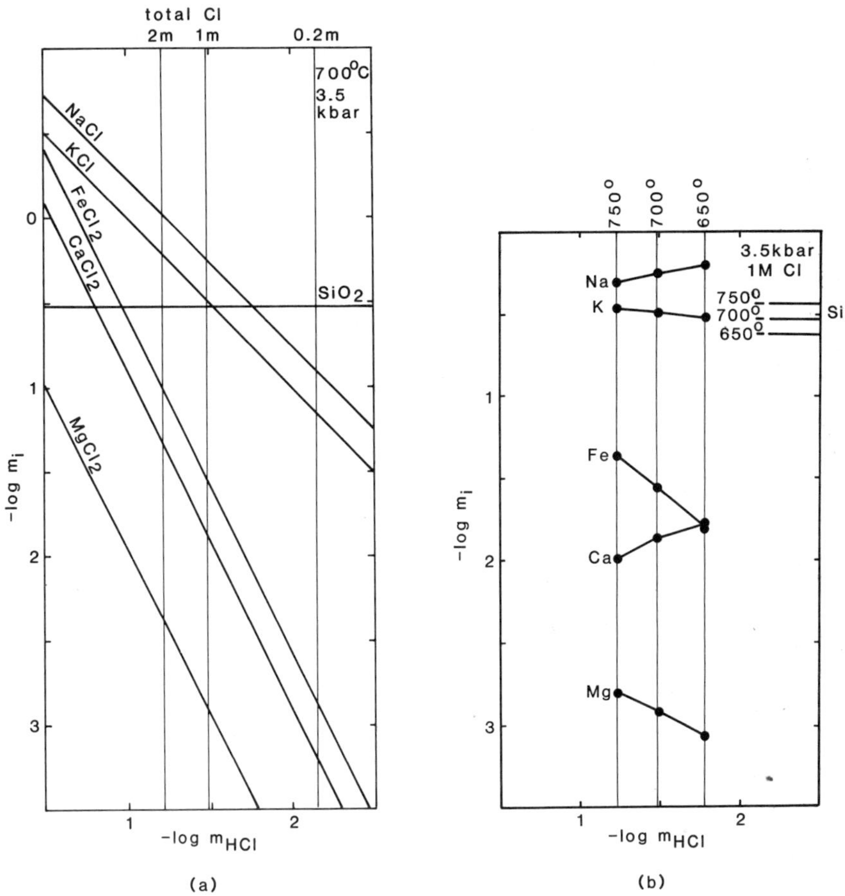

Figure 4.21*a* Estimated compositional range of the solution in equilibrium with qz–Kspar–plg–bio–mt(ms) at the *P–T* conditions of migmatization.
b Concentration gradients in the 1 M Cl solution due to temperature gradient.

closed-system migmatites and the migmatites formed by injection in amount of material introduced. This suggests that the Front Range migmatites contain introduced material which ranges in amount from nil to a large part of the leucosome and that the nature of the introduced material ranges from a water-rich fluid to a silicate melt. The closed-system migmatites, therefore, may only approximate closed systems and may have formed by a metasomatism and partial melting. If so, the amount of material introduced is too small to be detected by the method used.

Closed-system leucosomes in the Baltimore Gneiss and the Front Range migmatites tend to be thin layers up to a few centimetres at most, often planar with well-defined, thick (relative to leucosome) mafic selvages (Olsen, 1982,

Plate 1B). The open-system leucosomes thought to have formed by injection tend to be somewhat thicker and ptygmatically folded (Olsen, 1982, Plate 1A) and typically occur singly in a fairly large volume of homogeneous palaeosome. In the field, they may be classified as pegmatites or dykes. The difference is that the leucosomes are in most part conformable and have fairly well-defined mafic selvages. The other open-system migmatites thought to have formed by metasomatic infiltration of a fluid are more or less transitional between the two types above in size and appearance. These generalizations are based on very small number of samples and should not be relied upon in assessing mass balance in the field. They may, however, help in selecting samples to be studied in more detail.

4.5 Littleton Formation migmatites

Dougan (1979, 1981, 1983) used chemical (both major and minor elements) and textural evidence to assess mass balance in migmatitic metapelites in the Devonian Littleton Formation of New Hampshire and Maine. His samples came from layered metasedimentary rocks in which leucosomes one to several tens of centimetres thick had been developed, but his study samples contain leucosomes 1 to 5 cm thick. The leucosomes occur most commonly as lenses but also as irregular pods and discontinuous layers. Melanosomes, or mafic-rich zones 2 to 6 cm thick, foliated but homogeneous in appearance, border most of the leucosomes on both sides. Clear-cut mafic selvages are rare. Melanosomes, however, show systematic variations in composition, both chemical and modal, which are roughly symmetrical on both sides of the leucosome.

Although grossly similar, there are some significant petrochemical differences between the study samples of Dougan and those of Olsen discussed above. Each melanosome in Dougan's sample is a relatively thick zone a few times the thickness of the leucosome and there is a gradual change in mineral mode within the zone. On the other hand, mafic-rich selvages in Olsen's samples are generally much thinner than the leucosome and there is no discernible variation in mode. Leucosomes in both cases are similar in thickness.

4.5.1 *Mass-balance calculations: major elements*

The samples were cut normal to foliation in 1 to 2 cm slabs. Leucosomes were cut out in jigsaw fashion. Chips were cut from the melanosome with their lengths parallel to the leucosome (e.g. Dougan, 1979, Figure 4), and were used for thin sections and as samples for bulk rock and mineral (partial) analyses by X-ray fluorescence and atomic absorption methods. In general, variation in both chemical and modal compositions within the melanosome is almost linear. At a distance of 2 to 4 cm from the leucosome–melanosome contact, the composition becomes uniform (e.g. Dougan, 1979, Figure 4). The homogeneous portion of the melanosome was considered by Dougan to be the palaeosome or unaltered original rock.

There are also layerings in the melanosome and palaeosome which are interpreted to reflect original sedimentary layering: they show modal variations which are not systematic and unrelated to the distance to the leucosome. The unmigmatized homogeneous portion of the metasedimentary layer that contains a leucosome (approximately 1 m from the leucosome) has a composition identical to other similar metasedimentary layers without leucosomes at the same locality. The composition is also similar to the 'palaeosome' of the study sample containing the leucosome. This indicates that the leucosome and melanosomes formed from an originally homogeneous rock. The reason for selectiveness of the process which resulted in migmatization of one layer but not the other is not offered. There is a difference in composition between the palaeosome of migmatized samples and the unmigmatized rock, but it does not seem to be significant enough to indicate a compositional difference as the cause of selective migmatization (compare Johannes and Gupta, 1982).

The palaeosomes in these rocks are basically quartz–plagioclase–muscovite–biotite–sillimanite ± garnet schists. The melanosomes have basically the same assemblage with different modal proportions. The leucosomes typically have quartz–(muscovite or K-feldspar)–plagioclase ± sillimanite ± garnet assemblage. Roughly linear decreases in the leucosome phases within the melanosome suggest that segregation of these phases into a leucosome from the originally homogeneous palaeosome resulted in the melanosome. The melanosome, therefore, is a depletion zone. Compared to the palaeosome, the leucosome is most consistently depleted in SiO_2, Fe_2O_3 (= total Fe), MgO, and TiO_2 and enriched in Na_2O. Al_2O_3, K_2O, and CaO show trends which vary. Because a typical leucosome has little or no Fe_2O_3, the enrichment of Fe_2O_3 in the melanosome should be directly proportional to the removal of the leucosome material. MgO and TiO_2 can also be used as the reference. However, because their concentrations are low in the palaeosome, the results of calculations balanced on one of these components are likely to have large uncertainties.

In the simplest cases of the leucosomes, oxide wt % plotted on a variation diagram with wt % Fe_2O_3 along the abscissa show clearly linear variations (e.g. Dougan, 1979, Figure 7). Such variations indicate a simple extraction of some of the components that resulted in enrichment of Fe_2O_3 and remaining components. Modal variations can also be plotted against Fe_2O_3 (in wt %) after being recalculated as wt % from volume %. However, these variations are based on the assumption of constant volume or weight. If the leucosome material was extracted from the volume the melanosome now occupies, there must have been a fairly large volume decrease. Weight % of each mineral must therefore be recalculated to compensate for the volume decrease to obtain the true variation. Because

$$Fe_2O_3^P = X_L \cdot Fe_2O_3^L + (1 - X_L)Fe_2O_3^M$$

in which $Fe_2O_3^M$, $Fe_2O_3^P$, $Fe_2O_3^L$ are Fe_2O_3 contents in the melanosome, palaeosome, and leucosome respectively, and $Fe_2O_3^L \approx 0$,

$$X_L = \frac{Fe_2O_3^M - Fe_2O_3^P}{Fe_2O_3^M - Fe_2O_3^L} \approx \frac{Fe_2O_3^M - Fe_2O_3^P}{Fe_2O_3^M}$$

in which X_L is the weight proportion of the leucosome in the original palaeosome or the weight decrease due to extraction of leucosome material. Weight % of minerals in some of the melanosomes shows linear variations with Fe_2O_3 content (for example, Dougan 1979, Figure 8*b*) and the phases decreased are those present in the leucosome. The relationship is termed 'concordant' removal by Dougan (1979). This indicates that each phase was extracted as an entity which would be possible in each of the following cases: (1) mechanical transport of the mineral grain, which seems unlikely; or (2) congruent dissolution of the mineral into an intergranular phase which may be a fluid or silicate melt, and transport to the leucosome by either diffusion in or physical movement of the fluid or melt. Dougan concluded that the congruent melting and mechanical transport of the melt was the most likely mechanism.

The slope in the variation diagram such as oxides *v.* Fe_2O_3 is the rate by which the amount of each oxide is changing in the melanosome. Therefore relative slopes are relative proportions by which these components or phases are being extracted. If the migmatite has been a closed system and all the extracted material is now contained in the leucosome, the relative slopes should be identical to the relative proportions of components or phases in the leucosome. Such predicted leucosome compositions were calculated by Dougan for several samples. They agree well with the observed compositions indicating closed systems in these migmatites (e.g. Dougan, 1981, p. 339).

In some of the melanosomes, although there are linear variations in oxides, mineral variations are irregular (e.g. Dougan, 1979, Figure 11). Dougan concluded that the irregularities resulted from mineral reactions within the melanosome. The melanosome reactions are also indicated when the content of a mineral not present in the leucosome such as biotite decreases even if the recalculated variations are regular (Sample L3A, Dougan, 1981). Mineral reactions were deduced from the sympathetic or antipathetic relations between the amounts of phases, and presence of a phase which is present only in the melanosome or leucosome. A sample with the leucosome which contains a phase not present in the melanosome was termed a 'discordant' sample by Dougan (1979).

Basically two sets of reactions seem to be applicable to these rocks. The first is

$$\text{biotite} + 2\ \text{sillimanite} + \text{quartz} \rightleftarrows \text{garnet} + \text{muscovite} \qquad (4.7)$$

The second is a combination of two reactions:

$$\text{biotite} + \text{sillimanite} + 2\ \text{quartz} \rightleftarrows \text{garnet} + \text{K-feldspar} + H_2O \quad (4.8)$$

$$\text{muscovite} + \text{quartz} \rightleftarrows \text{K-feldspar} + \text{sillimanite} + H_2O \quad (4.9)$$

Because biotite is not present in the leucosome, a decrease in biotite content of the melanosome is assumed to be due to reaction 4.7 or 4.8, depending on whether assemblage (muscovite + garnet) or (K-feldspar + garnet) is present in the leucosome. An increase in garnet in the melanosome toward the leucosome with little or no garnet in the palaeosome was also interpreted as an indication that garnet was a reaction product. When the variations are recalculated to a constant biotite content in the melanosome, the variations in other minerals become linear and regular in some of the samples (e.g. Sample L2H-1, Dougan, 1979, Figure 11). If some irregularities still remain, the variations are recalculated using reaction 4.9. The results generally show more regular variations (e.g. Sample L2A2, Dougan, 1981).

Relative proportions of the phases removed, as determined after correcting the mineral proportions for reaction 4.7 or reactions 4.8 and 4.9, are the predicted leucosome compositions. Good agreements in most cases indicate validity of the assumption of a closed-system segregation of leucosome material and of the melanosome reactions postulated. Table 4.1 shows the

Table 4.1 Inferred reactions in two samples of Dougan (1981)

	Melanosome		Leucosome	Predicted leucosome	Observed leucosome
L3A					
	8.3 Bio		8.1 Gt		
'Discordant'	5.6 Sill	→	6.9 Ms		
removal	1.0 Qz			7.8% Gt	8.1% Gt
				22.5% Ms	22.4% Ms
	16.5 Ms	→	16.5 Ms	24.4% Pla	24.2% Pla
	25.4 Pla	→	25.4 Pla	45.2% Qz	44.9% Qz
'Concordant'	47.1 Qz	→	47.1 Qz	0.1% Sill	0% Sill
removal	0.1 Sill	→	0.1 Sill		
L2A2					
	2.0 Bio				
'Discordant'	0.7 Sill	→	1.2 K-spar		
removal	0.6 Qz				
	↓				
	2.0 Gt			13.9% K-spar	12.0% K-spar
				42.5% Pla	44.6% Pla
'Discordant'	0.4 Ms		0.3 K-spar	35.3% Qz	35.2% Qz
removal	0.1 Qz	→	0.2 Sill	8.3% Sill	8.2% Sill
	3.8 Qz	→	3.8 Qz		
'Concordant'	4.6 Pla	→	4.6 Pla		
removal	0.7 Sill	→	0.7 Sill		

mass transfer and reactions in Samples L3A and L2A2 as postulated by Dougan (1981).

It is interesting to note that if the reactions 4.8 and 4.9 progress at about equal rates, the net reaction can be reaction 4.7. In order for the net reaction to proceed to right as inferred by the data, reaction 4.9 must proceed from right to left, or it must be a retrograde reaction.

The slopes in variation diagrams *for oxides* of some samples are not linear but curvilinear even after the correction for volume decrease due to leucosome-component depletion (Dougan, 1979, Figure 15). Such changes in slope are postulated to be the indication for changes in composition of the material extracted from the melanosome. Relative proportions of the phases extracted from any point in the melanosome are equal to the ratios of slopes of tangents to the variation curve at the point. The compositions of the extracted material calculated for several melanosomes show changes away from the granite minimum (Dougan, 1979, Figure 20). The farther away the melanosome is from the leucosome, the closer the extracted material is to the initial melt in composition. The trend was interpreted to show the progressive melting of the melanosome and a more efficient removal of the melt from the zone closer to the leucosome.

4.5.2 *Mass-balance calculations: trace elements*

Trace-element analyses of the same migmatite samples also indicate closed-system segregation of the leucosome. Trace-element variations with Fe_2O_3 content in the melanosome are basically identical to those of the major elements: they are regular and systematic and are related to the degree of depletion of the leucosome material or increasing Fe_2O_3 contents (Dougan, 1981). In the case of the 'concordant' samples, in which all leucosome phases are present in the melanosome, trace-element concentrations are the same in each mineral, whether in the melanosome or leucosome. Such uniform concentration can be a result of equilibration by diffusion after the formation of leucosomes. However, the results from the 'discordant' samples, in which one of the leucosome phases is not present in the melanosome, rule out such re-equilibration. In the discordant samples, the trace-element concentrations in the minerals which are present in both melanosome and leucosome are different. For example, muscovite and plagioclase coexisting with quartz–sillimanite–biotite in the melanosome have much lower concentration of trace elements than those coexisting with quartz-garnet in the leucosome (Sample L3A, Dougan, 1979, Table 2; 1981, p. 341). The segregation process, therefore, had no effect on partitioning of the trace elements but the mineral reactions did. After the leucosome segregated and its minerals formed, it became a closed system and there was no equilibration by diffusion between the leucosome and melanosome.

Because trace-element variation in the melanosome is directly proportional to its degree of depletion in leucosome component, the trace element must have been removed in each leucosome phase as the latter was removed from the melanosome. A mechanical transport of the mineral grain is not likely. Dougan (1981) postulated the trace elements were transported by 'carrier' phases. He infers that such a 'carrier' phase was a partial melt.

Textural relationship in these Littleton migmatites also reinforces the reactions implied by the chemical and mineralogical analyses (Dougan, 1983). The grain contacts between the inferred reactant phases decrease as well as the number of grains per unit volume and grain size of the reactant phases (see also Ashworth and McLellan, this volume). The preferred reaction sites at mica-quartz and feldspar-quartz boundaries, implied by the textural analysis, agree with the preferred initial melting sites in the experiments (for example, Büsch *et al.*, 1974), indicating partial melting as the cause of leucosome segregation.

4.6 Discussion

Studies of mass balance in migmatites indicate that a migmatite can form in a closed system (= isochemically except for the volatile components) or in an open system (= non-isochemically). An external introduction of material into a migmatite may occur by a metasomatic introduction by a fluid or by an igneous injection. The results indicate the complex nature of migmatization processes, which can range widely in mechanism and in degree of departure from an isochemical system. It seems that the major driving force is partial melting and the gradients that can be created because of melting (Ashworth, 1976; Johannes and Gupta, 1982; Dougan, 1979; Olsen, 1983), although the process is not necessarily that of simple partial melting and filter pressing of the melt (Olsen, 1977, 1982, 1984).

The optimum size of a sample for a mass-balance study seems to be of the order 10 to 30 cm in diameter, although the closed-system migmatites seem to be small-scale features: these leucosomes range in thickness of only 5 to about 30 mm in diameter. Much larger systems than these sample sizes are probably mainly either open systems or contain original heterogeneities which cannot be studied by the type of method discussed above. Care must be taken to include 'palaeosomes' on both sides of the leucosomes far enough from the leucosome for the modal composition to become constant. In many cases, palaeosomes, or mesosomes, on two sides have significantly different compositions (Ljunggren, 1957; Kretz, 1966; Dougan, 1979; Olsen, 1984).

Although, ideally, mass-balance in migmatites should be studied in three dimensions, the tactical problem involved in such a study is formidable. It is rather interesting to see that all of the detailed, quantitative or semi-quantitative studies on mass-balance have been two-dimensional and yielded similar and reasonable results. This encourages one to assume that the results of a two-dimensional study can be justified as a reasonable approximation of three-dimensional mass-balance in migmatites. This also indicates that local mass transfer was mainly perpendicular to the leucosome-melanosome boundary and that any external introduction of the material along the strike of the leucosome becomes fairly uniformly distributed through a leucosome.

References

Ashworth, J.R. (1976) Petrogenesis of migmatites in the Huntly-Portsoy area, north-east Scotland. *Miner. Mag.* **40**, 661–682.

Babcock, R.S. (1973) Computational models of metasomatic processes. *Lithos* **6**, 279–290.

Barth, T.F.W. (1948) Oxygen in rocks: a basis for petrographic calculations. *J. Geol.* **56**, 50–60.

Brown, P.E. (1967) Major element composition of the Loch Coire migmatite complex, Sutherland, Scotland. *Contrib. Miner. Petrol.* **14**, 1–26.

Bryan, W.B., Finger, L.W. and Chayes, F. (1969) Estimating proportions in petrographic mixing equations by least-squares approximation. *Science* **163**, 926–927.

Burnham, C.W. (1979) Magmas and hydrothermal fluids. In *Geochemistry of Hydrothermal Ore Deposits*: second edition, ed. Barnes, H.L., Holt, Rinehart and Winston, New York, 71–136.

Büsch, W., Schneider, G. and Mehnert, K.R. (1974) Initial melting at grain boundaries. Part II: Melting in rocks of granodioritic, quartz-dioritic and tonalitic composition. *Neues Jb. Miner. Mh.* 345–370.

Carmichael, D.M. (1969) On the mechanism of prograde metamorphic reactions in quartz-bearing pelitic rocks. *Contrib. Miner. Petrol.* **20**, 244–267.

Cheng, Y.C. (1943) The migmatite area around Bettyhill, Sutherland. *Quart. J. Geol. Soc. Lond.* **99**, 107–154.

Dougan, T.W. (1979) Compositional and modal relationships and melting reactions in some migmatitic metapelites from New Hampshire and Maine. *Am. J. Sci.* **279**, 897–935.

Dougan, T.W. (1981) Melting reactions and trace element relationships in selected specimens of migmatitic pelites from New Hampshire and Maine. *Contrib. Miner. Petrol.* **78**, 337–344.

Dougan, T.W. (1983) Textural relations in melanosomes of selected specimens of migmatitic pelitic schists: Implications for leucosome-generating processes. *Contrib. Miner. Petrol.* **83**, 82–98.

Engel, A.E.J. and Engel, C.G. (1958) Progressive metamorphism and granitization of the major paragneiss, northwest Adirondack Mountains, N.Y. *Geol. Soc. Am. Bull.* **69**, 1369–1413.

Eugster, H.P. (1982) Rock-fluid equilibrium systems. In *Bochum Conference on High Temperature, High Pressure Equilibria*, ed. Schreyer, W., Stuttgart, pp. 501–518.

Eugster, H.P. and Gunter, W.D. (1981) The compositions of supercritical metamorphic solutions. *Bull. Mineral.* **104**, 817–826.

Ferry, J.M. (1982) A comparative geochemical study of pelitic schists and metamorphosed carbonate rocks from south-central Maine, U.S.A. *Contrib. Miner. Petrol.* **80**, 59–72.

Frantz, J.D. and Popp, R.K. (1979) Mineral-solution equilibria-I. An experimental study of complexing and thermodynamic properties of aqueous $MgCl_2$ in the system $MgO–SiO_2–H_2O–HCl$. *Geochim. Cosmochim. Acta* **43**, 1223–1239.

Gunter, W.D. and Eugster, H.P. (1980) Mica-feldspar equilibria in supercritical alkali chloride solutions. *Contrib. Miner. Petrol* **75**, 235–250.

Hedge, C.E. (1969) Petrogenetic and Geochronological Study of Migmatites and Pegmatites in the Central Front Range. Ph.D Thesis, Colorado School of Mines, Golden, Colorado (unpubl).

Henkes, L. and Johannes, W. (1981) The petrology of a migmatite. (Arvika, Värmland, Western Sweden). *Neues Jb. Miner. Abh.* **141**, 113–133.

Holmquist, P.J. (1921) Typen und Nomenklatur der Adergesteine. *Geol. Fören. Stockholm Förh.* **43**, 612–631.

Hopson, C.A. (1964) The crystalline rocks of Howard and Montgomery Counties. In *The Geology of Howard and Montgomery Counties*, Maryland Geol. Surv., 27–334.

Hughes, C.J. (1970) The significance of biotite selvedges in migmatites. *Geol. Mag.* **107**, 21–24.

Johannes, W. and Gupta, L.N. (1982) Origin and evolution of a migmatite. *Contrib. Miner. Petrol.* **79**, 114–123.

Kays, M.A. (1976) Comparative geochemistry of migmatized, interlayered quartzofeldspathic and pelitic gneisses: A contribution from rocks of southern Finland and northeastern Saskatchewan. *Precambrian Res.* **3**, 433–462.

Kenah, C. and Hollister, L.S. (1983) Anatexis in the Central Gneiss Complex, British Columbia. In *Migmatites, Melting and Metamorphism*, ed. Atherton, M.P. and Gribble, C.D., Shiva, Nantwich, 142–162.

King, B.C. (1965) The nature and origin of migmatites: metasomatism or anatexis. In *Controls of Metamorphism*, ed. Pitcher, W.S. and Flinn, G.W., Oliver and Boyd, Edinburgh, 219–234.

Kretz, R. (1966) Metamorphic differentiation at Einasleigh, N. Queensland: *J. Geol. Soc. Australia* **13**, 561–582.

Ljunggren, P. (1957) Banded gneiss from Gothenburg and their transformations: *Geol. Fören. Stockholm Förh.* **79**, 113–132.

Luth, W.C., Jahns, R.H. and Tuttle, O.F. (1964) The granite system at pressures of 4 to 10 kb. *J. Geophys. Res.* **69**, 759–773.

MacRae, N.D. and Nesbitt, H.W. (1980) Partial melting of common metasedimentary rocks: A mass balance approach. *Contrib. Miner. Petrol.* **75**, 21–26.

Mehnert, K.R. (1951) Zur Frage des Stoffhaushalts anatektischer Gesteine. *Neues Jb. Miner. Abh.* **82**, 155–198.

Mehnert, K.R. (1968) *Migmatites and the Origin of Granitic Rocks.* Elsevier, Amsterdam.

Mehnert, K.R. and Büsch, W. (1982) The initial stage of migmatite formation. *Neues Jb. Miner. Abh.* **145**, 211–238.

Misch, P. (1949) Metasomatic granitization of batholithic dimensions: Pt. 1, Synkinematic granitization in Nanga Parbat area, northwest Himalayas. *Am. J. Sci.* **247**, 209–245.

Misch, P. (1968) Plagioclase compositions and non-anatectic origin of migmatitic gneisses in N. Cascade Mountains of Washington State. *Contrib. Miner. Petrol.* **17**, 1–70.

Olsen, S.N. (1977) Origin of the Baltimore Gneiss migmatites at Piney Creek, Maryland. *Geol. Soc. Am. Bull.* **88**, 1089–1101.

Olsen, S.N. (1982) Open- and closed-system migmatites in the Front Range, Colorado. *Am. J. Sci.* **282**, 1596–1622.

Olsen, S.N. (1983) A quantitative approach to local mass balance in migmatites. In *Migmatites, Melting and Metamorphism*, ed. Atherton, M.P. and Gribble, C.D., Shiva, Nantwich, 201–233.

Olsen, S.N. (1984) Mass-balance and mass-transfer in migmatites from the Colorado Front Range: *Contrib. Miner. Petrol.* **85**, 30–44.

Peterman, Z.E., Hedge, C.E. and Braddock, W.A. (1968) Age of Precambrian events in the northeastern Front Range, Colorado. *J. Geophys. Res.* **73**, 2277–2296.

Roedder, E. (1972) The composition of fluid inclusions. *U.S. Geol. Surv. Prof. Paper* 440-JJ, 164.

Sederholm, J.J. (1967) *Selected Works. Granites and Migmatites.* Oliver and Boyd, Edinburgh.

Tuttle, O.F. and Bowen, N.L. (1958) Origin of granite in the light of experimental studies in the system $NaAlSi_3O_3$–$KAlSi_3O_8$–SiO_2–H_2O. *Geol. Soc. Am. Mem.* **74**.

White, A.J.R. (1966) Genesis of migmatites from the Palmer region of South Australia. *Chem. Geol.* **1**, 165–200.

Winkler, H.G.F., Boese, M. and Marcopoulos, T. (1975) Low temperature granitic melts. *Neues Jb. Miner. Mh.* 245–268.

Winkler, H.G.F. and Breitbart, R. (1978) New aspects of granitic magmas: *Neues Jb. Miner. Mh.* 463–480.

Wones, D.R. and Eugster, H.P. (1965) Stability of biotite: experiment, theory, and application. *Am. Miner.* **50**, 1228–1272.

Yoder, H.S. (1968) Albite–anorthite–quartz–water at 5 kb. *Yb. Carnegie Instn. Wash.* **66**, 477–478.

5 Textures

J.R. ASHWORTH* and E.L. McLELLAN**

5.1 Introduction

Textures are important as indicators of processes of migmatization, though they may be profoundly modified by post-migmatization events. Migmatization processes considered in this chapter are anatexis, metamorphic differentiation and segregation of aqueous fluid. The latter fluid is unlikely to segregate as large bodies, but leucosomes may grow gradually from aqueous fluids in fractures (Yardley, 1978). Anatexites usually segregate under the influence of stress (see section 5.6.2), leading to complex leucosome shapes. Segregation of melt may be incomplete or even non-existent: some rocks without discrete leucosomes are treated here, namely the ophthalmites (rocks with porphyroblastic 'eyes' of feldspar) of the anatexite suite of McLellan (1983*a*). Various textural properties will be treated in turn, with discussion of their bearing on the origin and subsequent evolution of leucosomes.

5.2 Grain size

5.2.1 *Primary grain sizes*

Compared with coexisting non-leucosome, a leucosome is usually coarser: this contributes to its pale colour (Johannes and Gupta, 1982). Leucosomes vary greatly in grain size, sometimes in parallel with mesosomes. Thus, within the Dalradian, at a shallow structural level Ashworth (1976) finds grain-sizes mostly ~ 0.1–1 mm in both mesosomes and leucosomes (e.g. Figure 5.1*a*), while McLellan (1983*a*) at a deeper level finds grain-sizes ~ 2–3 mm in mesosomes and ~ 5–25 mm in leucosomes. This background of variation from finer to coarser migmatite complexes should be borne in mind when considering the valuable

*Department of Geological Sciences, University of Aston, Gosta Green, Birmingham B4 7ET, UK.

**Department of Geology and Geophysics, Yale University, P.O. Box 6666, New Haven, CT 06511, USA. Present address: Department of Geology, University of Maryland, MD 20742, USA.

suggestion of Yardley (1978) that leucosome grain-size can indicate origin. *Pegmatitic* and *aplitic* leucosomes are distinguished, by analogy with the pegmatitic and aplitic dykes accompanying granite intrusions. Coarser (pegmatitic) leucosomes may grow from aqueous fluids, while water-saturated melts are expected to produce aplitic leucosomes (Yardley, 1978). It seems that pegmatitic leucosomes or veins might be defined by large grain-size *contrast* (> 1 order of magnitude) with metasedimentary mesosome (Figure 5.1*b*). The distinction should be easiest where both the hydrothermal pegmatites and melt-derived aplites were produced simultaneously in the same area; there are hints of this in the Moines (cf. Barr, this volume). More often, migmatization produces only one type in a given area. In the Dalradian of Connemara, Eire, the pegmatitic quartz–plagioclase veins (Yardley, 1975) are common in the staurolite + sillimanite zone, whereas aplitic migmatization is confined to higher grades with cordierite + K-feldspar + sillimanite (Barber, 1985): in this case, the genetic distinction is clear between metamorphic fluid segregation (pegmatitic) and anatexis (aplitic leucosomes).

Although anatexis produced some aplitic leucosomes in the Scottish Dalradian (Ashworth, 1976; McLellan, 1983*a*), these are not clearly distinct in grain-size (or grain-size contrast with mesosomes) from a suite having non-cotectic compositions and attributed to metamorphic differentiation (McLellan, 1983*a*). The process here was diffusion, under stress gradients which controlled the leucosome geometry (McLellan, 1984). The geometry is often similar to the rather regular banding produced, in finer-grained rocks at lower grade (Figure 5.1*c*), by similar processes associated with crenulation cleavage (Hobbs *et al.*, 1976, pp. 218–220).

In grain size (as in mineralogy), melanosomes usually grade into adjacent mesosome, whereas leucosomes often have sharp boundaries. In the anatexites of Ashworth (1976), quantitative studies (Figure 5.2) show that the range of median grain-size in leucosomes overlaps that in mesosomes, but individual leucosomes are coarser than *coexisting* non-leucosomes, and melanosomes are coarser than mesosomes. The statistical procedure here assumes approximately lognormal distributions (usually true of grain sizes), which seems reasonable as the data are approximately symmetrical when graphed on a logarithmic scale, but coarse-skewed when frequency is plotted against size on a linear scale (as is also found in other migmatite areas: Dougan, 1983; McLellan, 1983*a*). The spread of grain-size within a sample (measured by standard deviation) tends to increase with increasing median in Figure 5.2. A similar trend is seen in the altogether coarser anatexites of McLellan (1983*a*, Figure 11). The spread of the feldspar size distribution increases with increasing median from mesosomes (median approximately 4.6 mm), through modified mesosomes (with plagioclase blastesis) and ophthalmites, to leucosomes (median approximately 12 mm). In the modified mesosomes and ophthalmites, incipient melting is thought to have promoted coarsening, by growth of large grains at the expense of small ones (Ostwald ripening),

a

c

Figure 5.1 Migmatites and comparable rocks in sawn surface. The scale is the same throughout. (*a*) Typical stromatic migmatite, with leucosomes parallel to the foliation of garnet-mica-schist mesosome. 'Cowhythe Gneiss', Links Bay, E of Portsoy (Specimen 88C of Ashworth, 1979*a,b*). (*b*) Pegmatitic quartz-plagioclase vein in garnet–mica–schist, Connemara. Plagioclase is cloudy (white); quartz appears grey. This is the type of vein studied by Yardley (1975). (*c*) Layering produced by metamorphic differentiation under stress in Greenschist-facies semipelite. In this

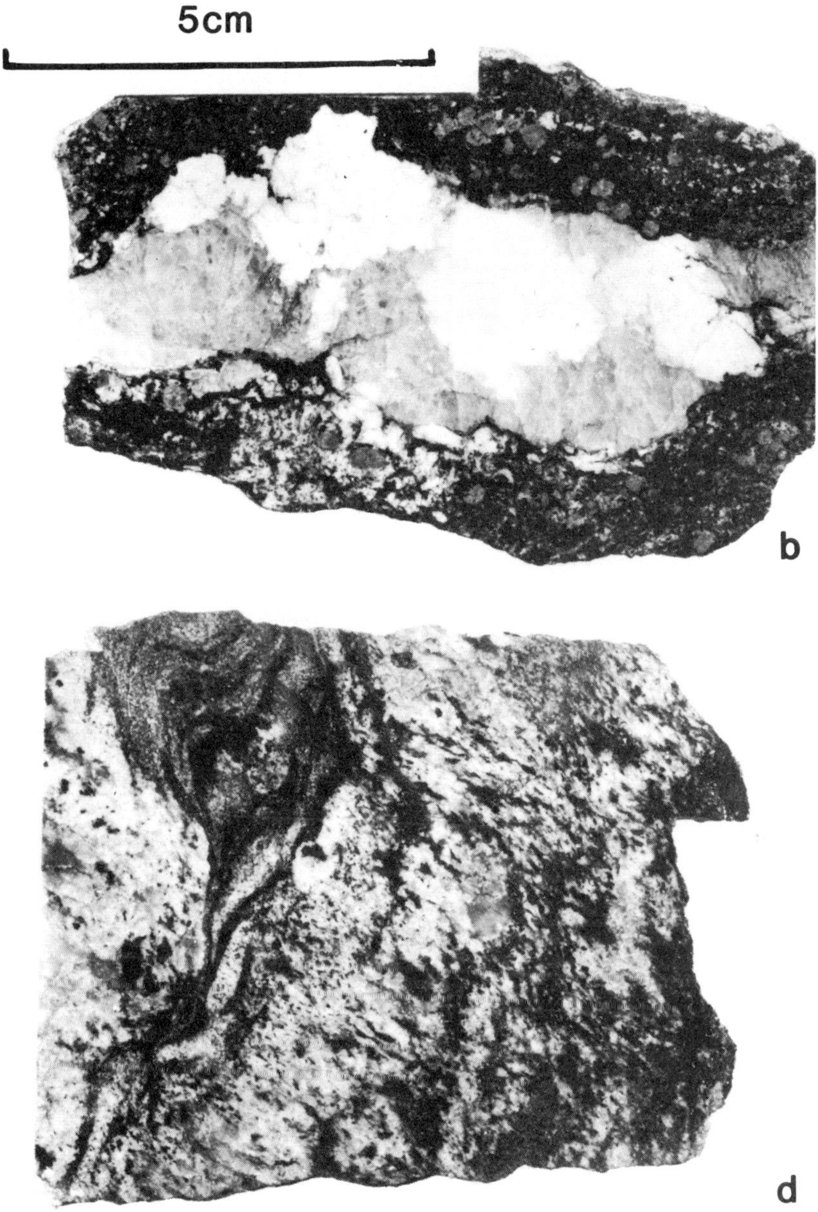

example, the layering has been subsequently folded and there are impersistent quartz veins parallel to the layering. New Harbour Group, Trearddur Bay, Anglesey. (*d*) Migmatite with complex structures, Redhythe Point, W of Portsoy. The mesosome is layered. The neosomes are coarser but have highly modified texture (see Figure 5.3*d*) and are foliated due to deformation after migmatization.

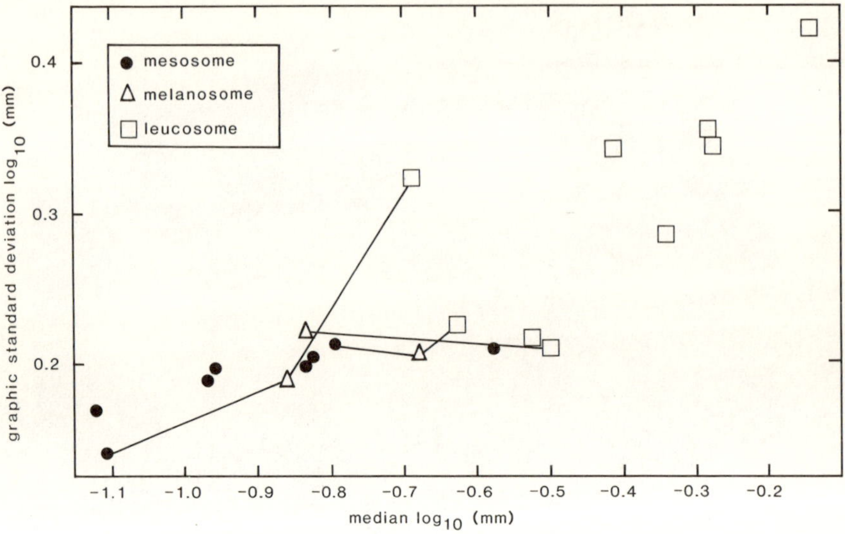

Figure 5.2 Plagioclase grain-size statistics for trondhjemitoid migmatites E of Portsoy ('Cowhythe Gneiss'). Tie-lines connect intimately coexisting bodies. Each point represents 100 grains measured by the maximum dimension in a direction perpendicular to the trace of (010) in thin-section. Statistical treatment is by the standard graphical method (Folk, 1968, p. 45).

accounting for the increasing spread (McLellan, 1983*a*). In a Precambrian paragneiss in Sweden, Johannes and Gupta (1982) conclude that melting in layers having fortuitously low-melting (granite-like) composition produced leucosomes by coarsening, with only minimal segregation of melt from residue. Grain growth in the presence of dispersed melt is also invoked in a late (Caledonian) aureole in the Moines (Tyler and Ashworth, 1982), and explains coarsening of the melanosomes in Figure 5.2 (though the melt was eventually extracted from these). The coarsest mesosomes in this area (Figure 5.2) have probably also been affected by this process.

Dougan (1983) reports *decrease* in grain-size from mesosomes to melanosomes. In a meticulous study of 3 anatexite specimens from Maine and New Hampshire, he finds that some minerals systematically decrease in size with increasing proportion of melt extracted from the melanosomes; the latter is inferred from mass-balance considerations (Olsen, this volume) and tends to be highest for melanosome samples closest to a leucosome. Statistical significance of grain-size trends was tested by analysis of variance. The results are summarized in Table 5.1. Different behaviour by different minerals can largely be explained by the melting reaction inferred for the particular rock. In specimen L2D1 muscovite is a major reactant while biotite is not, in L2A2 the reverse is true, and the grain-size effects reflect this distinction. Dougan (1983) also recorded the minerals in contact with each measured grain in thin-section, and considered the probabilities of unrecorded contacts in three dimensions.

Table 5.1 Summary of the main results of Dougan (1983) on grain-size decreases with progressive extraction of melt from melanosomes

Specimen number	Minerals reacting to produce melt, in approximate order of decreasing importance (inferred from modal mass balance)	Minerals (and particular contact associations) showing significant grain-size decreases	Minerals studied that do not show significant grain-size decreases
L2D1	Quartz, plagioclase, muscovite, sillimanite	Muscovite (especially where in contact with quartz) Plagioclase (especially where in contact with quartz) Quartz (only where in contact with plagioclase or muscovite)	Biotite
L2A2	Plagioclase, quartz, biotite, sillimanite, muscovite	Biotite (only where in contact with quartz) Plagioclase	Muscovite Quartz
L7G2	Plagioclase, quartz, biotite, muscovite	Muscovite (only where in contact with quartz) Biotite (only where in contact with quartz)	Garnet

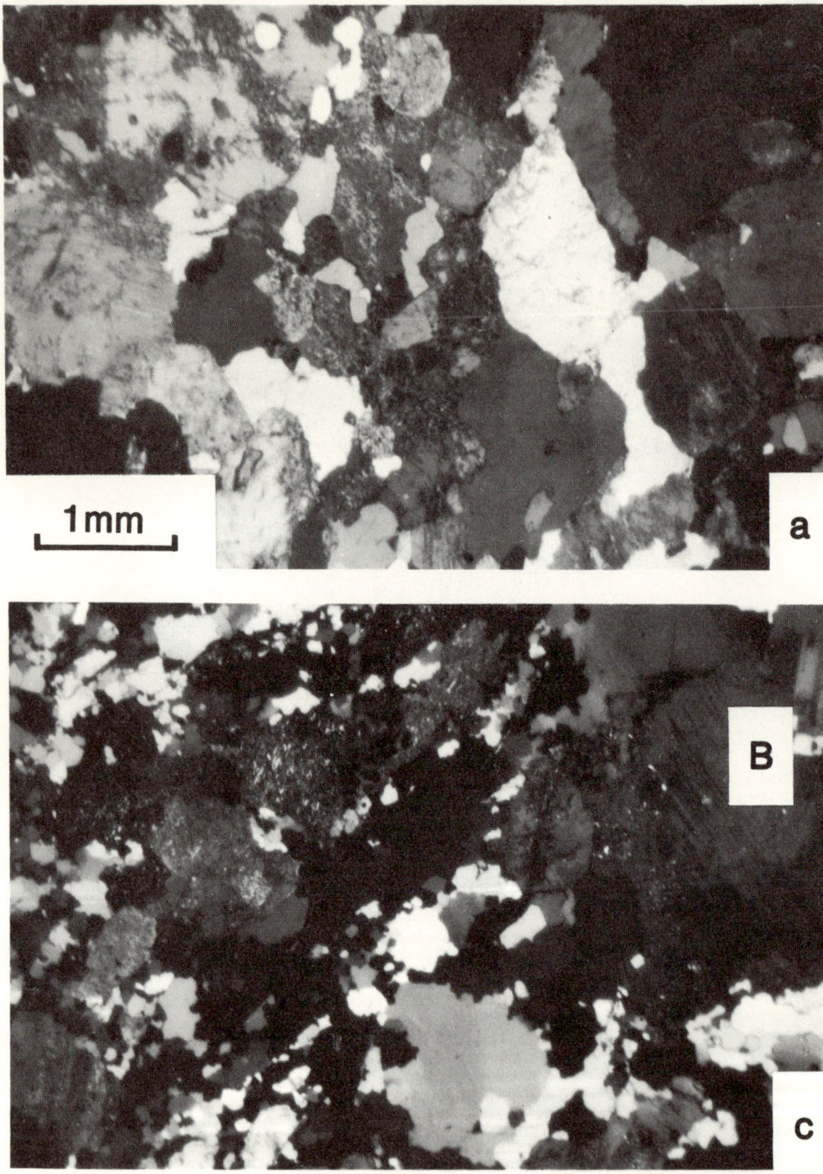

Figure 5.3 Comparison of textures in undeformed (*a*) and deformed (*b–d*) leucosomes. Slightly uncrossed polars. The scale is the same throughout. (*a*) Late, anatectic granitoid leucosome, Strontian area (Specimen A8 of Tyler and Ashworth, 1982). (*b*) Deformed leucosome with bent plagioclase (B), subgrain boundaries in quartz (e.g. at S), and aggregate of recrystallized quartz (A). Upper Findhorn area of Ashworth (1979*a,b*). (*c*) More intensely deformed leucosome from the

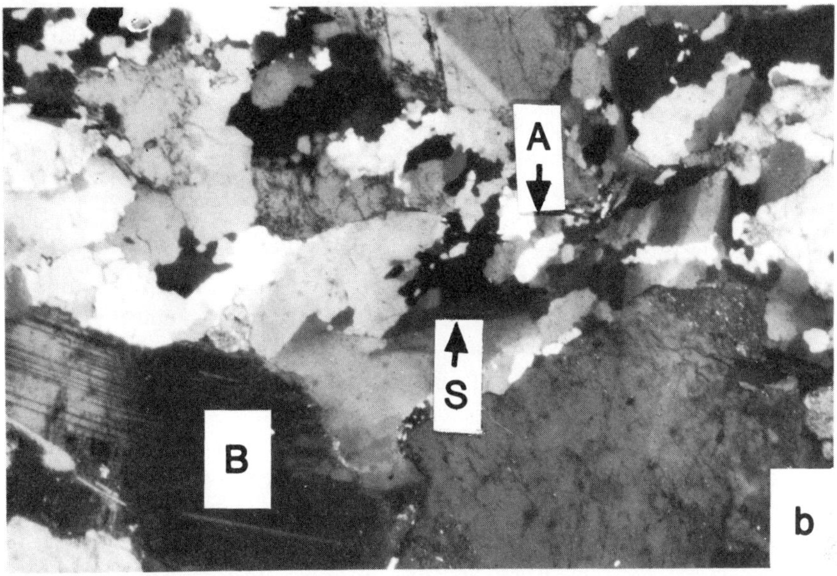

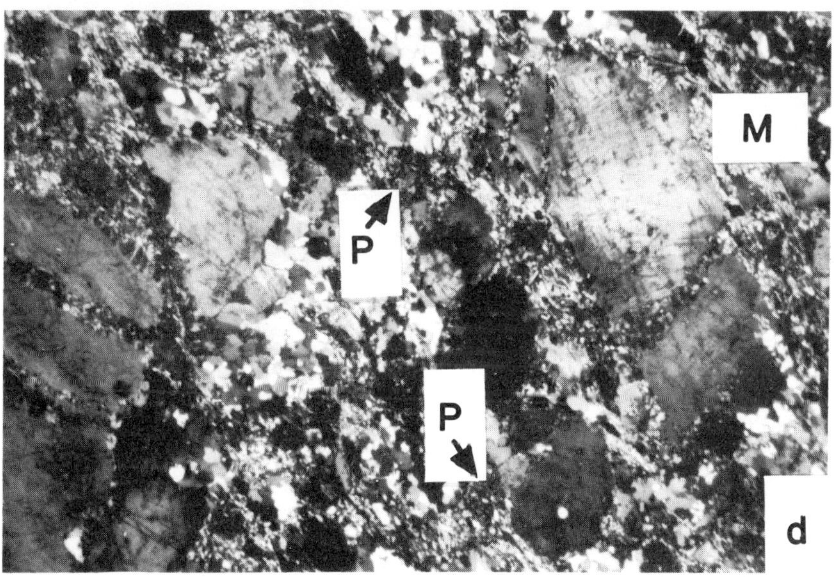

same area, with bent and broken plagioclase (B), and extensive recrystallization of quartz. (*d*) Intensely deformed leucosome with complete recrystallization of quartz, and also of plagioclase (P). Fine-grained plagioclase also occurs along cracks in the large grains, which are K-feldspar. Fine-grained micaceous material (M) defines a tectonic foliation. Redhythe Point, W of Portsoy.

He shows that the grain-size effect is statistically greatest for grains in contact with other reactant minerals, mica-quartz and feldspar-quartz contacts showing particularly notable effects. Thus the grain-size decreases are attributed directly to grain-boundary melting. In L7G2, the results are used to infer reaction at mica-quartz contacts in preference to muscovite-biotite contacts (Dougan, 1983, p. 96). Also in this rock, garnet increases in size with progressive melt extraction, presumably being a product of a dehydration-melting reaction. Dougan (1983) further infers that some grains of some minerals have been consumed completely (muscovite and quartz in L2D1; plagioclase in L2A2): the method is to count grains per unit area and apply a correction factor for bulk shrinkage during leucosome extraction.

These results of Dougan (1983) contrast strikingly with the general coarsening, in the presence of melt, inferred in the Scottish non-leucosomes. Dougan (1983, p. 97) points out the close comparability between his inferred grain-boundary reactions and incipient melting effects actually observed, in experiments of short duration. It seems that melt was efficiently extracted from Dougan's grain-boundaries, an extreme opposite case from its total failure to segregate in McLellan's (1983*a*) ophthalmites. Segregation is probably controlled by an interplay of deformation and extent of melting (section 5.6.2).

5.2.2 *Recrystallization after migmatization*

Coarse rocks such as leucosomes can recrystallize to finer ones (Figure 5.3), particularly after tectonic strain (Ashworth, 1979*a,b*). Even in severe cases, the primary grain-size contrast with non-leucosomes may remain visible (Figure 5.1*d*). Quartz recrystallizes more readily than feldspars, which may, however, fracture to produce a fine 'tail' in their size distribution (fine-skewness when plotted on a logarithmic scale; Ashworth, 1979*b*). Recrystallization tends to produce small crystals as aggregates, an effect which has been studied mostly by grain-contact frequency measurements (section 5.6). After slight strain, the original quartz grains may remain unrecrystallized, but with undulose extinction or with relatively sharp (through quantitatively small) changes in extinction position between *subgrains*. Subgrain boundaries are regular arrays of dislocations, produced by annealing during or after deformation (Hobbs *et al.*, 1976, pp. 93–96); their presence does not obscure the pre-existing grain boundaries. Recrystallization, on the other hand, produces new strain-free grains from the old strained ones, usually with strong changes in orientation and a reduction in grain-size; the primary texture is progressively eliminated (Figure 5.3*b*–*d*). The Upper Findhorn rocks (Figure 5.3*b,c*) are 'old Moines', from the Central Highland Division as defined by Piasecki (1980), the deformation probably being of Caledonian age following pre-Caledonian migmatization. The extremely modified migmatite of Figure 5.3*d* (the same rock is shown macroscopically in Figure 5.1*d*) comes from Redhythe Point, W of Portsoy, an occurrence near (but not contiguous with) the outcrop of simple-textured migmatites described by Ashworth (1976). The unit containing the latter is often called the 'Cowhythe Gneiss' (Read, 1923), and these 'Cowhythe' migmatites E of Portsoy can now be contrasted with the evolved 'Redhythe' migmatites. Ashworth (1976) erroneously failed to recognize the 'Redhythe' rocks as migmatites (because of their textural anomalies), but they are undoubtedly migmatites strongly recrystallized as a result of tectonic events which have not affected the 'Cowhythe' migmatites. In or near the 'Redhythe' unit may be the location of the major tectonic discontinuity (thrust) sought by Sturt *et al.* (1977).

5.3 Textures attributed to crystallization from a melt

Two effects are regarded as particularly characteristic of melt origin.

5.3.1 *Idiomorphic zoning in plagioclase*

Occasionally, in leucosomes, an idiomorphic zone-boundary is found in normally-zoned plagioclase. In Figure 5.4, the interior of the grain has almost constant composition; outwards from the idiomorphic boundary there is normal zoning. Mehnert (1968, p. 261) finds this effect in diatexites. Dougan (1979) figures idiomorphic zone-boundaries and mentions oscillatory zoning. These observations would be difficult to reconcile with solid-state growth of a grain in contact with neighbouring grains. The implication is unimpeded growth in an isotropic medium, presumably a melt (prior to final impingement on the surrounding grains). Ashworth (1976) and McLellan (1983*a*) use this argument to support an anatectic origin for some leucosomes lacking K-feldspar (trondhjemitoid, i.e. predominantly quartz + plagioclase, leucosomes).

5.3.2 *'Primary replacement' textures*

An aspect of the complexity of some leucosome textures is the replacement of some grains by aggregates of minerals that also occur as primary minerals outside the aggregates. K-feldspar is particularly often replaced, by a muscovite-quartz intergrowth together with myrmekite (plagioclase-quartz intergrowth; Figure 5.5). Characteristically, some muscovite and plagioclase grains in the aggregate are continuous with grains of the same minerals outside it, so that the boundary of the pseudomorph is 'blurred'; also, the compositions of these minerals are very similar in the aggregate and elsewhere in the leucosome (Ashworth, 1972, 1975). This texture is interpreted in terms of reaction between the cooling melt and the minerals crystallizing from it. The reaction is the retrograde equivalent of a major prograde dehydration-melting

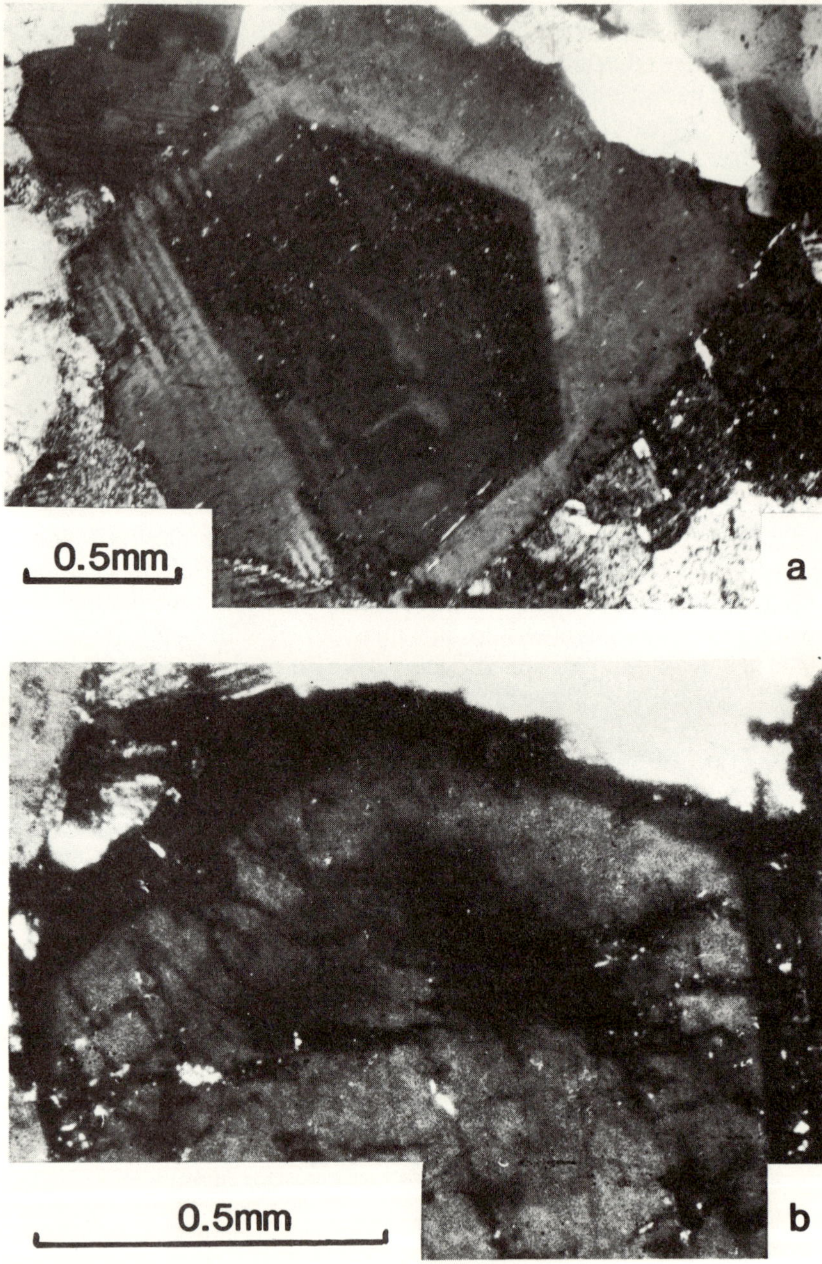

Figure 5.4 Idiomorphic zone boundaries in plagioclase. Crossed polars. Specimens 105753 (left) and 105799/2 (right) of Ashworth (1976).

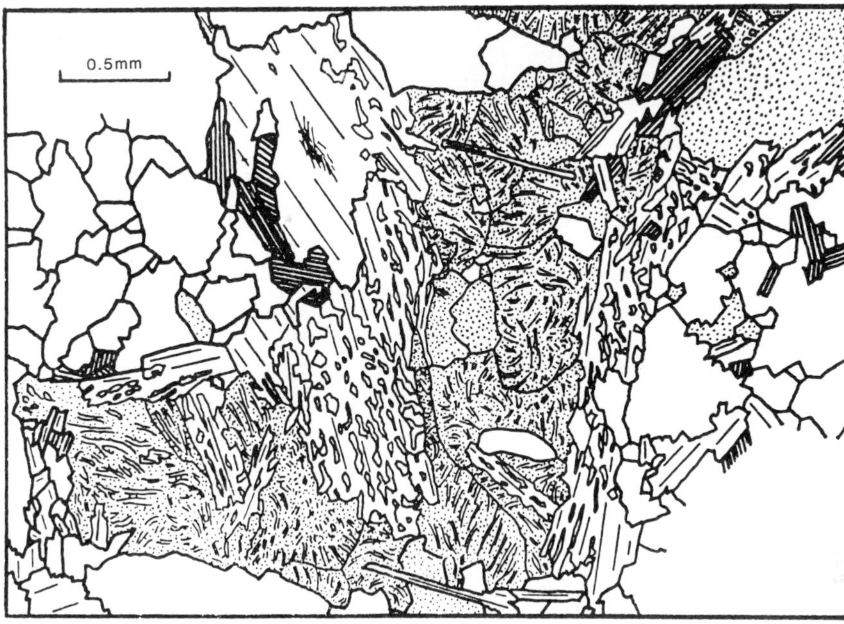

Figure 5.5 Complete replacement of K-feldspar by myrmekite and muscovite + quartz intergrowths (from Ashworth, 1972, Figure 1 *b*).

reaction, which generated much of the leucosome melt. The well-known dehydration reaction

$$\text{muscovite} + \text{quartz} \rightarrow \text{sillimanite} + \text{K-feldspar} + H_2O$$

when it occurs in the presence of the melt, is coupled with granite melting to produce a continuous reaction approximately

$$\text{muscovite} + \text{quartz} + \text{plagioclase}_1$$
$$\rightarrow \text{sillimanite} + \text{K-feldspar} + \text{plagioclase}_2 + \text{melt}.$$

During crystallization *in situ* of melts produced by this reaction, the reverse reaction is almost inescapable, and explains the observed textures. Symplectic intergrowths (notably myrmekite) form because the crystal which is being replaced imposes its Al:Si ratio on the local products (Ashworth, 1972), but the reaction involves the melt still present around this crystal, and involves other nearby crystals.

Analogous replacement of cordierite in higher-grade leucosomes, by biotite + sillimanite ± quartz (Ashworth, 1976), is attributed to the retrograde equivalent of the major prograde reaction

$$\text{biotite} + \text{sillimanite} + \text{quartz} + \text{plagioclase}_1$$
$$\rightarrow \text{cordierite} + \text{K-feldspar} + \text{plagioclase}_2 + \text{melt}.$$

These 'primary' replacement textures contrast strongly with the replacement of K-feldspar by simple, granular aggregates of muscovite, quartz and plagioclase, found in some leucosomes that are known to have been re-metamorphosed. An example is figured by Tyler and Ashworth (1982, Figure 4*a*) from a Moine leucosome that is probably pre-Caledonian ('early' Moine migmatite in the sense of Barr, this volume). Incompatibility of K-feldspar with sillimanite during Caledonian regional metamorphism has produced a more equilibrated texture than the retrograde reaction with melt. The essential distinction is probably the time scale of reaction, rather than the presence or absence of melt. The texture of Figure 5.5 (found in a shallow-level migmatite) might not have been preserved if cooling had been very protracted, though coarsened and degraded varieties of it are still recognizable in some deep-level and polymetamorphic migmatites (Ashworth, 1979*a*).

5.4 Grain shape

Dougan (1983) and McLellan (1983*a*) studied the aspect ratio (length/width) of some minerals. Dougan (1983) defines length as the maximum dimension of a grain in thin-section, and width as the short side of the rectangle which can be visualised enclosing the grain, the long side being the length of grain. He finds that significant decreases in mica grain-sizes, due to leucosome extraction, are always accompanied by significant increases in aspect ratio. It is not clear why grain-boundary melting should make the residual micas more elongate; if the efficient melt extraction (section 5.2.1) is due to deformation (section 5.6.2), the elongation effect may also be deformational, produced by diffusive mass transfer as in the lower-temperature examples of Elliott (1973).

McLellan (1983*a*) is able to discriminate between anatectic and non-anatectic trondhjemitoids by the aspect ratio of plagioclase. This is lower in the partially melted rocks than in the non-anatectic leucosomes (where values are similar to those in mesosomes). The greater elongation of plagioclase grown in the solid state is attributed to the influence of pre-existing fabric (McLellan, 1983*a*) combined with anisotropic diffusive mass transfer (McLellan, 1984).

5.5 Grain orientation (petrofabric)

In favourable cases, petrofabric studies can elucidate the relative timing of migmatization in the structural evolution of a terrain. This is an important role for quantitative petrography, because the macroscopic morphology of leucosomes can be misleading: stromatites formed after a deformation can be *mimetic*, mimicking the structures due to the deformation so that their shapes may be indistinguishable from those of layers that really are deformed (Figure 5.6). The internal, microscopic features of the leucosomes should always be consulted, in the hope of escaping from endless cycles of unedifying structural

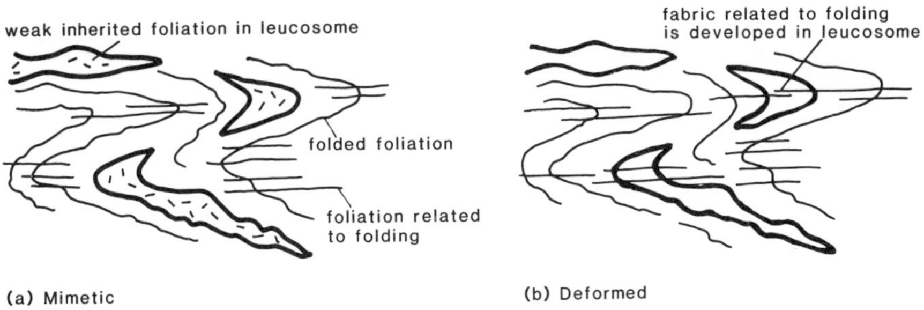

Figure 5.6 Idealized, schematic comparison of stromatic, anatectic leucosomes having different times of origin with respect to deformation of the mesosome. In both cases the leucosomes have complex shapes (pinch-and-swell, concentration in fold cores), but in (*a*) this is envisaged as mimetic on pre-existing fold structure, melt segregation being post-tectonic or syntectonic with respect to folding; in (*b*) the leucosome has been deformed in the solid state (is pretectonic relative to the folding).

controversy. Where leucosomes show no fabric, the case is perfectly clear. On the other hand, leucosomes with fabric (*gneissose* leucosomes of Mehnert, 1968, pp. 78–80) have either inherited a pre-existing fabric or been deformed during or after migmatization. In the case of anatectic migmatites, grains growing from a *segregated melt* are not expected to develop much preferred orientation, and Ashworth (1979*b*) shows that undeformed leucosomes from the 'Cowhythe Gneiss' have very little preferred orientation of biotite. Some of this biotite probably grew from the melt, but some may have been mechanically incorporated from melanosomes, and unsurprisingly there is some fabric derived from the pre-existing non-leucosome foliation (responsible also for stromatic geometry), but much weaker (Ashworth, 1979*b*, Figure 1, Portsoy specimens). The results are consistent with other indications of lateness of the 'Cowhythe' migmatization relative to local deformations. In such work, it is worthwhile to search for leucosomes with disseminated biotite; if clots or schlieren of biotite (i.e. small inclusions of melanosome) had been counted in the study, the inherited fabric would misleadingly appear much stronger. In the Mid-Strathspey specimen of Ashworth (1979*b*), where structural control of segregation indicates that migmatization was syntectonic, there is some biotite fabric (girdle of (001) poles, due to rotational deformation) and a quartz shape fabric; if an anatectic origin is accepted, then the deformation must have outlasted the crystallization of the melt. Quartz shape fabric is stronger, and a true biotite foliation (point maximum of (001) poles) is developed, in a highly deformed pre-Caledonian leucosome (Ashworth, 1979*b*, Upper Findhorn specimen). Foliation is also obvious in the 'Redhythe' leucosomes (Figures 5.1*d*, 5.3*d*).

In the case of post-tectonic anatexis with weak segregation, McLellan (1983*a*) describes the progressive destruction of an existing foliation. This effect neatly distinguishes her ophthalmites from the superficially similar rock-type called *augen-gneisses*, whose large feldspars are relics in a foliated matrix produced by deformation *after* feldspar growth (Mehnert, 1968, p. 36). In the *in-situ* anatexis described by Johannes and Gupta (1982), a pre-existing gneissose foliation is found to be reduced or eliminated in the developing leucosomes. In the syntectonic case studied by Tyler and Ashworth (1982), the coarsened mesosomes are gneissose but the segregated leucosomes are not (Figure 5.3*a*).

In *metamorphic differentiation*, grain growth in already foliated rocks can produce a mimetic shape fabric in leucosomes (McLellan, 1983*a*). McLellan (1984) distinguishes two suites of metamorphic migmatites, a syntectonic suite with plagioclase shape fabric and a pretectonic suite which additionally has a crystallographic fabric in quartz (preferred *c*-axis orientation) and recovery and recrystallization effects consequent on deformation of the quartz. The syntectonic suite formed by a deformation process, but this was intergranular mass transfer ('diffusional flow' or 'grain boundary diffusion creep' of Elliott, 1973) rather than intragranular dislocation motion which would have produced a crystallographic fabric.

5.6 Grain contact relations

The contrast between simple metamorphic textures and the relatively intricate textures of some leucosomes was noted by Ashworth (1976) and Dougan (1979), with such descriptions as 'textural immaturity' and 'textural disequilibrium' applied to the leucosomes. Ashworth (1976) sought to quantify this effect, as a possible indicator of anatectic origin. Frequencies of contacts among minerals were used, following work on metamorphic textures by Kretz (1969) and Flinn (1969). Since the method has given interesting results (e.g. McLellan, 1983*a*) but is still not widely familiar, the principles will now be reviewed. The basis is counting the numbers of contacts of different types; in a rock consisting of two minerals A and B, there will be AA, BB and AB contact-types. In a *random* spatial distribution (Kretz, 1969), A would have no preference for being in contact with either A or B. There are two simple kinds of departure from randomness: if like contacts (AA and BB) are more frequent than in a random distribution, at the expense of unlike (AB) contacts, then we have an *aggregate* distribution; the opposite case is a *regular* distribution (Kretz, 1969; sometimes called a *dispersed* distribution—e.g. McLellan, 1983*a*). Mature or equilibrated metamorphic textures tend to have regular distributions, while granite textures approach randomness (Flinn, 1969; McLellan, 1983*a*). Thus the possibility arises of distinguishing melt-grown (anatectic or igneous) textures from others. Effects of post-migmatization annealing can also be studied. As far as possible, *homogeneous* rocks should be sampled, or results may be anomalous. For example, aggregate spatial relations in the mesosomes of Dougan (1983) may be due to the rocks being banded (e.g. Figure 1 of Dougan, 1983) so that the aggregate behaviour of a mineral such as biotite merely reflects the fact that it is concentrated in mafic bands.

5.6.1 *Methods*

Observation. Of the several approaches described by Kretz (1969), two have been found useful in migmatites. Ashworth (1976, 1979*a*), Dougan (1983) and McLellan (1983*a*) use the *line-transect* method (Figure 5.7*a*). This records the sequence of minerals encountered in thin-section, along a line laid out using a mechanical stage. In theory, long traverses are connected by short line-segments (Figure 5.7*a*). In practice, the short segments can be omitted, and it is not necessary to make long traverses in opposite directions alternately. The distance between adjacent long traverses should exceed the maximum grain-size, to avoid counting a contact more than once on different traverses.

In intricate textures, some contacts are encountered repeatedly on the same long traverse; Ashworth (1976) identified these *recurrences* and omitted them from the totals, which are arrived at as follows. From the recorded sequence of minerals the contact-types are readily identified (Figure 5.7) and these are usually tallied to build up totals for Q → P, P → K etc. (rather than written down

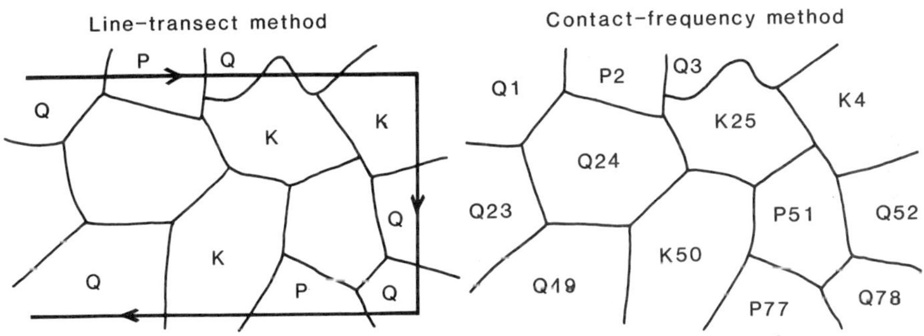

Figure 5.7 Observational methods used in grain contact studies. The minerals are supposed to be K-feldspar (K), plagioclase (P) and quartz (Q).
(a) Line-transect method. Sequence of grains recorded (parentheses indicate recurrence): ...QPQK(Q)KQQPKQ... Contacts counted: ...Q→P, P→Q, Q→K (K→Q recurrence omitted), Q→K, K→Q, Q→Q, Q→P, P→K, K→Q...
(b) Contact-frequency method. Data array:

⋮
24 Q 25 49 50
⋮
51 P 52 77 78
⋮

Contacts counted: ...QK, QQ, QK; KK, KP; ...PQ, PP PQ; ...

individually as in Figure 5.7). To avoid totals < 5 (which can perturb the statistical analysis discussed below), accessory minerals are counted with one of the major minerals. For unlike contacts the method naturally yields two totals ($A \rightarrow B$ and $B \rightarrow A$). Omitting recurrences can reveal any tendency for one mineral to enclose another; if B tends to be included by A, then $A \rightarrow B$ contacts tend to be followed by $B \rightarrow A$ recurrences, omission of which produces a $B \rightarrow A$ total lower than the $A \rightarrow B$ total. However, in studies so far, inclusive tendencies are weak. McLellan (1983*a*) ignores small quartz grains included in plagioclase. Subgrain boundaries should not be counted (McLellan, 1983*a*). True grain boundaries arising from recrystallization must be counted; it may not be possible to distinguish them from primary grain boundaries. Consistently large numbers of contacts should be counted (~ 1000 per sample).

Karlsson and Wahlgren (1982) note the inefficiency of the line-transect method where there is a large spread of grain-size (due to the restriction on spacing of the long traverses). They adopt a method, called the *contact-frequency method* by Kretz (1969), in which *all* contacts are counted. Tracings are made from photomicrographs of the thin-sections. All grains are identified and numbered (Figure 5.7). A data array is then built up in lines, each line beginning with the grain number n, and its identification, followed by the numbers m of grains in contact with it (omitting grains with $m < n$, to avoid counting contacts twice). Karlsson and Wahlgren (1982) suggest a rough method of removing effects attributed to recrystallization, by ignoring contacts within aggregates of small, like grains. This is rather subjective, and if it is attempted, the result of counting all contacts should also be reported.

Calculation of expected contact frequencies. The *expected* frequency of a contact-type is that which would occur according to some hypothesis, usually the hypothesis of randomness. For the *line-transect method*, the calculation of expected frequencies follows Kretz (1969). The observed totals for each contact-type are written in the cells of a *contingency table* (Figure 5.8). Here $C_{Q \rightarrow P}$ is the total number of quartz $\rightarrow$ plagioclase contacts, etc. The row totals are written at the right, e.g. $(\Sigma C)_{Q \rightarrow}$ for all transitions from quartz, and the column totals are written at the bottom. $\Sigma\Sigma C$ is the grand total of all contacts counted (Figure 5.8). Denoting the ith row total $(\Sigma C)_{i \rightarrow}$ and the jth column total $(\Sigma C)_{\rightarrow j}$, the expected frequency (on the hypothesis of randomness) for the contact-type $i \rightarrow j$ is

$$E_{ij} = \frac{(\Sigma C)_{i \rightarrow} \cdot (\Sigma C)_{\rightarrow j}}{\Sigma\Sigma C},$$

i.e. the expected value in each cell is given by the product of the corresponding row and column totals, divided by the grand total. The expected value is often written under the observed value in each cell (e.g. Kretz, 1969, Ashworth, 1976).

For the *contact-frequency method*, the treatment by Karlsson and Wahlgren

Line-transect method

	Q	P	K	
Q	$C_{Q\to Q}$	$C_{Q\to P}$	$C_{Q\to K}$	$(\Sigma C)_{Q\to}$
P	$C_{P\to Q}$	$C_{P\to P}$	$C_{P\to K}$	$(\Sigma C)_{P\to}$
K	$C_{K\to Q}$	$C_{K\to P}$	$C_{K\to K}$	$\Sigma C)_{K\to}$
	$(\Sigma C)_{\to Q}$	$(\Sigma C)_{\to P}$	$(\Sigma C)_{\to K}$	$\Sigma\Sigma C$

Contact-frequency method

	Q	P	K	
Q	$C_{Q\ Q}$	$1/2C_{P\ Q}$	$1/2C_{K\ Q}$	$(\Sigma C)_Q$
P	$1/2C_{P\ Q}$	$C_{P\ P}$	$1/2C_{K\ P}$	$(\Sigma C)_P$
K	$1/2C_{K\ Q}$	$1/2C_{K\ P}$	$C_{K\ K}$	$(\Sigma C)_K$
	$(\Sigma C)_Q$	$(\Sigma C)_P$	$(\Sigma C)_K$	$\Sigma\Sigma C$

Figure 5.8 Contingency tables constructed from data of the two methods in Figure 5.7.

(1982) is defective. They use the number of grains of minerals *i* and *j*, rather than summations of their contacts, in estimating E_{ij}. This follows an exercise by Kretz (1969) who, however, notes that it is unsatisfactory because different minerals may have different average numbers of contacts per grain (the most obvious reason being that the minerals may have different grain-sizes). Karlsson and Wahlgren (1982) observe plagioclase with more than twice as many contacts per grain as quartz, and attempt a correction, but only for one mineral of a pair, i.e. for *i* but not *j*. The proper procedure (Figure 5.8) is simply analogous to that for the line-transect method. Since the method does not distinguish $i \to j$ from $j \to i$ contacts, for $i \to j$ the observed totals are placed half in cell *ij* and half in cell *ji*. The row totals then equal the corresponding column totals, and

$$E_{ij} = E_{ji} = \frac{(\Sigma C_i)\cdot(\Sigma C)_j}{\Sigma\Sigma C}.$$

Observed/expected ratios are often used to display the pattern of results. In purely aggregate distributions, *like* ratios (e.g. for quartz–quartz and plagioclase–plagioclase contact-types) are > 1; in purely regular distributions they are < 1. Ratios based on small expected frequencies are liable to statistical fluctuation. In granitoid (K-feldspar-bearing) leucosomes, plagioclase–plagioclase expected frequencies are smaller than in trondhjemitoid leucosomes, and plagioclase–plagioclase observed/expected ratios may fluctuate randomly: Ashworth (1979*a*) counted plagioclase together with K-feldspar when comparing granitoid with trondhjemitoid migmatites.

Statistical tests. These seek *significant differences*, from some *null hypothesis*, at some specified *level* of significance, the weakest level commonly used being 5% (i.e. 5% risk of being wrong in rejecting the null hypothesis, assuming that a valid procedure has been used). To test a *single specimen* for *difference from randomness*, χ^2 is calculated from the contingency table:

$$\chi^2 = \Sigma\left[\frac{(C_{ij} - E_{ij})^2}{E_{ij}}\right]$$

where the summation is over all the cells. The value obtained is compared with standard tables of percentage points of the χ^2 distribution for the appropriate number of degrees of freedom, $v = (n - 1)^2$ where n is the number of cells in a row or column. The method is invalid if any C_{ij} is < 5, and more generally the result may depend on sample size. On the basis that (at constant sample-size and constant v) larger χ^2 values mean stronger deviations from randomness, the use in practice of χ^2 (or χ^2/v; McLellan, 1983*a*) has been to make comparisons between suites of specimens that yield obviously different ranges of χ^2 (e.g. Ashworth, 1976). χ^2 does not assess the *direction* of departure from randomness (regular or aggregate), and different minerals can behave differently in this respect, which suggests testing one contact-type at a time. Using the ratio R = *observed/expected ratio* for the contact-type, we can test (1) one suite of specimens for difference from randomness, or (2) two suites for difference between them. Noting that a non-parametric method is preferable to the Student *t*-test with its unjustified assumption of a normal distribution, Karlsson and Wahlgren usefully introduce Wilcoxon statistics for case (2); the calculations are easy and can easily be extended to case (1). The basis is ranking of the R values, which need not be accurately known, since it is necessary only to judge the extent of overlap with some hypothesized value.

(1) A test for significant *difference from randomness* in one contact-type in *one suite of specimens* is the Wilcoxon signed-rank test (Hollander and Wolfe, 1973). The null hypothesis is that the population sampled has a median $R = 1$. It is necessary that the data have been treated validly to obtain R values. These are then ranked from smallest, 1, to largest, n (Table 5.2). The ranks are multiplied by S where $S = 1$ if $R > 1$ and $S = 0$ if $R < 1$, i.e. the ranks are set to zero where $R < 1$ (Table 5.2). These modified ranks are summed to obtain the statistic T^+. It is appropriate to perform a two-tailed test (i.e. assume no prior knowledge of the direction of departure from $R = 1$). The null hypothesis is accepted if T^+ falls in the range

$$\left[\frac{n(n+1)}{2} - t\right] < T^+ < t,$$

where the critical value t depends on n and on the chosen significance level α (t here is not the same as Student's t.) Interpolating in Hollander and Wolfe's (1973) tables of $\alpha/2$ (for the two-tailed test) as a function of n and t, for $n = 9$ and $\alpha/2 = 0.025$ we find $t = 39.3$, so the quartz–quartz data (Table 5.2) differ significantly from randomness at the 0.05 (5%) level. In the plagioclase–plagioclase data, two R values are not measurably larger or smaller than 1 and are dropped from the analysis (Siegel, 1956), leaving $n = 7$. Table 5.2 shows that the null hypothesis cannot be rejected for these data.

Table 5.2 Wilcoxon signed-rank test applied to data of McLellan (1983*a*) for ophthalmites

Contact-type:	quartz–quartz				plagioclase–plagioclase			
	observed/ expected R	rank	S	rank $\times S$	observed/ expected R	rank	S	rank $\times S$
	1.11	5.5	1	5.5	1.41	7	1	7
	1.20	7	1	7	1.00	—	—	—
	1.27	9	1	9	1.00	—	—	—
	1.23	8	1	8	1.27	6	1	6
	1.11	5.5	1	5.5	0.89	3	0	0
	1.07	4	1	4	0.98	5	0	0
	0.78	1	0	0	0.95	4	0	0
	1.01	3	1	3	0.87	2	0	0
	0.87	2	0	0	0.84	1	0	0
T^+				42				13
				$n=9$				$n=7$

At 5% significance level $t = 39.3$. Significant difference from $R = 1$ if $T^+ > 39.3$ or $T^+ < 5.7$.

At 5% significance level $t = 25.9$. Significant difference from $R = 1$ if $T^+ > 25.9$ or $T^+ < 2.1$.

Table 5.3 Wilcoxon rank-sum test for two suites, applied to data of Ashworth (1976)

Contact-type:	plagioclase–plagioclase		plagioclase–quartz	
	R	rank	R	rank
leucosomes				
	0.92	9	1.14	5
	0.84	5	1.17	6.5
	0.89	8	1.17	6.5
	0.87	7	1.20	8
	0.85	6	1.13	3.5
$m = 5$				
mesosomes				
	0.50	1	1.11	2
	0.51	2	1.13	3.5
	0.60	3.5	0.94	1
	0.60	3.5	1.23	9
W		10		15.5
$n = 4$	$n(m+n+1) = 40$			

At 5% significance level $w = 28.4$. Significant difference between leucosomes and mesosomes if $W > 28.4$ or $\bar{W} < 11.6$.

(2) To compare *two suites*, the Wilcoxon rank-sum test is used (Karlsson and Wahlgren, 1982). Hollander and Wolfe (1973) show that it is equivalent to the Mann–Whitney U test, which is more familiar to geologists. It should be valid *within a study* even if the method of calculating R is suspect. A suite comprising m values of R is compared with one comprising n values, where $m \geqslant n$ (Table 5.3). The ranks of the n values are summed to obtain the statistic W. The null hypothesis of equal medians is accepted at significance level α, if

$$\left[n(m+n+1)-w\right] < W < w,$$

where $\alpha/2$ is tabulated as a function of w, m and n (Hollander and Wolfe, 1973). In Table 5.3, only the plagioclase–plagioclase data pass the test.

5.6.2 *Interpretations of results*

Ashworth (1976) treats a simple case of migmatites with only three major minerals (quartz, plagioclase and biotite). A gross distinction is found between mesosomes (strongly regular spatial distributions) and leucosomes (much weaker tendency to be regular). This result is used to support an anatectic origin for trondhjemitoid leucosomes. Similarly, in trondhjemitoid leucosomes regarded as anatectic (with discordant field relations and cotectic compositions), McLellan (1983*a*) reports R values close to 1 for both like and unlike contacts: a general randomization of spatial relations is traced from gneissose mesosomes with regular distributions, through ophthalmites to segregated leucosomes. A tendency to $R < 1$ for plagioclase–plagioclase contacts in ophthalmites and gneissose leucosomes is attributed to textural inheritance in these incompletely segregated rocks. On the other hand, a tendency to $R > 1$ for the same contact-type in segregated leucosomes may be a flow effect.

In comparison with anatectic trondhjemitoid leucosomes, McLellan (1983*a*) finds that those produced by metamorphic differentiation have a greater tendency to aggregate relations, having $R > 1$ for all like contacts among major minerals. This is attributed to nucleation of new grains at existing boundaries of the same mineral. In another example (McLellan, 1984), as material moved into the leucosome, new grains grew from subgrains formed at an earlier stage of the migmatization. Grain contact distributions produced by such a process should be distinct from melt-grown ones.

However, the above results are for migmatites that post-date most of the regional deformation. Earlier leucosomes tend to have aggregate distributions due to recrystallization, particularly in quartz. In migmatites from Deeside (whose age is controversial), McLellan (1983*a*) adduces unevolved textures in support of late origin. Even among late-Caledonian (nominally post-tectonic) Grampian leucosomes, McLellan (1983*a*) finds that the deeper-level ones have aggregate quartz–quartz relations (Figure 5.9), due to recrystallization. Evolution towards regular distributions (like those in mesosomes), which might naively be expected during static annealing, is not found. A trend (Figure 5.9) from weakly regular original distributions, to aggregate distributions in deformed leucosomes (Ashworth, 1979*a*) is also consistent with the data of Karlsson and Wahlgren (1982) after reprocessing (see section 5.6.1). 'Lumping' the two feldspars is sensible, not only because of low plagioclase–plagioclase frequencies (section 5.6.1), but also because *R* tends to be < 1 for plagioclase–K-feldspar if it is < 1 for plagioclase–plagioclase and K-feldspar–K-feldspar, i.e, the two feldspars behave somewhat like a single mineral. For mesosomes, Karlsson and Wahlgren (1982) give insufficient data for reprocessing, but their result that mesosomes are similar to leucosomes seems valid since they treated all the data consistently. It is not clear whether this similarity reflects incipient migmatization in the mesosomes (McLellan, 1983*a*) or the granodioritic nature of the rocks, which might respond to metamorphism (and deformation) more like a leucosome than a pelite.

The melanosomes of Dougan (1983) evolve complexly, showing some tendency towards aggregate relations, with progressive melt extraction. The inferred grain-boundary melting reactions (Table 5.1) can be correlated with decreases in *R* for some contact-types: muscovite–quartz and plagioclase–quartz in L2D1; biotite–quartz, biotite–sillimanite and plagioclase–quartz in

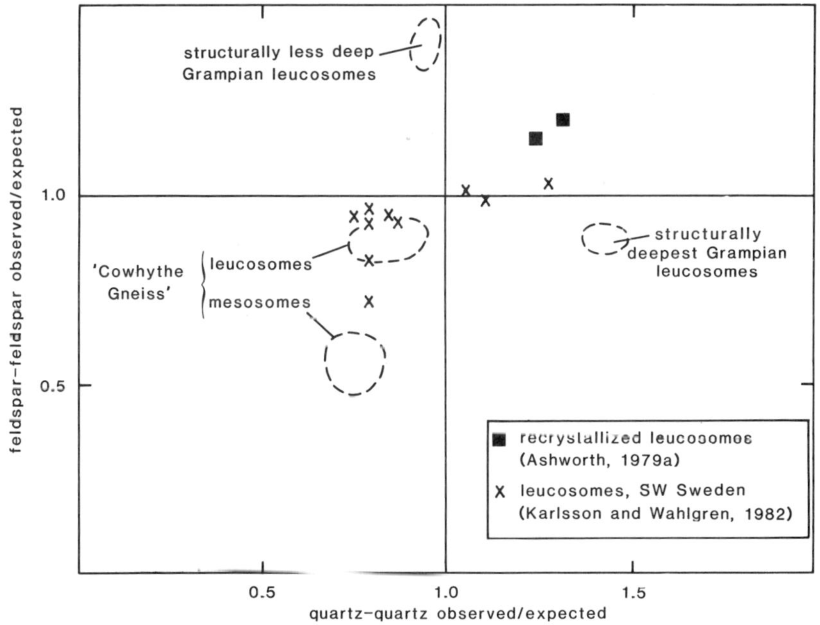

Figure 5.9 Observed/expected ratios for the major contact-types quartz–quartz and feldspar–feldspar in some migmatites. Ratios > 1 represent aggregate distributions of these minerals, < 1 regular distributions. Data of Karlsson and Wahlgren (1982) have been reprocessed (see text). Grampian data are from McLellan (1983*a*).

L2A2; and biotite–quartz, muscovite–quartz and plagioclase–quartz in L7G2. The efficient extraction of melt implied by all Dougan's (1983) results probably requires some deformation. Under static (no-strain) conditions, unsegregated ophthalmites are produced (McLellan, 1983*a*) even where randomization of contact relations (and of petrofabrics) suggests melting beyond the 'rheologically critical melt percentage' (RCMP) of Arzi (1978), which is ~ 30%. Efficient drainage of melt, even at such large proportions, is unlikely under static conditions because felsic melts show little tendency to wet grain boundaries (Jurewicz and Watson, 1984). Below the RCMP, limited deformation may not cause drainage, but rather disperse the melt through the rock (van der Molen and Paterson, 1979; Paquet *et al.*, 1982; cf. Tyler and Ashworth, 1982). Segregation is favoured by large strains at low degrees of melting, or by > 30% melting at lower strains (see also Barr, this volume). McLellan (1983*b*) distinguishes drained from undrained systems by the extent of melt-restite re-equilibration.

5.7 Conclusions

Textures, used sensibly in conjunction with other petrological and field data, can help to elucidate the origin of those migmatites with little subsequent metamorphic history, and the post-migmatization history of others. Grain contact relations show promise for distinguishing anatectic from non-anatectic leucosomes, but will often be obscured by recrystallization. They are easiest to study in rocks with simple mineralogy (e.g. trondhjemitoid), and so complement the use of reaction textures to infer anatexis in more complex assemblages. Petrofabrics, crucial in deducing the timing of migmatization, may also reflect process: anatexis tends to destroy fabrics, while metamorphic differentiation under stress can enhance grain elongation and preferred orientation. Grain size may help distinguish leucosomes grown from aqueous fluids and melts, and grain-size evolution in anatexites is sensitive to the degree of melt segregation.

References

Arzi, A.A. (1978) Critical phenomena in the rheology of partially melted rocks. *Tectonophysics* **44**, 173–184.

Ashworth, J.R. (1972) Myrmekites of exsolution and replacement origins. *Geol. Mag.* **109**, 45–62.

Ashworth, J.R. (1975) The sillimanite zones of the Huntly-Portsoy area in the north-east Dalradian, Scotland. *Geol. Mag.* **112**, 113–136.

Ashworth, J.R. (1976) Petrogenesis of migmatites in the Huntly-Portsoy area, north-east Scotland. *Mineral. Mag.* **40**, 661–682.

Ashworth, J.R. (1979*a*) Textural and mineralogical evolution of migmatites. In *The Caledonides of the British Isles—Reviewed*, eds. Harris, A.L., Holland, C.H. and Leake, B.E., *Geol. Soc. Lond. Spec. Publ.* **8**, 357–361.

Ashworth, J.R. (1979*b*) Comparative petrography of deformed and undeformed migmatites from the Grampian Highlands of Scotland. *Geol. Mag.* **116**, 445–456.

Barber, J.P. (1985) Conditions of high grade metamorphism in the Dalradian of Connemara, Ireland. *J. Geol. Soc. Lond.*, in press.

Dougan, T.W. (1979) Compositional and modal relationships and melting reactions in some migmatitic metapelites from New Hampshire and Maine. *Am. J. Sci.* **279**, 897–935.

Dougan, T.W. (1983) Textural relations in melanosomes of selected specimens of migmatitic pelitic schists: implications for leucosome-generating processes. *Contrib. Miner. Petrol.* **83**, 82–98.

Elliott, D. (1973) Diffusion flow laws in metamorphic rocks. *Geol. Soc. Am. Bull.* **84**, 2645–2664.

Flinn, D. (1969) Grain contacts in crystalline rocks. *Lithos* **3**, 361–370.

Folk, R.L. (1968) *Petrology of Sedimentary Rocks.* Hemphill's, Austin, Texas.

Hobbs, B.E., Means, W.D. amd Williams, P.F. (1976) *An Outline of Structural Geology.* Wiley, New York.

Hollander, M. and Wolfe, D.A. (1973) *Nonparametric Statistical Methods.* Wiley, New York.

Johannes, W. and Gupta, L.N. (1982) Origin and evolution of migmatite. *Contrib. Miner. Petrol.* **79**, 114–123.

Jurewicz, S.R. and Watson, E.B. (1984) Distribution of partial melt in a felsic system: the importance of surface energy. *Contrib. Miner. Petrol.* **85**, 25–29.

Karlsson, G. and Wahlgren, C.H. (1982) A statistical investigation of grain contacts in a migmatite. *Neues Jb. Miner. Mh.* 348-360.

Kretz, R. (1969) On the spatial distribution of crystals in rocks. *Lithos* **2**, 39–66.

McLellan, E.L. (1983*a*) Contrasting textures in metamorphic and anatectic migmatites: an example from the Scottish Caledonides. *J. Metamorphic Geol.* **1**, 241–262.

McLellan, E.L. (1983*b*) Plagioclase compositions in anatectic migmatites. *Geol. Soc. Am. Abstr. Programs* **15**, 641.

McLellan, E.L. (1984) Interrelations of deformation and metamorphism in the formation of metamorphic migmatites. To be submitted to *Geol. Mag.*

Mehnert, K.R. (1968) *Migmatites and the Origin of Granitic Rocks.* Elsevier, Amsterdam.

Paquet, J., Francois, P. and Nédélec, A. (1981) Effect of partial melting on rock deformation: experimental and natural evidences on rocks of granitic composition. *Tectonophysics* **78**, 545–565.

Piasecki, M.A.J. (1980) New light on the Moine rocks of the Central Highlands of Scotland. *J. Geol. Soc. Lond.* **137**, 41–59.

Read, H.H. (1923) The geology of the country around Banff, Huntly and Turriff. *Mem. Geol. Surv. Scotl.* for Sheets 86 and 96.

Siegel, S. (1956) *Nonparametric Statistics for the Behavioral Sciences.* McGraw-Hill, New York.

Strut, B.A., Ramsay, D.M., Pringle, I.R. and Teggin, D.E. (1977) Precambrian gneisses in the Dalradian sequence of northeast Scotland. *J. Geol. Soc. Lond.* **134**, 41–44.

Tyler, I.M. and Ashworth, J.R. (1982) Sillimanite-potash feldspar assemblages in graphitic pelites, Strontian area, Scotland. *Contrib. Miner. Petrol.* **81**, 18–29.

van der Molen, I. and Paterson, M.S. (1979) Experimental deformation of partially-melted granite. *Contrib. Miner. Petrol.* **70**, 299–318.

Yardley, B.W.D. (1975) On some quartz–plagioclase veins in the Connemara schists, Ireland. *Geol. Mag.* **112**, 183–190.

Yardley, B.W.D. (1978) Genesis of the Skagit Gneiss migmatites, Washington, and the distinction between possible mechanisms of migmatization. *Geol. Soc. Am. Bull.* **89**, 941–951.

6 Migmatite occurrences in New England

ROBERT J. TRACY*

6.1 Introduction

The high-grade metamorphic belts of the northern Appalachians in New England comprise one of the classic migmatite terrains of the world. Migmatites occur within the highest-grade metamorphic zones in roughly linear north–south belts running from Long Island Sound to near the Canadian border—a distance of about 500 km (Figure 6.1). These linear high-grade belts have been interpreted to be related to a succession of convergent tectonic events throughout the Palaeozoic (Robinson, 1983; Hall and Robinson, 1982). Ages of metamorphism and migmatization generally decrease from west to east. The westernmost belt of upper amphibolite–lower granulite facies metamorphism occurs in Proterozoic Y rocks exposed in the Green Mountains–Berkshires–Housatonic Highlands–Hudson Highlands massifs; highest-grade metamorphism of these rocks is generally considered to be of Grenville age (approximately 1100 Ma) although they may be overprinted by later, lower-grade metamorphism (Robinson, 1983).

Coming eastward, the next belt of migmatitic rocks is exposed only sporadically in domes or windows through lower-grade cover-rocks, especially in southern New England. The migmatized rocks are Cambro-Ordovician and probably reached their highest metamorphic grade in the Middle Ordovician (Taconian), although this interpretation is complicated by superposed Devonian amphibolite facies metamorphism (Robinson, 1983). Little petrological work has been done on these rocks, and their relationship to more typical Taconian metamorphic rocks is poorly known.

The so-called eastern Acadian Metamorphic High (Robinson, 1983), shown in Figure 6.1, is the locus of the easternmost well-developed belt of migmatitic rocks, and the best studied. This zone runs from Connecticut to northwestern Maine and encompasses the eastern part of the Bronson Hill anticlinorium and the western part of the Merrimack synclinorium. In much of this belt the Devonian (Acadian) metamorphic grade is sillimanite + K-feldspar or higher,

*Department of Geology and Geophysics, Yale University, New Haven, CT 06520, USA.

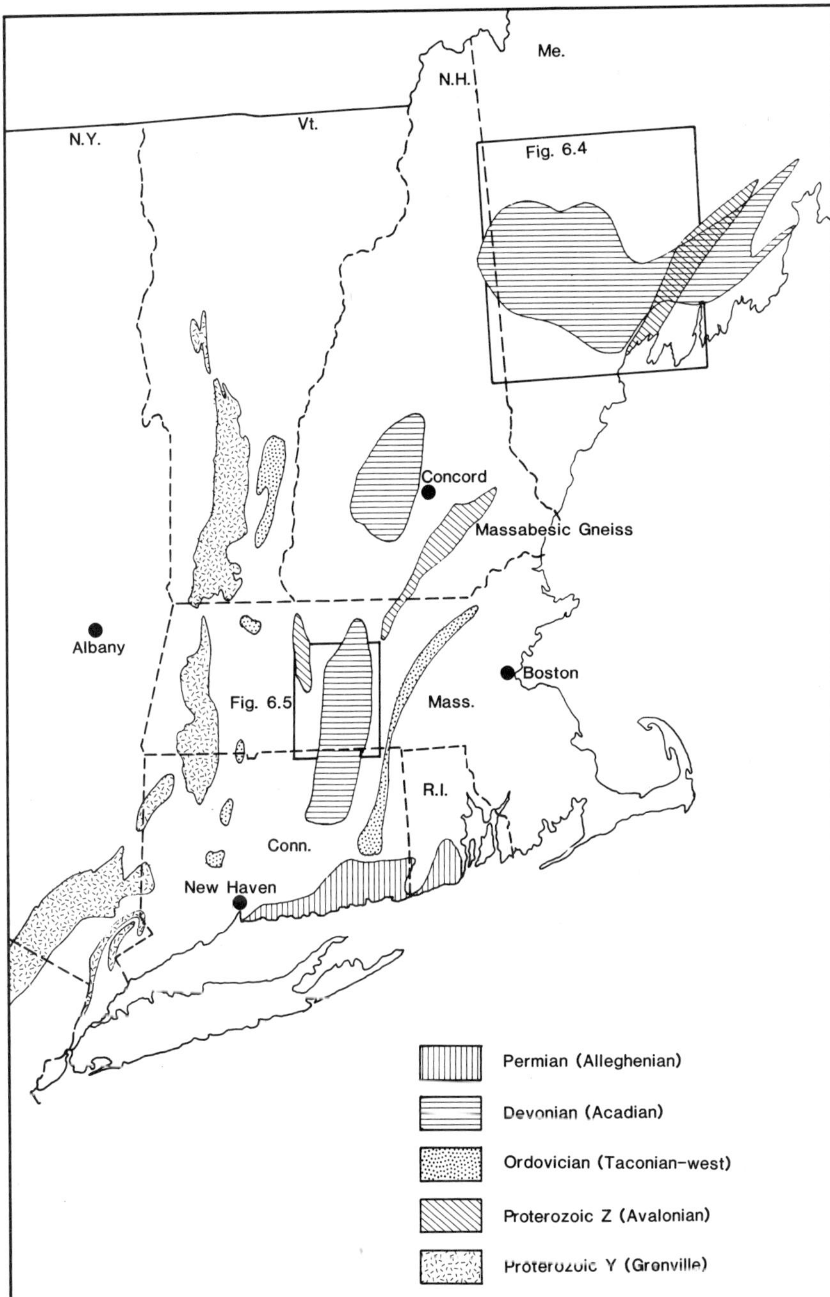

Figure 6.1 Map of southern and central New England showing the distribution and current age assignment of high-grade metamorphic and migmatitic terrains. Each terrain is typically indicated by sillimanite + K-feldspar grade or higher. Areas of more detailed maps of Maine and Massachusetts are shown.

and the migmatites have been well established to be the result of partial fusion of the deeper levels of a thickened crust. There has been ongoing debate about whether this high-grade metamorphism requires extra heat input from arc magmatism above an active Acadian subduction zone, or simply resulted from crustal thickening and thermal relaxation. Considerable *in-situ* melting seems to have occurred, however, regardless of the gross regional heat sources.

Finally, of more problematic origin are the sporadic occurrences of migmatites east of the Acadian Metamorphic High. This region presents a problem because of the recently postulated existence of a number of exotic or suspect terrains which may have experienced their own individual metamorphic histories before being assembled in eastern New England in the late Palaeozoic. Several of these terrains contain Proterozoic rocks and so may have had long and complicated histories before accretion. The occurrence of local high-grade metamorphism and melting, as for example in southern Rhode Island, may be due to regional thermal relaxation following Permian (Alleghenian) collisional tectonics and crustal thickening or may have occurred much earlier.

The purpose of this chapter is to review the occurrence and characteristics of migmatites in the northern Appalachians in New England. In order to accomplish this, the succeeding sections will discuss each of the above-mentioned belts in detail, especially noting the relation of migmatite occurrences to metamorphic grade. As will become obvious, in order to understand the distribution and characteristics of migmatites in New England (or in any terrain) it is vital to understand the tectonic and metamorphic evolution of the terrain. Because readers of this book may not be familiar with the tectonic framework of New England, the first section below is devoted to a summary of the tectonic setting.

6.2 Tectonic evolution of New England

Over the past fifteen or twenty years, significant advances in our understanding of detailed geological and geochronological relations in New England have made possible the development of reasonable plate-tectonic models for the region. These advances were to some extent sparked by the classic paper by Bird and Dewey (1970) and hastened by a broader application of relatively unambiguous radiometric dating techniques.

New England is a relatively young North American terrain—the oldest known rocks are those of the Chain Lakes Massif in Maine, about 1600 Ma, apparently the southernmost part of a Helikian terrain (Lyons *et al.*, 1982). More widely exposed in New England are the 1100 Ma old Proterozoic Y (Grenville) rocks of the western New England massifs and the 600–650 Ma old Proterozoic Z (Avalonian) rocks of eastern and southern New England, both shown in Figure 6.2. Geographically between the two are a diverse sequence of Cambro-Ordovician to Devonian metasedimentary and metavolcanic rocks

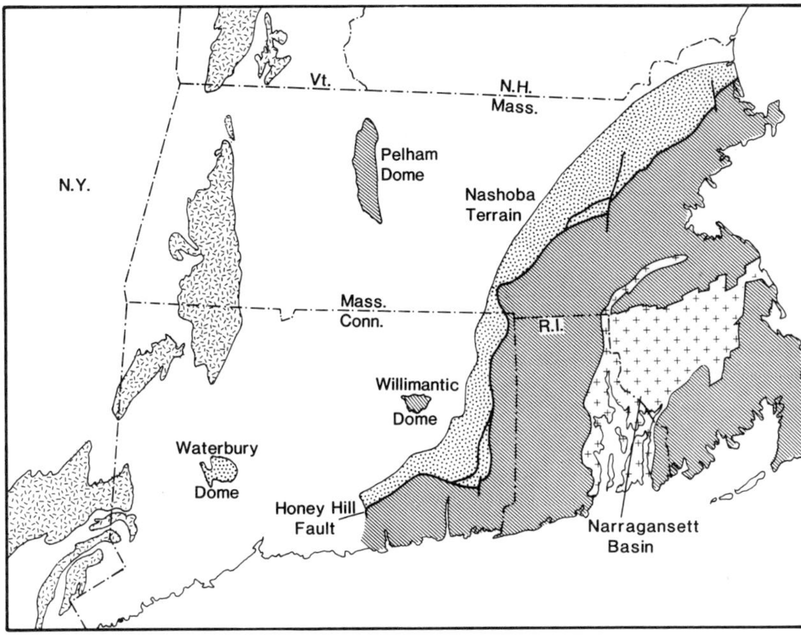

Figure 6.2 Map of southern New England indicating locations of a number of geographical and geological locales mentioned in the text. Grenville age rocks are shown by hachured symbol, Avalonian by diagonal ruling, Ordovician by stipple, and Pennsylvanian–Permian by crosses.

that represent various oceanic basins that closed throughout the Palaeozoic. This sequence has been cut through by plutons ranging in age from Ordovician to Permian, and has two rift basins developed in it, the Carboniferous Narragansett Basin of southeastern New England and the Mesozoic graben of central New England.

Figure 6.3 shows the major tectonic subdivisions of New England (Osberg, 1978; Robinson and Hall, 1980; Lyons *et al.*, 1982). The Green Mountains–Berkshire anticlinorium is the axis of exposed Grenville massifs in western New England, uncomformably overlain by a Cambro-Ordovician sequence. The next major structure to the east is the Connecticut Valley–Gaspe synclinorium in which metasedimentary and metavolcanic rocks from Cambrian to Devonian are metamorphosed and deformed (Hall and Robinson, 1982). The west flank of the Connecticut Valley synclinorium is penetrated by a number of gneiss domes of apparent Devonian age which expose core rocks of uncertain age, though likely either Precambrian or Cambro-Ordovician.

Next to the east is the Bronson Hill anticlinorium, a classic mantled gneiss dome terrain in which a sequence of late middle Ordovician to Devonian rocks form a nappe sequence (Thompson *et al.*, 1968) that was later deformed by gneiss domes which bring to the surface metamorphosed Ordovician felsic volcanics (Zartman and Naylor, 1984) or more rarely, anomalous Avalonian

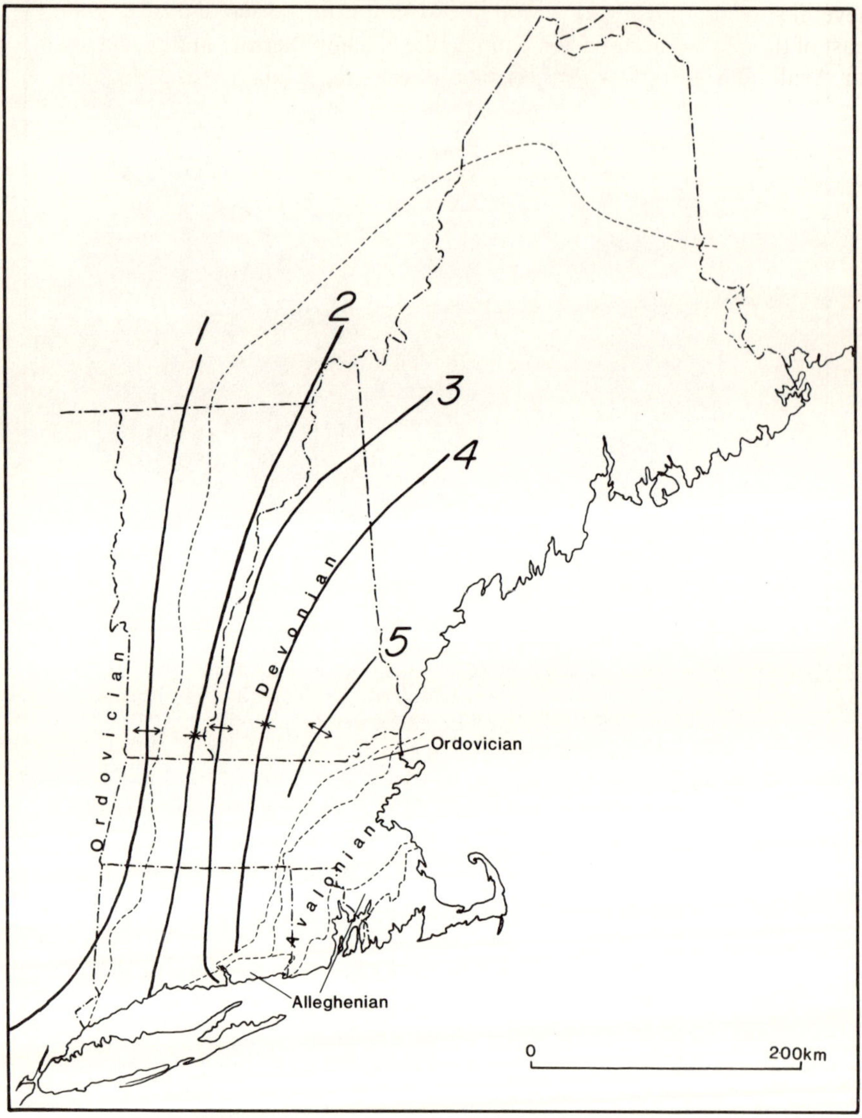

Figure 6.3 Major tectonic zones of New England, and distribution of metamorphisms by age. (1) Manhattan–Berkshire–Green Mountains anticlinorium; (2) Connecticut Valley–Gaspe synclinorium; (3) Bronson Hill anticlinorium; (4) Merrimack synclinorium; (5) Massabesic anticlinorium. Ages of most recent documented metamorphism are shown.

basement. East of the Bronson Hill anticlinorium is the Merrimack synclinorium, a rapidly eastward-thickening wedge of Ordovician to Devonian metasediments. Lyons *et al.* (1982) have suggested some redefinition of the Merrimack synclinorium in northern New England, notably introducing several smaller anticlinoria within it that split it into discrete basins. Finally, east of the Merrimack synclinorium lie the Nashoba terrain and the Dedham, or Avalonian, terrain whose tectonic development may be independent of the Palaeozoic events which shaped the tectonic zones to the west.

Ages of metamorphism in New England range from Precambrian to late Palaeozoic and fit into a chrono-geographical scheme which Robinson (1983) refers to as 'the seven realms of metamorphism'. (1) Grenville-age (Proterozoic Y—approximately 1100 Ma) upper amphibolite to lower granulite facies metamorphism in North American basement of western New England. (2) Avalonian-age (Proterozoic Z—approximately 650 Ma) low-grade and contact metamorphism in eastern New England which rises to amphibolite facies at the exposed western edge of the terrain and locally to lower granulite facies in isolated occurrences of the Pelham and Willimantic Domes (see Figure 6.2). (3) Ordovician upper amphibolite to lower granulite facies metamorphism in the Nashoba zone (Figure 6.2), probably unrelated to Taconian Ordovician metamorphism in western New England. (4) Ordovician (Taconian) metamorphism, with early high-pressure and later high-temperature phases, localized along the Berkshire–Green Mountains anticlinorial axis and widely overprinted by later (Devonian) metamorphism of similar grade. (5) Western Acadian metamorphic high, locally reaching kyanite and sillimanite grade, and bordered to the east by the Connecticut Valley metamorphic low. (6) Eastern Acadian metamorphic high, centred just east of the Bronson Hill anticlinorium, which runs from southern Maine to the Connecticut coast and has high-grade culminations along it that reach lower granulite facies. (7) Alleghenian (Pennsylvanian–Permian) metamorphism best documented in the Narragansett Basin, but now thought to extend westward along the Connecticut coast and northward at least to southeastern New Hampshire; this metamorphism may in large part be contact metamorphism associated with Alleghenian granitic plutons.

6.3 Grenville massifs of western New England

Rocks of Grenville age are exposed in several massifs: the Green Mountains of Vermont, the Berkshires of Massachusetts, the Housatonic Highlands of Connecticut and the Hudson Highlands of Connecticut and New York. These rocks were metamorphosed at upper amphibolite or lower granulite facies in the Grenville orogenic event of about 1100 Ma ago, as indicated by Rb/Sr ages on plutonic rocks in the Hudson Highlands that were intruded more or less contemporaneously with the intense deformation and metamorphism (Helenek and Mose, 1984). K/Ar ages of these Grenville rocks, as of those from the central and southern Appalachians (Rankin *et al.*, 1983) reflect some later thermal event which has reset the ages to 850–950 Ma.

Migmatites in the New England Grenville massifs are both stromatic and ophthalmitic, similar to migmatites in the classic Grenville terrain to the north and west, such as the Adirondacks (Weiner *et. al.*, 1984; Engel and Engel, 1953) and adjacent Canada (Wynne-Edwards, 1967). There is very little in the literature concerning the petrology of Precambrian migmatites in New England; one recent study is the discussion by Helenek and Mose (1984) of migmatitic rocks adjacent to the Canada Hill Granite in the Hudson Highlands. The Canada Hill Granite is a massive ternary granite, weakly foliated in places, which appears to be intrusive into Proterozoic Y paragneiss. The paragneiss consists of a variety of lithic types including rusty gneisses and schists, amphibolites, calc-silicates and ferruginous metasediments of lower granulite facies. Along contacts with the Canada Hill Granite plutons the paragneiss is migmatitic, consisting of alternating light and dark bands of medium-grained granitic leucosome and garnet-, biotite- and sillimanite-enriched melanosome (Helenek and Mose, 1984).

Helenek and Mose argue for an origin of these migmatites by a 'closed-system metasomatic process,' based upon leucosome bulk compositions that diverge from expected minimum melt compositions and upon the similarity of plagioclase compositions between leucosome and palaeosome. They do, however, note that metamorphic P, T and P_{H_2O} are permissive of an origin by anatexis, as is the localization of migmatitic gneisses around the main Canada Hill granite body. Migmatitic paragneiss is not restricted to the margins of Canada Hill Granite, however; this author has observed, and D. Murray (pers. comm.) confirms, that major fold hinges in the paragneiss are commonly strongly migmatitic whereas the fold limbs are rarely so. This observation is an argument for the migration of locally derived partial melts to the low-shear environment of fold hinges.

6.4 Ordovician migmatites in the western high grade belt

East of the exposed Precambrian massifs lies a metamorphic terrain composed of rocks ranging in age from Cambro-Ordovician to Devonian which have been metamorphosed and deformed in both Taconian and Acadian events (Robinson, 1983). In this terrain, the Taconian metamorphism is typically of lower grade than the Acadian, but most exposed pre-Silurian rocks show relatively superficial levels of a Taconian metamorphic section at greenschist or lower amphibolite facies. However, structural domes of presumed Acadian age expose the deeper levels of the Taconian metamorphic pile. The 'core' rocks in the domes are of problematic age, but in the southernmost and largest of the domes in this belt, the Waterbury Dome in Connecticut (Figure 6.2), there is evidence that rocks no older than Cambro-Ordovician are exposed (Dietsch, 1983).

The core rocks of the Waterbury Dome are much more intensely deformed and more highly metamorphosed than surrounding rocks of similar age (O'Connor, 1973; Dietsch, 1983), and it has been proposed that they are in thrust contact with the overlying rocks (Dietsch, 1983). Metamorphic grade in the core rocks is as high as kyanite + K-feldspar, and the metamorphism seems to have been at high pressure because kyanite is ubiquitous in aluminous bulk compositions and sillimanite and cordierite are unknown in any rocks of the dome. Thermobarometry by Dietsch (1984) has shown that metamorphic conditions for the kyanite + K-feldspar schist were 700–750 °C at 7–7.5 kbar. Under these conditions, at reasonably high P_{H_2O}, pelitic rocks would be expected to undergo anatectic melting, and many of the pelitic and semipelitic schists and gneisses in the Waterbury Dome are migmatitic. There are as yet no detailed petrological studies of the migmatites in the Waterbury Dome or other domes to the north in this belt, so it is not possible to argue for either a metasomatic or anatectic origin of the migmatites.

The age of the high-grade metamorphism in the core rocks of the Waterbury Dome is at present ambiguous, but Dietsch (1983) has argued that it is most likely Taconian. He bases this on the involvement only of Ordovician or older rocks in the highest grade metamorphism, and on the patchy retrogression of the highest-grade schists in a later event, presumably Acadian. If the interpretation is correct that the Acadian structural domes have exposed the deep levels of an imbricated early Palaeozoic metasedimentary pile, then these rocks deserve a considerably more intense petrological scrutiny than they have received.

6.5 Migmatites of the eastern Acadian metamorphic high

This terrain is certainly the most thoroughly studied high-grade metamorphic belt in New England. Extensive development of sillimanite + K-feldspar grade or higher occurs from east central Connecticut to northwestern Maine (Thompson and Norton, 1968; Robinson, 1983; Guidotti *et al.*, 1983). The highest reported metamorphic grades in this terrain occur in southern Massachusetts and northeastern Connecticut where grades observed in pelites and metabasites approach or reach lower granulite facies (Tracy *et al.*, 1976; Tracy, 1978; Tracy and Robinson, 1980). Widespread migmatite development has been reported from Maine (Evans and Guidotti, 1966), south-central New Hampshire (Heald, 1950; Dougan, 1979), central Massachusetts and northeastern Connecticut (Tracy and Robinson, 1980, 1983) and east-central and southeastern Connecticut (Lundgren, 1966; Wintsch, 1975). Because of the detail in which migmatites from this terrain have been described, each of the above areas will be discussed separately.

6.5.1 *Northwestern Maine*

The timing and relative intensity of metamorphic events in Maine have recently been reviewed by Guidotti *et al.* (1983) and Holdaway *et al.* (1982). They recognize four phases of Acadian metamorphism in that part of the Merrimack synclinorium occurring in southern Maine, three regional and one contact. Their M3 metamorphism, the youngest regional event, is the only one which produces grades as high as sillimanite + K-feldspar. As they note (Guidotti *et al.*, 1983, p. 242), an 'enormous area' of highest-grade rocks occurs in association with apparently synmetamorphic large granite plutons, but nowhere do the pelitic schists contain sillimanite and K-feldspar without muscovite (see Figure 6.4). The above authors argue that this limitation on assemblages may have been caused by thermal buffering due to extensive

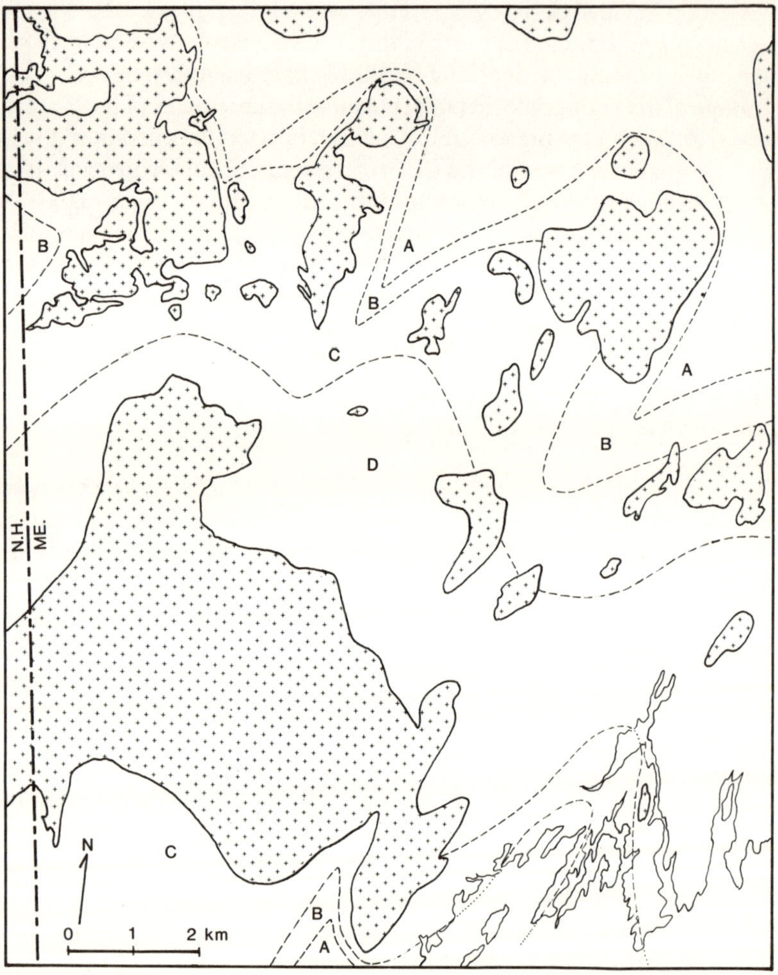

Figure 6.4 Isograd map of western Maine, also showing the distribution of Devonian granite plutons in relation to metamorphic isograds. (*A*) Epidote amphibolite grade; (*B*) lower amphibolite, characterized by andalusite or kyanite, with staurolite, in pelites; (*C*) medium amphibolite, characterized by sillimanite; (*D*) upper amphibolite characterized by sillimanite + K-feldspar in pelites and widespread development of migmatites (adapted from Guidotti *et al.*, 1983).

partial melting and the presence of 'abundant small granitic plutons and pegmatites.' It seems equally likely to this author that ubiquitous muscovite occurrence might be due to pervasive retrogression of sillimanite + K-feldspar assemblages by the hot aqueous fluids given off during crystallization of the abundant small plutons (Figure 6.4).

Researchers working on the high-grade rocks of southern Maine have concentrated their published efforts on discussing upper amphibolite-facies subsolidus metamorphic reactions, and have not concentrated attention on migmatite-forming processes. Guidotti *et al.* (1983) and Evans and Guidotti (1966) suggest that partial melting may have played a significant part in forming what they describe as a broadly migmatitic high-grade zone, but they do not discuss possible melting reactions. They do note that the M3 metamorphism appears to be post-tectonic and static, and give outcrop descriptions which imply that many of the migmatitic rocks are blastomylonitic. If melting actually is post-deformational in Maine, this would be a very different pattern from that observed in other Acadian migmatites to the south (see below).

6.5.2 *New Hampshire*

One of the earliest published reports of sillimanite + K-feldspar grade metamorphism in New England is that of Heald (1950) which describes the geology of the Lovewell Mountain quadrangle in southwestern New Hampshire. He notes that the highest-grade rocks contain leucocratic streaks or layers that are orthoclase-bearing, and ascribes their origin to either partial fusion or solution segregation. As in the case of the southern Maine migmatites, there is an intimate intermingling of apparently plutonic granitic and quartz monzonitic rocks with the migmatitic schists and gneisses, implying a role for magmatic heating in the migmatite-forming process.

More recently, Dougan (1979, 1981, 1983) has reported on an investigation of migmatites from central New Hampshire, in the vicinity of Concord. Dougan (1979) provides a thorough description of the migmatites and concludes that an origin by anatexis is most likely. He describes the leucosomes as roughly lenticular, coarse-grained and conformable with the foliation. They contain quartz, plagioclase, K-feldspar and a variety of aluminous accessory minerals, and are texturally immature (cf. Ashworth, 1976). In contrast, melanosomes were finer-grained and much more mature texturally. Dougan took the leucosome–melanosome textural contrasts to indicate solid-state annealing in the melanosome and crystallization from a melt in the leucosome.

In his papers, Dougan also dealt with major and trace element geochemistry of leucosomes and melanosomes, and concluded that segregation occurred under closed-system conditions for non-volatile elements (see Olsen, this volume). He further noted that systematic compositional and modal profiles could be used to deduce segregation-forming mineral reactions. Dougan

(1983) used a textural study to deduce that the textures of melanosomes were consistent with melt reactions having generated the leucosome–melanosome separation (Ashworth and McLellan, this volume). He points out that all but one of the possible melting reactions for these rocks (Dougan, 1979, Table 1) involves either muscovite or biotite, and that locally higher f_{H_2O} may be required for melting involving quartz and feldspar. This observation of the potential importance of mica decomposition as a supplier of the H_2O needed for melting has also been made by Lundgren (1966) and others.

6.5.3 *Central Massachusetts*

Migmatitic rocks in the Bronson Hill anticlinorium and Merrimack Synclinorium have been discussed in a number of papers (Robinson, 1967; Tracy, 1978; Tracy and Robinson, 1980, 1983; Thompson and Tracy, 1979; Tracy *et al.*, 1976). The area in which migmatites occur is generally restricted to the higher metamorphic grade portions of the two adjacent major tectonic belts of central New England. The Bronson Hill anticlinorium, a gneiss dome terrain, shows late Acadian (Devonian) uplift immediately following high-grade metamorphism whereas the adjacent Merrimack synclinorium appears to have been downwarped at the same time (Robinson and Hall, 1980). Tracy and Robinson (1980) pointed out the apparent differences in $P-T$ paths taken by the two terrains during metamorphism, and discussed the tectonic and metamorphic implications.

Estimated peak metamorphic conditions in this terrain are 6 to 6.5 kbar and 700 to 750 °C based upon various thermobarometers (Tracy *et al.*, 1976; Robinson *et al.*, 1982), and the temperature appears to have exceeded 650° over a very wide area (Tracy and Robinson, 1983, Figure 2). Much of the terrain has therefore achieved metamorphic conditions appropriate to melting in pelites, and the case has been made that many of the migmatitic rocks originated through melting (Tracy, 1978).

Interpretation of migmatites in this terrain is complicated, however, by the existence of obvious solid-state quartzofeldspathic segregations at grades as low as staurolite. These segregations are rarely vein-like but more commonly conformable with compositional layering and foliation. Mineralogically, they are composed of various proportions of quartz and plagioclase, commonly approaching virtually pure end-members of each. Their stromatic appearance and interlayering with biotite, aluminosilicate and garnet-rich schists gives rise to outcrops that are certainly descriptively 'migmatitic' even though no melting has taken place, as the estimated low temperatures (550 to 625 °C) and absence of K-feldspar indicate. Solid-state segregations of this type are relatively common in many metamorphic terrains (e.g. Yardley, 1978; Olsen, 1977, 1982).

An obvious complication of having pre-existing segregations of this type in rocks that are beginning to melt is that mechanical heterogeneities such as segregations may act as loci for concentration of locally derived melts (McLellan, 1983), which may in turn mobilize or recrystallize the earlier segregation minerals. The result, as suggested by Tracy (1978), may be a modified leucosome with a bulk composition that does not necessarily closely approximate an expected minimum melt composition. In fact, compositional data on central Massachusetts migmatites (Tracy and Thompson, in preparation) indicate that although there is a concentration of leucosome compositions near the minimum in a quartz–flagioclase–orthoclase diagram, most of the discrepant compositions are enriched in quartz or plagioclase or both, as would be expected in the assimilation of quartz–plagioclase segregations.

Tracy and Robinson (1983) noted that there are two distinct migmatite types in central Massachusetts, a stromatic type associated with metamorphic grades near the sillimanite + K-feldspar isograd and a more nebulous vein-type occurring more commonly in near-granulite facies pelites (Figure 6.5). Associated with this latter variety are highly aluminous layers not obviously related to migmatites but which have compositions most easily explained as residua from partial fusion of pelites. These rocks are anomalously enriched in garnet, cordierite, sillimanite and biotite and depleted in silica and alkalis, and have been described by Tracy and Robinson (1983), Robinson *et al.* (1982) and Lasaga *et al.* (1977).

The distinction between the two migmatite types has been ascribed by Tracy and Robinson (1983) to the difference in melting reactions and in melt extraction efficiency between higher-pressure and lower-pressure metamorphic facies series (see also Thompson and Tracy, 1979). The stromatic migmatites of the central Massachusetts terrain occur relatively close to the sillimanite + K-feldspar isograd where temperatures did not greatly exceed those required for initial melting and where any melt formed would have remained saturated, or nearly so, with H_2O. As pointed out by Thompson (1982), a melt of this type would have had relatively little mobility and would likely have remained near its site of origin.

On the other hand, the vein-type migmatites occur in pelitic rocks at sillimanite + K-feldspar + garnet + cordierite grade (near-granulite) and show no evidence of having originated through melting reactions involving muscovite. In fact, the evidence, including presence of large garnet and cordierite crystals in leucosomes, points toward melting having occurred on a vapour-absent reaction at temperatures well above the breakdown of muscovite (Tracy and Robinson, 1983). The enhanced mobility of melts under these conditions (Arzi, 1978; Thompson, 1982) helps to explain the locally intrusive vein morphology of these migmatites and the occurrence of apparent aluminous residua without paired leucosome.

The central Massachusetts migmatites are useful for evaluating the contention by Guidotti *et al.* (1983) that partial melting of muscovitic schists may provide a thermal buffering effect that prevents the temperature rising

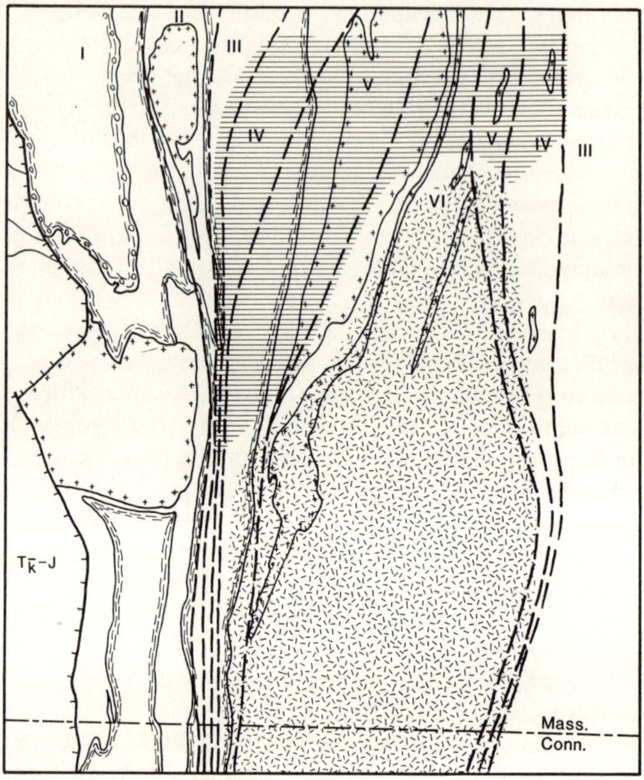

Figure 6.5 Generalized geological–metamorphic map of central Massachusetts showing the distribution of dome gneisses, Devonian plutons, metamorphic isograds and migmatite types. Precambrian (Avalonian) and Ordovician dome gneisses are shown with a short-line pattern, with and without circles, respectively. Devonian intermediate to felsic plutons, typically foliated, are shown with crosses. Metamorphic zones are characterized by assemblages in pelites: (I) kyanite–staurolite; (II) sillimanite–staurolite; (III) sillimanite–muscovite; (IV) sillimanite–muscovite–orthoclase; (V) sillimanite–orthoclase; (VI) sillimanite–orthoclase–garnet–cordierite. The horizontal ruled pattern shows the distribution of stromatic migmatites only, and the hachured pattern shows the extent of vein-type migmatites, commonly coexisting with minor stromatic migmatites.

sufficiently for the disappearance of muscovite, i.e., the achievement of a true sillimanite + K-feldspar zone. In Massachusetts, stromatic migmatites occur in both muscovite-bearing and muscovite-free palaeosomes, although almost all of the leucosomes contain muscovite, perhaps reflecting the suspected higher a_{H_2O} in melt layers. Abundant K-feldspar-bearing and muscovite-free schist layers may be found just above the sillimanite + K-feldspar isograd, with or without adjacent leucosome layers. These observations suggest that in Maine it is the presence of large granite or adamellite plutons (absent in Massachusetts) that produces the huge extent of sillimanite + K-feldspar + muscovite schists, possibly through the retrogressive action of the water exsolved by crystallizing plutons. In the absence of retrograde reaction, it might also be possible to generate sillimanite + muscovite + K-feldspar assemblages of regional extent by having systematic regional variation in a_{H_2O} (Tyler and Ashworth, 1982).

In Massachusetts, attention has also been paid to the timing of the migmatite formation and its relation to the structural history of the terrain. The Bronson Hill anticlinorium has been interpreted to have undergone three major deformational events that produced outcrop-scale structures corresponding to regional nappe formation, backfolding and dome formation (Robinson, 1967). High-grade metamorphism and melting have been placed in the sequence as predating the isoclinal second phase folding because migmatite leucosomes are isoclinally folded in this fold generation. It has also been presumed (Tracy, 1978) that the melts had solidified by this time because the leucocratic layers define the wavelength and amplitude of the large folds, and are therefore the most competent layers. Oddly, however, there is no evidence of cataclasis, annealing or recrystallization in the leucosomes as should be expected if these rocks had undergone such intense deformation in the solid state. Ashworth (1976) has posed similar points for migmatites in the Huntly-Portsoy area of Scotland, and has suggested the possibility that granitic melts were intruded conformably into pre-existing isoclinal folds, and crystallized in their present morphology. An obvious objection to this explanation concerns the high viscosity of such melts and the difficulty of such delicate injection.

There are several explanations that avoid the above difficulty. One is that the folds were formed in a layered sequence containing earlier-formed metamorphic differentiation layers whose mechanical properties defined the scale of the folds. If melting followed the folding, collection of locally derived melt into the pre-existing segregation would greatly aid in the annealing of the deformed fabric. Another scenario is that melting followed folding, but occurred in a layer whose composition was appropriate for nearly complete melting within relatively few degrees above minimum melting temperature (Johannes, 1983). A third possibility is that a layered sequence containing both micaceous schists and melt was folded, with the melt layer being the most mechanically competent and determining the fold geometry. This might happen if the melt was of sufficiently high viscosity and if the deformation occurred at high strain rates; such a phenomenon is currently being investigated and modelled by the author.

6.5.4 *East central Connecticut*

The Acadian high-grade belt continues south into Connecticut where the tectonic style becomes somewhat different (Rodgers, 1981; Robinson and Tucker, 1982), complicating the relation between the Bronson Hill anticlinorium and the Merrimack synclinorium. As far south as central Connecticut, upper amphibolite to lower granulite facies rocks of the Acadian metamorphism locally contain stromatic migmatites that appear to have resulted from melting. The characteristics of this area are in general similar to those of central Massachusetts, described above.

Robinson (1983, p. 254) shows the southward termination of highest-grade Acadian metamorphism to be in central Connecticut, with middle to upper amphibolite facies rocks continuing south to the Honey Hill fault. There are abundant sillimanite + K-feldspar grade rocks in southern Connecticut, but the ages of metamorphism and migmatization of these rocks are matters of some dispute. Robinson (1983, p. 252) shows a narrow belt of high-grade Ordovician metamorphism extending southward from the Nashoba terrain of eastern Massachusetts and culminating in granulite facies rocks in southeastern Connecticut. Ordovician high-grade metamorphism has been well dated in Massachusetts, but not in Connecticut where dating must be regarded as somewhat speculative. This Ordovician metamorphism is apparently not related to Taconian metamorphism in western New England.

The Willimantic Dome in eastern Connecticut contains high-grade schists and gneisses that are abundantly migmatitic, and thermobarometry (unpublished) by the author on these rocks yields temperatures in the melting range (approx. 650–700 °C). The Pelham Dome, to the north and west in Massachusetts, contains rocks of very similar character and possible pre-Acadian metamorphic grade (Robinson, 1983), although the metamorphism of the Pelham Dome may be somewhat older than that of the Willimantic Dome. Basement in both domes appears to be of Avalonian age, and the structural relationships between core gneisses and mantling Ordovician metavolcanics are not entirely clear. There is clear evidence in the core of the Pelham Dome for a pre-Acadian sillimanite + K-feldspar (or higher) grade metamorphism (Robinson *et al.*, 1975) which may be either Avalonian or Taconian. As pointed out by Lyons *et al.* (1982), the presence of Avalonian basement with a cover of metamorphosed Cambro-Ordovician rocks this far west may indicate a larger extent of the eastern cratonic platform of Iapetus in early Ordovician time.

To recapitulate, some migmatitic rocks of eastern Connecticut that were thought to have been metamorphosed in the Acadian are now ascribed by Robinson to Ordovician metamorphism. Because these migmatites are developed in Avalonian basement, it is also possible that their metamorphism is late Precambrian, probably reworked in both the Ordovician and Devonian.

To further complicate the issue, Wintsch (1975; Wintsch and Fout, 1982) has suggested that although the rocks of the Willimantic Dome underwent amphibolite-facies metamorphism in the Acadian or before, they were migmatized by solid-state processes in an Alleghenian (Permian) shearing event. In his model, Wintsch argues that the chemical and physical processes accompanying ductile shearing decreased the stability of muscovite by reducing a_{H_2O}, thereby producing coarse-grained K-feldspar-bearing leucosomes at temperatures well below those typically required. Specifically,

Wintsch suggested that during ductile shear of micaceous rocks, the constant grinding-off of surface layers of the mica could change the chemistry of intergranular fluids, effectively reducting a_{H_2O} and destabilizing muscovite. He further suggested that this process operated in discrete conformable zones and could produce both large rotated K-feldspar porphyroblasts and continuous K-feldspar-bearing layers that might mimic melt leucosomes. Wintsch clearly notes that he views his process as a retrograde one operating at temperatures well below those required for melting—for example, he believes that retrograde production of chlorite along shear zones is contemporaneous with ‘feldspathization’, and that kyanite was the stable aluminous phase at these low temperatures and H_2O activities.

Because the rocks of the Willimantic Dome are of radiometrically undocumented, though possibly Avalonian, age (Robinson, 1983) and because they closely resemble the rocks of the Pelham Dome, it is possible that the metamorphism of both is at latest Precambrian, as Robinson (1983, p. 250) claims for the Pelham Dome. We are therefore left with possible migmatization ages of these rocks of late Precambrian, Ordovician, Devonian or Permian, the first three probably by melting and the last by differentiation, according to Wintsch's model. It is clear that many of the migmatized rocks of eastern Connecticut deserve further study, preferably combining geological and isotopic techniques.

6.6 Avalonian terrains overprinted in the Alleghenian

Much of eastern and southeastern New England is made up of apparently small individual terrains which may have been accreted on to the North American continental margin at various times in the Palaeozoic, typically in the Permian. Several of these have been studied in some detail in recent years, and a pattern is emerging of tectonic and metamorphic events that may not be represented in rocks farther to the west.

6.6.1 *Southern Connecticut*

Migmatitic rocks along the Connecticut shoreline from just east of New Haven into southern Rhode Island present another possible case of ambiguity of age for migmatization. This terrain appears to be the southern and eastern extension of the Bronson Hill anticlinorium, but it has recently been suggested that domes along the Connecticut coast may be Alleghenian rather than Acadian as is true farther to the north (Losh and Bradbury, 1984; Bradbury, pers. comm.; Hall and Robinson, 1982). Lundgren (1963, 1966) argued that the high-grade metamorphism and migmatization of this area occurred in the Alleghenian (approx. 265 Ma) based on radiometric dating of pegmatites that he assumed were contemporaneous with high-grade metamorphism and on the occurrence of metamorphism of Carboniferous rocks up to sillimanite grade in the nearby Narragansett Basin (Grew and Day, 1972). The Connecticut coastal belt is also characterized by numerous small and intermediate-size peraluminous granitic plutons well-dated as Permian (Zartman *et al.*, 1970), and a strong Permian thermal disturbance of U-Pb systematics in granites of Avalonian age has been documented in southern Rhode Island (Hermes *et al.*, 1981).

The petrology of the high grade migmatitic schists in southeastern Connecticut has been well described by Lundgren (1966). He mapped a sillimanite + K-feldspar isograd that crosscuts the regional structural trends, and observed that formation of migmatites largely coincided with the isograd. He ascribed the migmatite formation to partial fusion and suggested that the melting event followed a subsolidus muscovite dehydration reaction in which the release of aqueous fluid allowed melting to occur. A more modern view could be that melting occurred on a fluid-present or fluid-absent 'dehydration-melting' curve (Huang and Wyllie, 1974; Thompson and Tracy, 1979). Unpublished thermobarometric evidence of the author indicates that the metamorphic P–T trajectory of these rocks probably passed above the invariant point generated by the intersection of the muscovite dehydration and water-saturated granite solidus curves.

As noted above, the age of migmatization of the Connecticut coastal belt has not been definitively established. Lundren argued for, and successive authors (e.g. Robinson, 1983) have accepted, the Permian age. The absence of metamorphosed rocks younger than Ordovician south of the Honey Hill Fault (Rodgers, 1981) gives the possibility of one or more older metamorphisms. Robinson (1983) recognized this possibility and commented on the likelihood of an Alleghenian overprinting on both Ordovician and Devonian metamorphisms. Evidence cited above indicates that there was an important thermal disturbance here in the late Palaeozoic; there is no definite evidence, however, that temperatures high enough to cause melting were achieved. Because of the possibility that the Honey Hill fault was a zone of major horizontal transport, it is hazardous to infer that these rocks south of it had a similar pre-Alleghenian metamorphic history to those north of it.

6.6.2 *The Massabesic gneiss*

Over the past ten years, considerable research, much of it geochronological, has clarified the origin and history of this migmatitic gneiss terrain in southeastern New Hampshire (Figure 6.1), the best studied of a number of older migmatitic terrains that occur in the cores of small anticlinoria in eastern New England (see Lyons *et al.*, 1982, Figure 2). As summarized in Aleinikoff *et al.* (1979), the metasedimentary and metavolcanic units of the Massabesic Gneiss yield fairly well constrained ages of about 650 Ma (U–Pb zircon). The above paragneiss units are cross-cut by an orthogneiss which has yielded a similar age. The intimate intermingling of the two gneiss types in complexly folded outcrops apparently reflects a high-grade metamorphism and melting event in the late Precambrian, immediately followed by intense deformation (Sriramadas, 1966; Aleinikoff, 1978).

Evidence for an Alleghenian overprint on the Massabesic Gneiss is provided by a granite near Milford, New Hampshire, which has recently been dated as 275 Ma old, or Permian (Aleinikoff *et al.*, 1979). This granite is clearly intrusive into the Gneiss and contains intimately intruded roof pendants and xenoliths. An Ordovician overprint is also indicated by the occurrence of cross-cutting orthogneiss that has been dated as 480 Ma old (Lyons *et al.*, 1982). These authors suggest that the migmatization event is of uncertain age, but might be Ordovician or younger, i.e. at one of the times of formation of anatectic granites in New Hampshire.

An important point made by Lyons *et al.* (1982) is that the granite at Milford is quite similar to the Permian Narragansett Pier Granite of southern Rhode Island, also intrusive into late Precambrian gneisses. Both reworked Avalonian terrains are of Pan-African affinity, suggesting that major sutures may exist between them and more typical North American basement to the west. The timing of earlier metamorphic events in these terrains may therefore bear only coincidental relation to the classic Taconian and Acadian events of the western belts.

6.7 Summary

The distribution of tectonic and metamorphic events in New England, both areally and in time, is especially complex and reflects the operation of a variety of collisional tectonic processes over hundreds of millions of years. The occurrence of migmatites in New England is similarly, and relatedly, complicated and the study of migmatites in this terrain has included their use and implications for tectonic reconstructions as well as the petrological details of their origin.

Many metamorphic studies in New England have shown that high-grade metamorphism, with temperatures and pressures high enough to cause crustal anatexis, has occurred repeatedly in this orogenic belt. Petrological examination of many pelitic migmatites, notably those of the eastern Acadian metamorphic high, has unambiguously established that partial melting was important in the development of these rocks (Tracy, 1978; Dougan, 1979). Other New England migmatites are of more ambiguous origin, although the case has been made for the operation of solid-state metamorphic segregation processes in the genesis of these migmatites (e.g. Wintsch, 1975; Helenek and Mose, 1984). Quite a few of the reported migmatites have not been described in enough detail to be able to make much of an argument one way or the other.

One of the most significant problems in fully characterizing some migmatites is that the age of migmatization is not as well constrained as one would like, commonly because of uncertainty over the mechanism of migmatite formation, or because of the overlap of high-grade metamorphisms of different ages which makes different migmatization ages possible. Migmatites from the Massabesic Gneiss of New Hampshire, from the Willimantic Dome of Connecticut and from the southern coast of Connecticut all fall into this category.

Despite the fact that quite a few good studies have been done on migmatites in New England, the above review makes it clear that many important questions remain, among them:

(1) Are there particular tectono-metamorphic regimes that better lend themselves to the production of migmatites at higher metamorphic grade? As a corollary, which regimes (accretionary wedge, continental subduction, obduction, etc.) are capable of reaching the temperatures and pressures required for melting or large-scale metamorphic segregation?
(2) On the scale of the whole of the Grenville-Caledonide Orogen, can one make sense of the distribution of highest-grade metamorphism and migmatization both in space and time?
(3) What are the roles of purely conductive *v.* magmatic heat transport in the production of near-granulite grade orogenic metamorphic rocks and the initiation of crustal anatexis, and how do these two heat transport mechanisms operate in various metamorphic regimes?

Acknowledgements

The author would like to acknowledge many stimulating discussions on this topic with Peter Robinson, Alan Thompson and Eileen McLellan. Financial support was provided by the U.S. National Science Foundation through grant EAR 81–20670.

References

Aleinikoff, J.N. (1978) Structure, Petrology, and U-Pb Geochronology in the Milford (15′) Quadrangle, New Hampshire. Unpublished Ph.D. Thesis, Dartmouth College.

Aleinikoff, J.N., Zartman, R.E. and Lyons, J.B. (1979) U–Th–Pb geochronology of the Massabesic Gneiss and the granite near Milford, south-central New Hampshire: New evidence for Avalonian basement and Taconic and Alleghenian disturbances in eastern New England. *Contrib. Miner. Petrol.* **71**, 1–11.

Arzi, A.A. (1978) Fusion kinetics, water pressure, water diffusion and electrical conductivity in melting rock, interrelated. *J. Petrol.* **19**, 153–169.

Ashworth, J.R. (1976) Petrogenesis of migmatites in the Huntly-Portsoy area, north-east Scotland. *Mineral. Mag.* **40**, 661–682.

Bird, J.M. and Dewey, J.F. (1970) Lithosphere plate-continental tectonics and the evolution of the Appalachian orogen. *Geol. Soc. Am. Bull.* **81**, 1031–1060.

Dietsch, C.W. (1983) Preliminary correlation of polyphase folding of rocks in the Waterbury Gneiss Dome, west-central Connecticut (abstract). *Geol. Soc. Am. Abstr. Programs* **15**, 558.

Dietsch, C.W. (1984) Character and evolution of structural domes in western Connecticut (abstract). *Geol. Soc. Am. Abstr. Programs.* **16**, in press.

Dougan, T.W. (1979) Compositional and modal relationships and melting reactions in some migmatitic metapelites from New Hampshire and Maine. *Am. J. Sci.* **279**, 897–935.

Dougan, T.W. (1981) Melting reactions and trace element relationships in selected specimens of migmatitic pelites from New Hampshire and Maine. *Contrib. Miner. Petrol.* **78**, 337–344.

Dougan, T.W. (1983) Textural relations in melanosomes of selected specimens of migmatitic pelitic schists: implications for leucosome-generating processes. *Contrib. Miner. Petrol.* **83**, 82–98.

Engel, A.E. and Engel, C.G. (1953) Grenville Series in northwest Adirondack Mountains, New York. Part II: Origin and metamorphism of the Major Paragneiss. *Geol. Soc. Am. Bull.* **64**, 1049–1098.

Evans, B.W. and Guidotti, C.V. (1966) The sillimanite–potash feldspar isograd in western Maine, U.S.A. *Contrib. Min. Petrol.* **12**, 25–62.

Grew, E.S. and Day, H.W. (1972) Staurolite, kyanite and sillimanite from the Narragansett Basin, Rhode Island. *U.S. Geol. Survey Prof. Paper* **800-D**, 151–157.

Guidotti, C.V., Trzcienski, W.E. and Holdaway, M.J. (1983) A northern Appalachian metamorphic transect–eastern townships, Quebec to the central Maine coast. In *Regional Trends in the Appalachian-Caledonian-Hercynian-Mauritanide Orogen,* ed. Schenk, P.E., NATO A.S.I. Series, 235–247.

Hall, L.M. and Robinson, P. (1982) Stratigraphic-tectonic subdivisions of southern New England. In *Major Structural Zones and Faults of the Northern Appalachians*, ed. St. Julien, P. and Beland, J., *Geol. Assoc. Canada Spec. Paper* **24**, 15–41.

Heald, M.T. (1950) Structure and petrology of the Lovewell Mountain Quadrangle, New Hampshire. *Geol. Soc. Am. Bull.* **61**,43–89.

Helenek, H.L. and Mose, D.G. (1984) Geology and geochronology of Canada Hill Granite and its bearing on the timing of Grenvillian events in the Hudson Highlands, New York. In *The Grenville Event in the Appalachians and Related Topics*, ed. Bartholomew, M.J., *Geol. Soc. Am. Spec. Paper* **194**, 57–73.

Hermes, O.D., Gromet, L.P. and Zartman, R.E. (1981) Zircon geochronology and petrology of plutonic rocks in Rhode Island. In *New England Intercoll. Geol. Conf., 73rd Annual Mtg. Guidebook*, ed. Boothroyd, J.C. and Hermes, O.D., 315–338.

Holdaway, M.J., Guidotti, C.V., Novak, J.M. and Henry, W.E. (1982) Polymetamorphism in medium- to high-grade pelitic metamorphic rocks, west-central Maine. *Geol. Soc. Am. Bull.* **93**, 572–584.

Huang, W.L. and Wyllie, P.J. (1974). Melting relations of muscovite with quartz and sanidine in the $K_2O–Al_2O_3–SiO_2–H_2O$ system to 30 kilobars and an outline of paragonite melting relations. *Am. J. Sci.* **274**, 378–395.

Johannes, W. (1983) On the origin of layered migmatites. In *Migmatites, Melting and Metamorphism*, eds. Atherton, M.P. and Gribble, C.D., Shiva, Nantwich, 234–248.

Lasaga, A.C. Richardson, S.M. and Holland, H.D. (1977) The mathematics of cation diffusion and exchange between silicate minerals during retrograde metamorphism. In *Energetics of Geologic Processes*, ed. Saxena, S.K. and Bhattacharji, S., Springer-Verlag, New York, etc. 353–388.

Losh, S. and Bradbury, H.J. (1984) Late Paleozoic deformation within the Honey Hill-Lake Char Fault zone (abstract). *Geol. Soc. Am. Abstr. Programs* **16**, 48.

Lundgren, L.W. (1963) The bedrock geology of the Deep River Quadrangle. *Quadrangle Report* **13**, Conn. State Geological and Natural History Survey, Hartford.

Lundgren, L.W. (1966) Muscovite reactions and partial melting in southeastern Connecticut. *J. Petrol.* **7**, 421–453.

Lyons, J.B., Boudette, E.L. and Aleinikoff, J.N. (1982) The Avalonian and Gander Zones in central eastern New England. In *Major Structural Zones and Faults of the Northern Appalachians*, ed. St. Julien, P. and Beland, J., *Geol Assoc. Canada Spec. Paper* **24**, 43–66.

McLellan, E.L. (1983) Contrasting textures in metamorphic and anatectic migmatites: an example from the Scottish Caledonides. *J. Metamorphic Geol.* **1**, 241–262.

O'Connor, B.J. (1973) A Petrologic and Electron Microprobe Study of Pelitic Mica Schists in the Vicinity of the Staurolite-disappearance Isograd in Philadelphia, PA and Waterbury, CT. Unpublished Ph.D. Dissertation, The Johns Hopkins University, Baltimore.

Olsen, S.N. (1977) Origin of the Baltimore Gneiss migmatites at Piney Creek, Maryland. *Geol. Soc. Am. Bull.* **88**, 1089–1101.

Olsen, S.N. (1982) Open- and closed-system migmatites in the Front Range, Colorado. *Am. J. Sci.* **282**, 1596–1622.

Osberg, P.H. (1978) Synthesis of the geology of the Northern Appalachians, U.S.A. *Geol. Surv. Canada Paper* **78–13**, 137–147.

Rankin, D.W., Stern, T.W., McLelland, J., Zartman, R.E. and Odom, A.L. (1983) Correlation chart for Precambrian rocks of the eastern United States. *U.S. Geol. Surv. Prof. Paper* 1241–E.

Robinson, P. (1967) Geology of the Quabbin Reservoir area, central Massachusetts. In *New England Intercoll. Geol. Conf., 59th Annual Mtg. Guidebook*, 114–127.

Robinson, P. (1983) Realms of regional metamorphism in southern New England, with emphasis on the eastern Acadian metamorphic high. In *Regional Trends in the Geology of the Appalachian-Caledonian-Hercynian-Mauritanide Orogen*, ed. Schenk, P.E., NATO A.S.I. Series, 249–258.

Robinson, P. and Hall, L.M. (1980) Tectonic synthesis of southern New England. In *Proceedings, The Caledonides in the U.S.A.*, ed. Wones, D.R., *Virginia Polytechnic Inst. and State Univ. Memoir* **2**, 73–82.

Robinson, P. and Tucker, R.D. (1982) Discussion: The Merrimack synclinorium in northeastern Connecticut. *Am. J. Sci.* **282**, 1735–1744.

Robinson, P. Tracy, R.J. and Ashwal, L.D. (1975) Relict sillimanite-orthoclase assemblage in kyanite-muscovite schist, Pelham Dome, west-central Massachusetts (abstract). *Trans. Am. Geophys. Union* **56**, 466.

Robinson, P. Tracy, R.J., Hollocher, K.T. and Dietsch, C.W. (1982) High grade Acadian regional metamorphism in south-central Massachusetts. In *New England Intercoll. Geol. Conf., 74th Annual Mtg. Guidebook*, eds. Joesten, R.A. and Quarrier, S.S., 289–339.

Rodgers, J. (1981) The Merrimack synclinorium in northeastern Connecticut. *Am. J. Sci.* **281**, 176–186.

Sriramadas, A. (1966) The geology of the Manchester quadrangle, New Hampshire. *N.H. Dept of Resources and Econ. Dev. Bull.* **2**.

Thompson, A.B. (1982) Dehydration melting of pelitic rocks and the generation of H_2O-undersaturated granitic liquids. *Am. J. Sci.* **282**, 1567–1595.

Thompson, J.B., Jr. and Norton, S.A. (1968) Paleozoic regional metamorphism in New England and adjacent areas. In *Studies of Appalachian Geology, Northern and Maritime*, eds. Zen, E-An *et al.* Interscience, New York, 319–327.

Thompson, A.B. and Tracy, R.J. (1979) Model systems for anatexis of pelitic rocks: II. Subsolidus and melting reactions in the system $CaO–KAlO_2–NaAlO_2–Al_2O_3–SiO_2–H_2O$. *Contrib. Miner. Petrol.* **70**, 429–438.

Thompson, J.B., Jr., Robinson, P. Clifford, T.N. and Trask, N.J. (1968) Nappes and gneiss domes in west-central New England. In *Studies of Appalachian Geology, Northern and Maritime*, eds. Zen, E-An *et al.* Interscience, New York, 203–218.

Tracy, R.J. (1978) High grade metamorphic reactions and partial melting in pelitic schist, west-central Massachusetts. *Am. J. Sci.* **278**, 150– 178.

Tracy, R.J. and Robinson, P. (1980) Evolution of metamorphic belts: information from detailed petrologic studies. In *Proceedings, The Caledonides in the U.S.A.*, ed. Wones, D.R., *Virginia Polytechnic Inst. and State Univ. Memoir*, **2**, 189–195.

Tracy, R.J. and Robinson, P. (1983) Acadian migmatite types in central Massachusetts. In *Migmatites, Melting and Metamorphism*, eds. Atherton, M.P. and Gribble, C.D. Shiva, Nantwich, 163–173.

Tracy, R.J., Robinson, P. and Thompson, A.B. (1976) Garnet composition and zoning in the determination of temperature and pressure of metamorphism, central Massachusetts. *Am. Miner.* **61**, 762–775.

Tyler, I.M. and Ashworth, J.R. (1982) Sillimanite-potash feldspar assemblages in graphitic pelites, Strontian area, Scotland. *Contrib. Miner. Petrol.* **81**, 18–29.

Weiner, R.W., McLelland, J.M., Isachsen, Y.W. and Hall, L.M. (1984) Stratigraphy and structural geology of the Adirondack Mountains, New York: Review and synthesis. In *The Grenville Event in the Appalachians and Related Topics*, ed. Bartholomew, M.J., *Geol. Soc. Am. Spec. Paper* **194**, 1–56.

Wintsch, R.P. (1975) Feldspathization as a result of deformation. *Geol. Soc. Am. Bull.* **86**, 35–38.

Wintsch, R.P. and Fout, J.S. (1982) Structure and petrology of the Willimantic Dome and the Willimantic Fault, eastern Connecticut. In *New England Intercoll. Geol. Conf., 74th Annual Mtg. Guidebook*, ed. Joesten, R.A. and Quarrier, S.S., 465–482.

Wynne-Edwards, H.R. (1967) Westport map-area, Ontario, with special emphasis on the Precambrian rocks. *Geol. Surv. Canada Memoir* **346**.

Yardley, B.W.D. (1978) Genesis of the Skagit Gneiss Migmatites, Washington, and the distinction between possible mechanisms of migmatization. *Geol. Soc. Am. Bull.* **89**, 941–951.

Zartman, R.E. and Naylor, R.S. (1984) Structural implications of some radiometric ages of igneous rocks in southeastern New England. *Geol. Soc. Am. Bull.* **95**, 522–539.

Zartman, R.E., Hurley, P.M., Krueger, H.W. and Giletti, B.J. (1970) A Permian disturbance of K-Ar radiometric ages in New England: its occurrence and cause. *Geol. Soc. Am. Bull.* **81**, 3359–3373.

7 Migmatites in the Moines

DAVID BARR*

7.1 Introduction

7.1.1 *Introductory remarks*

The Moines (summaries in Johnstone, 1975; Johnson *et al.*, 1979; Johnson, 1983) are a series of late Precambrian metasediments which have undergone both 'Grenvillian' (*c.* 1000 Ma—Brook *et al.*, 1976) and Caledonian (490–440 Ma—Fettes, 1979) amphibolite facies metamorphism. The effects of tectonic reworking and renewed migmatization on rocks of an earlier migmatite complex are particularly well displayed. The conditions of pressure, temperature, etc. under which the early and late migmatites formed may well have been different, giving rise to differing mechanisms of migmatization. Rather than seek a single mechanism for a set of very different leucosomes of varying age within a single outcrop, it will often be more appropriate to seek a common mechanism for a widely-distributed suite of petrographically-similar migmatites of the same age. To this end, the various migmatites must first be classified according to their relationships with widely-correlatable structures, from which inferences may be drawn about their temporal relationships.

7.1.2 *Regional geology*

The Moines crop out in northern Scotland between the Caledonian front and the Great Glen fault (Figure 7.1) and inliers have been recognized south of the fault where they appear to be unconformably overlain by the post-Grenvillian 'young Moines' (Grampian Division of Piasecki and van Breemen, 1983). These latter metasediments pass stratigraphically into rocks of the Eocambrian Dalradian Supergroup (Treagus and King, 1978; Harris *et al.*, 1978) which have suffered only Caledonian deformation and metamorphism. Only the 'old Moines' will be considered further. They were deposited unconform-

*British Geological Survey, 19 Grange Terrace, Edinburgh EH9 2LF, UK.

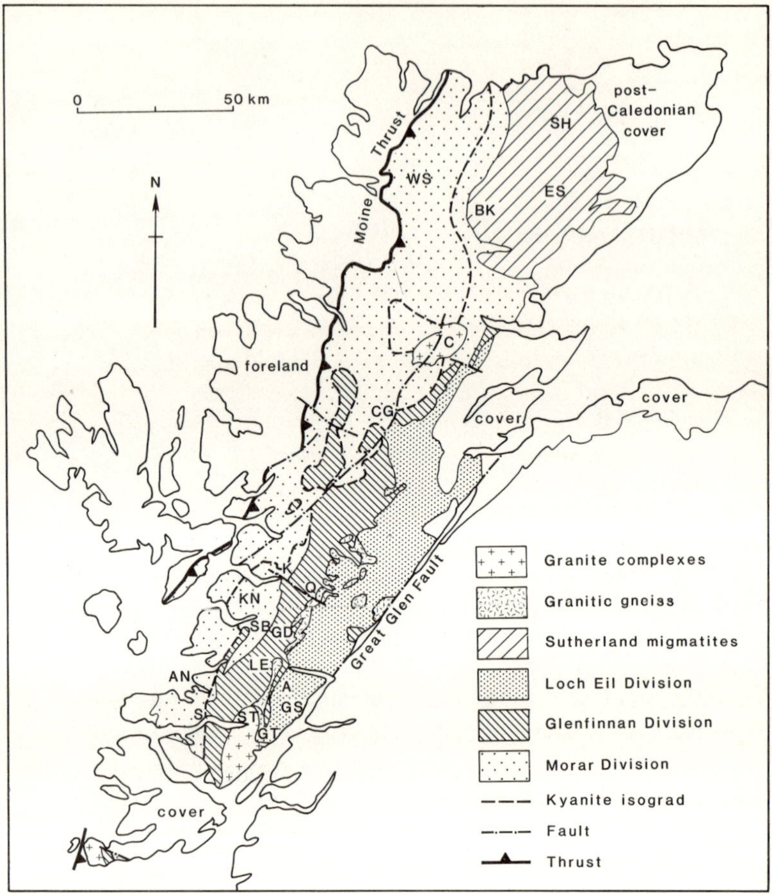

Figure 7.1 The Northern Highland Moines. Lithostratigraphic boundaries based on Johnstone (1975); kyanite isograd from Winchester (1974). Lewisian basement inliers not distinguished. A Ardgour; AN Ardnish; BK Ben Klibreck; C Carn Chuinneag; CG Carn Gorm; ES East Sutherland; GD Glen Dessary; GS Glen Scaddle; GT Glen Tarbert; K Kinlochourn; KN Knoydart; LA Loch Arkaig; LE Loch Eilt; Q Loch Quoich; S Salen; SB Sgurr Breac; SC Sguman Coinntich; SH Strath Halladale; ST Strontian; WS West Sutherland.

ably on to early Proterozoic Lewisian basement (Peach *et al.*, 1910; Ramsay, 1958; Moorbath and Taylor, 1974) comparable to that which forms the Caledonian foreland.

The Northern Highland Moines have been subdivided into three major units, the Morar, Glenfinnan and Loch Eil divisions (Figure 7.1; Johnstone *et al.*, 1969). The Glenfinnan Division comprises interbanded psammites and pelites and passes up stratigraphically into the dominantly psammitic Loch Eil Division (Roberts and Harris, 1983), but has been tectonically emplaced on to the dominantly psammitic Morar Division by the Sgurr Beag slide, an early Caledonian ductile thrust (Tanner *et al.*, 1970; Rathbone *et al.*, 1983). Parts of the Central Highland Division south of the Great Glen are lithologically comparable to the Glenfinnan Division (Piasecki and van Breemen, 1983), but its exact status is not known. Absolute ages can be assigned to most of the major tectonic events recognized in the Moines, mainly by reference to deformed igneous intrusions and pegmatites. In the following summary, the numbering (Dn = deformation, Fn = fold, Ln = lineation, MPn = post-Dn metamorphism) is for descriptive convenience only, e.g. D2 in two separate areas need not be contemporaneous, and neither need be the second event in an absolute sequence.

In Sutherland, the Morar Division is deformed in a major imbricate fan or duplex of Caledonian age which extends westwards to the Moine thrust (Soper and Barber, 1982; unpublished work by R.E. Holdsworth and the author), but the east Sutherland migmatites (migmatized Glenfinnan and Loch Eil division lithologies containing Lewisian inliers and granitic intrusions) record little or no Caledonian strain. Further south, much of the E–W shortening has been accommodated in tight, NNE–SSW–trending folds with steep axial planes (D3). These structures define a 'steep belt' which extends westwards from the Glenfinnan/Loch Eil division boundary (Roberts and Harris, 1983). In the eastern Morar Division, where several major Caledonian slides are folded by 'steep belt' structures, the slides have been assigned to D3 and the steep belt to D4 (e.g. Powell *et al.*, 1981). In all three divisions, two early deformation events (D1 and D2) can be recognized in areas of low Caledonian strain. In the Morar Division, they are associated with major E-verging fold nappes (Powell, 1974) and with an early episode of regional metamorphism and local migmatization. In the Glenfinnan and Loch Eil divisions, major and minor N–S-trending recumbent sheath folds (D2) deform an earlier D1 fabric and migmatitic banding (Holdsworth and Roberts, 1984). In view of the large Caledonian displacement on the Sgurr Beag slide, detailed D1/D2 correlations should not be made between the Morar and Glenfinnan/Loch Eil divisions.

Moine pelites generally lack index minerals other than garnet (due mainly to their high K/Al ratios Barr, 1983*a*) and so regional metamorphic maps have traditionally been based on mineral assemblages in calc-silicates (early diagenetic concretions) with additional control from rare aluminosilicate occurrences (Kennedy, 1949; Soper and Brown, 1971; Winchester, 1974; Tanner, 1976). All of the Glenfinnan Division and most if not all of the Loch Eil Division is kyanite grade or higher, with a small area of sillimanite + K-feldspar assemblages occurring in western Ardgour (Figure 7.1). Metamorphic grade in the Morar Division rises rapidly from garnet in the west to sillimanite + muscovite in the east. This is due both to Caledonian foreshortening of Precambrian isograds and to a smooth eastwards increase in Caledonian metamorphic grade resulting from progressively deeper burial beneath the overthrust Glenfinnan and Loch Eil divisions. Migmatites of one form or another are widespread in the kyanite and sillimanite zones, particularly in the Glenfinnan Division.

7.1.3 *Classification of migmatites*

Two main categories of migmatite can be recognized in the Moines: early migmatites and late migmatites. The early migmatites are invariably deformed and primary textures are rarely preserved; the late migmatites are only weakly deformed (except in certain readily-recognized high-strain zones) and often retain primary textures. Except where otherwise indicated, the descriptions which follow are based on the work of Barr (1983*a*). The early migmatites are widespread and have been subdivided lithologically, while the late migmatites are restricted to particular geographical areas or tectonic settings and are described on that basis. The Sutherland migmatites are somewhat isolated from the remainder of the Moine migmatites and will be discussed separately.

7.2 Early migmatites

7.2.1 *Regional trondhjemitoid migmatites*

Trondhjemitoid migmatites are restricted to the kyanite and sillimanite zones of the Moines, where they correspond to the 'permeation gneisses' of early Geological Survey workers (e.g. Read, 1931). They are typically represented by stromatic (lit-par-lit) pelitic gneisses with K-feldspar-free leucosomes consisting mainly of plagioclase (An_{25-30}) and quartz. Coarse-grained biotite + muscovite + garnet screens (former selvages) occur within the larger veins. In undeformed migmatites, plagioclase forms tabular, subidioblastic grains up to 2 mm in length enclosed by larger quartz plates of several centimetres' grain size. With deformation, quartz is recrystallized to a few millimetres' grain size and plagioclase suffers marginal polygonization. At high strains, all grains are reduced to 1–2 mm, with quartz defining a shape fabric. Leucosomes carry K-feldspar only in those rare cases where it is also present in the host rock. The mesosome consists of quartz, biotite, muscovite, garnet, plagioclase and ilmenite; accessory apatite, zircon and monazite are common, as is retrogressive sphene and/or haematite. Its texture is typically schistose (Figure 7.2), with grain size varying from 0.1–1 mm (quartz) to 0.2–2 mm (biotite) according to the intensity of subsequent deformation. The quartzofeldspathic portion of the mesosome is often segregated into small lenticles, and a coarser-grained (*c.* 5 mm) biotite-rich selvage or melanosome is often, but not invariably, present at the margin of the leucosome (Figures 7.2, 7.3).

Migmatitic banding generally lies parallel to bedding and is often harmonically folded. The folds are concordant with those of bedding and have dominant wavelengths which are controlled by the thickness of the leucocratic veins: this last feature confirms that the veins were solid at the time of folding and more rigid than the micaceous mesosomes or melanosomes. The stromatic layering has itself been accentuated by deformation. Both in

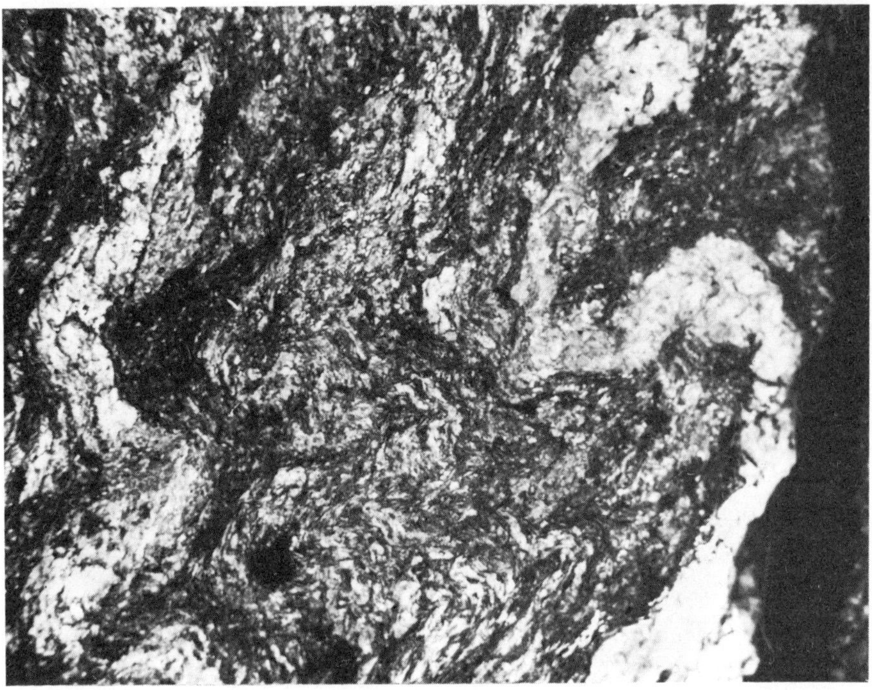

Figure 7.2 Migmatitic pelite crenulated by F3 folds; field of view 20 cm. Sguman Coinntich, UK National Grid reference [NG 969 292]. In all other photographs, the hammer shaft is 40 cm long and the lens cover 5 cm in diameter.

Caledonian slide zones (Figure 7.4) and on the highly-strained limbs of F2 (Precambrian) sheath folds, resistant feldspars are wrapped by plastically-deformed quartz ribbons. Where the leucosome appears at first sight to have segregated in the S2 fabric, close inspection invariably reveals intrafolial folding and transposition of thin veins (Figure 7.3*b*). Where subsequent strain is low, the leucosomes are more irregular and pod-like, although they still lie within a bedding-parallel S1 fabric (Figure 7.3*a*) and appear to be of broadly MP1 age.

The regular, small-scale geometry of the leucosomes with their distinct selvages, the coincidence of mesosome and leucosome plagioclase compositions, the absence of intrusive relationships and the simple metamorphic textures all favour a subsolidus origin for these migmatites (cf. Yardley, 1978; McLellan, 1983). This conclusion is supported by experimental and theoretical data on the partial melting of model pelite systems: a K-feldspar-free Moine pelite should not undergo partial melting until the sillimanite + K-feldspar isograd, where dehydration melting at 650–700 °C (depending on plagioclase composition and a_{H_2O}) would yield a granitoid leucosome and an

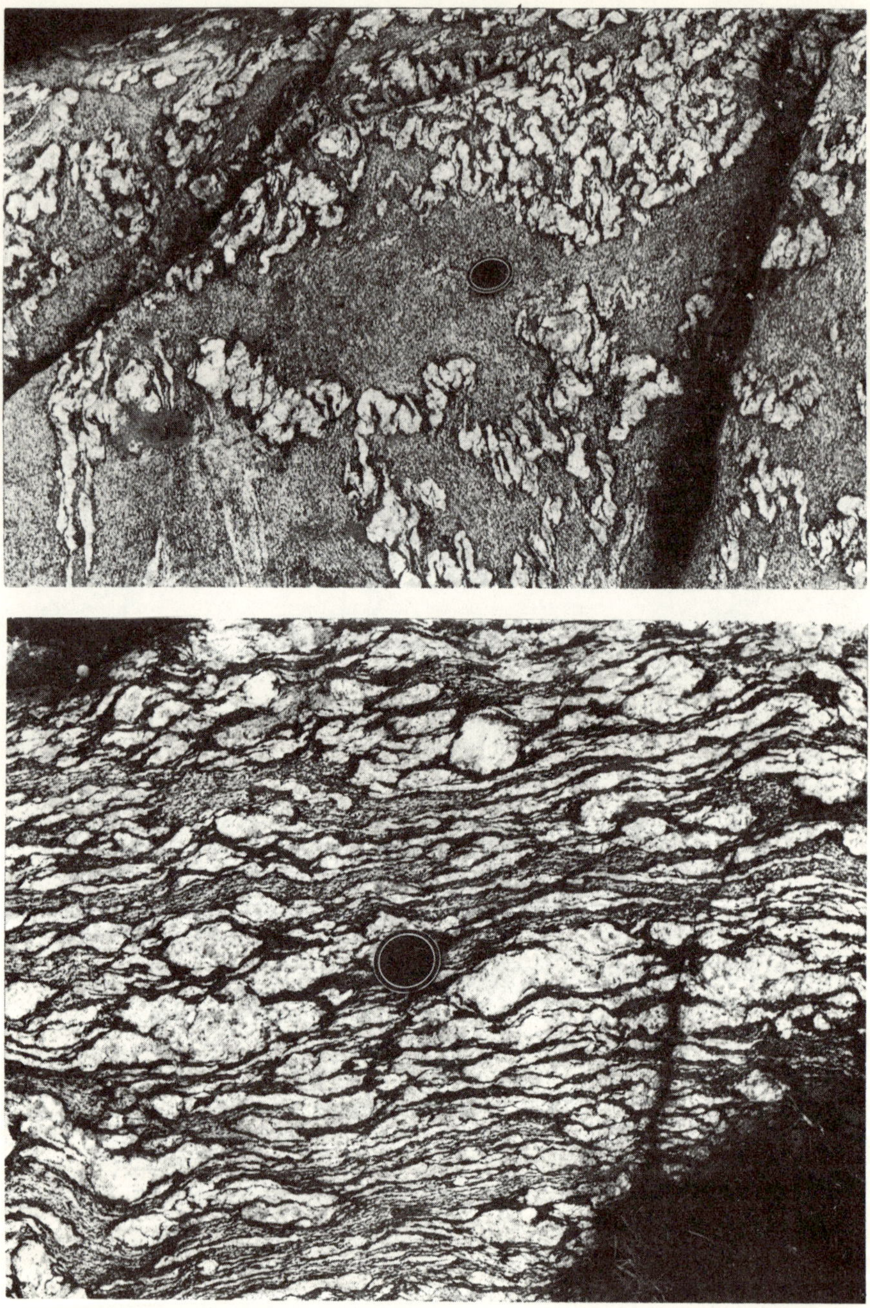

(a)

(b

Figure 7.3 Migmatitic pelites from Loch Ruthven in the Central Highland Division [NH 640 302]. (*a*) Bedding trends left to right, alternate beds are variably migmatized. (*b*) Deformed trondhjemitoid leucosomes with augened feldspars define intrafolial folds; note biotite selvages.

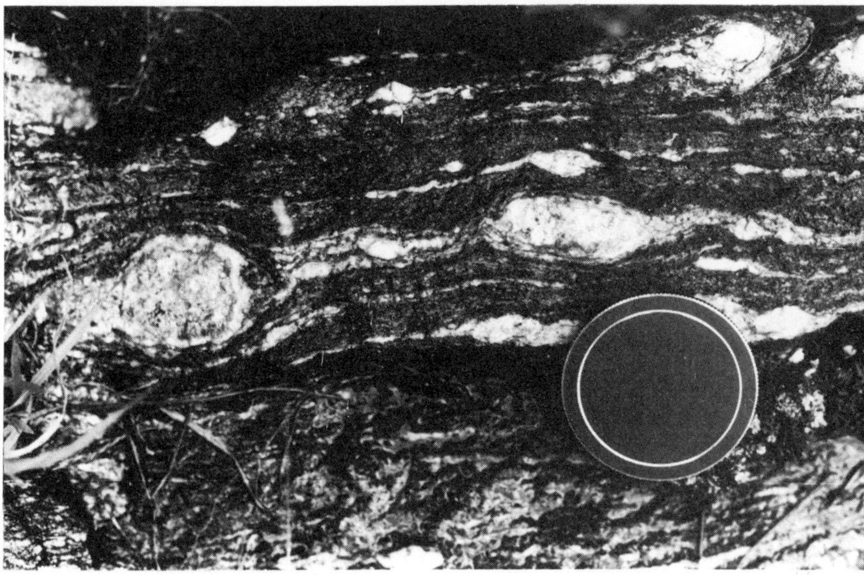

Figure 7.4 Deformed trondhjemitoid leucosomes in the hanging-wall shear zone of the Sgurr Beag slide at Kinlochourn [NG 946 067]. Note quartz ribbons wrapping feldspar porphyroclasts.

aluminosilicate-bearing palaeosome (e.g. Thompson and Tracy, 1979). *P, T* estimates obtained by Barr (1983*a*) from mineral assemblages broadly coeval with the trondhjemitoid leucosomes range from 580°C (kyanite zone) to 640 °C (sillimanite zone) at 5.5–6.5 kbar, and lie to the left of the dehydration melting curves (temperatures were based on Thompson's (1976) garnet-biotite thermometer and pressures on a variety of barometers calculated from the dataset of Helgeson *et al.* (1978): calibration against Newton and Haselton's (1981) garnet–plagioclase–Al_2SiO_5–quartz barometer in rare Al_2SiO_5-bearing pelites suggests that the pressures are *c.* 0.5 kbar too low).

Steveson (1968, 1971) showed in a rigorous statistical study that there is no significant difference in bulk chemical composition between the garnet-grade Morar Pelite and the same stratigraphic unit where it crops out as a sillimanite-grade migmatitic gneiss. The distribution of migmatites within the kyanite and sillimanite zones is controlled primarily by lithology and not, for example, by proximity to major tectonic features or igneous bodies. Certain stratigraphic units are invariably highly migmatized, while on a smaller scale, individual beds display contrasting degrees of migmatization (Figure 7.3*a*). This compositional control was investigated further by separating 21 new analyses and 48 others culled from the literature into migmatites and non-migmatites and comparing them by statistical means (Barr, 1983*a*); preliminary results were communicated in Barr (1983*b*). Various groupings were tested (new analyses, pelites only, corrected for SiO_2 dilution, etc.) but all show

Table 7.1 Statistical analysis of migmatitic and non-migmatitic Moines.

	non-migmatites		migmatites					recalculated SiO_2–free		
	mean	st. dev.	mean	st. dev.	F	t	U	F	t	U
SiO_2	62.23	10.78	66.32	7.85	*	*	84			
TiO_2	0.88	0.37	0.83	0.28						
Al_2O_3	18.69	6.91	15.67	3.53	*	**	84	*	**	92
Fe_2O_3	1.21	0.75	0.93	0.45	*	*	83	*		
FeO	4.54	2.08	4.39	1.94						
MnO	0.09	0.04	0.09	0.05						
MgO	1.99	0.94	1.85	0.78						
CaO	1.68	0.94	1.92	0.66	*			*	*	79
Na_2O	2.39	1.07	3.05	0.68	*	**	98		*	98
K_2O	4.09	1.79	3.53	0.89	*	*	61			
P_2O_5	0.18	0.11	0.14	0.08						
Tot Fe as Fe_2O_3	6.25	2.53	5.81	2.40						
$Na_2O/Na_2O + Al_2O_3$	12.53	5.94	16.65	3.44	*	**	99			
$Na_2O/Na_2O + CaO$	59.41	8.38	61.86	7.64						
$Fe_2O_3/Fe_2O_3 + FeO$	22.24	12.79	18.75	7.03	*					
Ba	1052	611	880	254	*					
Ce	73	33	75	60	*					
La	31	20	36	29						
Nd	37	20	44	43	*					
Pb	23	5	24	5						

Rb	150	64	130	30	*
Sc	16	7	16	7	
Sr	237	96	284	92	
Th	17	7	15	8	
Ti	5335	1984	4949	1212	
V	102	50	88	41	
Y	35	15	48	55	**
Zn	100	42	96	49	
Zr	245	59	299	290	**
K/Rb	257	42	249	65	
K/Ba	38	9	36	7	
Ba/Rb	7.1	1.3	7.4	3.8	*
Rb/Sr	1.24	2.21	0.51	0.22	**
Ca/Y	377	181	498	455	*
Ca/Sr	52	16	50	17	

Notes

1. *Number of samples per group*

	majors	traces
non-migmatites	16	10
migmatites	53	11

2. *F and t tests*

*significant at 95% level (i.e. there is a 95% probability that the difference in mean or standard deviation did not arise by chance)

**significant at 99% level

3. *U tests*

The level at which the test is significant is quoted

essentially the same features. Data for the largest groups are presented in Table 7.1. The *F* test shows that the non-migmatites have significantly higher variance in most elements (except those associated with ferromagnesian minerals), while the migmatites have a very restricted range of compositions. The *t* test identifies Na_2O, SiO_2 and CaO as higher in the migmatites and K_2O, Al_2O_3 and Fe_2O_3 as higher in the non-migmatites. The *t* test can be over-optimistic in cases where the two distributions have significantly different variances, particularly if the sample sizes are also different. The Mann-Whitney *U* test (Siegel, 1956, pp. 116–127) is more reliable in such situations, and accepts only the differences in Na_2O and $Na_2O/Na_2O + Al_2O_3$ as significant. Al_2O_3 comes close to 95% significance on an SiO_2-free basis; other oxides (SiO_2, Al_2O_3, Fe_2O_3) show only a slight correlation (there is a one in six probability that the differences arose by chance), but since the *U* values improved consistently as larger samples were taken, they may reflect real but small differences. In Sutherland, Brown (1967) found high Na_2O and a restricted range of oxidation ratios to be features of migmatitic pelites but not of low-grade pelites, and attributed this to the action of a metasomatizing fluid. In fact, these are characteristics of high-grade migmatites *v.* high-grade non-migmatites and of the low-grade garnetiferous pelite *v.* other low-grade pelites (cf. Butler, 1965) and must be inherited from their original sedimentary compositions. Thus any high-grade sample which is restricted to migmatites is bound to be severely biased.

The *t* and *F* tests were also applied to trace-element data from the new analyses; no elements or ratios thereof yielded a significant *t* test, so the *U* test (which serves mainly to reject differences wrongly accepted by the *t* test) was not applied. Ba, Rb, Ba/Rb and Rb/Sr have significantly higher variance in the non-migmatites, reflecting their wide range of mica and plagioclase contents. Zr, Y, Ca/Y and the rare earths show greater variance in the migmatites, but this is probably spurious since the data are dominated by one heavy mineral-rich specimen.

In terms of their metamorphic mineralogies, these lithogeochemical differences indicate that the migmatites tend to be plagioclase-rich and the non-migmatites micaceous (Figure 7.5*a*). Quartzose psammites are quite common in the Moines and are clearly under-represented in Figure 7.5*a*. Steveson (1968) randomly sampled the sillimanite-grade Morar Division north of Loch Eilt, where migmatites are volumetrically minor and are restricted to the garnetiferous pelite and some feldspathic psammites. He also specifically collected migmatites and a suite of samples from the garnetiferous pelite. His migmatites (Figure 7.5*b*) are virtually identical to those of Figure 7.5*a*. The randomly-sampled group is less feldspathic and consists mainly of non-migmatitic pelites and semipelites (equivalent to Butler's (1965) Types and 1 and 2) and quartzites. The garnetiferous pelites are identical to the pelitic migmatites of the Glenfinnan Division and correspond to Butler's (*op. cit.*) Type 3 pelites (of 'metagreywacke' composition). The overlap of non-

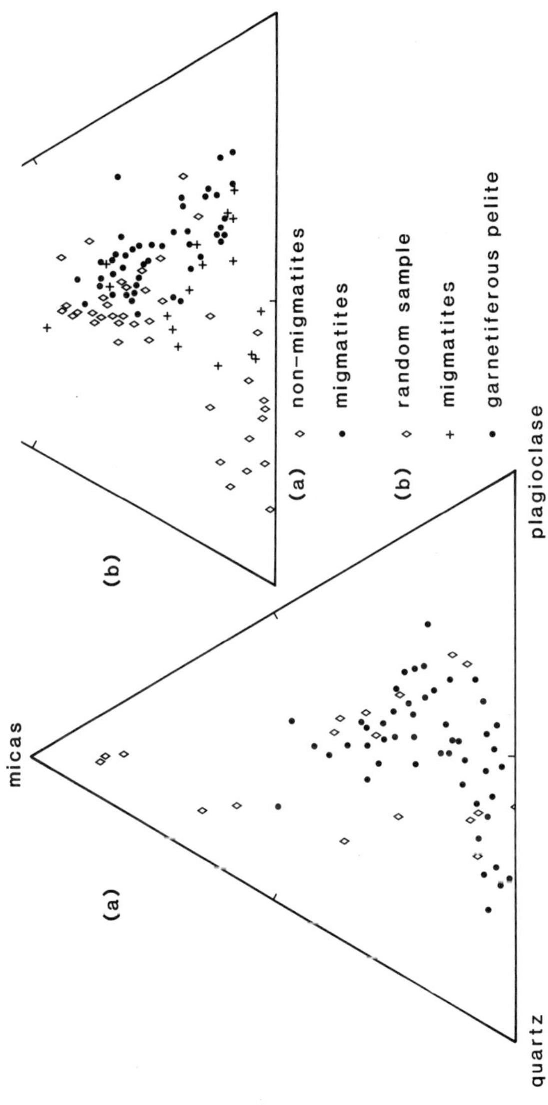

Figure 7.5 Mesonorm quartz:plagioclase:micas diagrams, kyanite and sillimanite zone rocks. (*a*) New and previously-published analyses. (*b*) Steveson's (1968) analyses.

migmatites on to the migmatite field in Figure 7.5*a* may be a result of the indiscriminate sampling procedure, since field observations indicate that a wider range of lithologies is migmatized in the sillimanite zone than in the kyanite zone. In addition, factors such as grain size, nature and distribution of micas and/or graphite, degree of deformation and composition of the fluid phase may exert a secondary control once the initial lithogeochemical requirements are satisfied.

The nature of the compositional control suggests that partial melting was not the dominant mechanism of migmatization: if it were, one would expect to find that rocks with albitic plagioclase (high $Na_2O/Na_2O + CaO$) or K-feldspar were migmatized, rather than simply plagioclase-rich rocks. Since the distinguishing feature of a migmatite is the presence of coarse, plagioclase-rich (rather than quartz-rich) veins, it may simply be that a large plagioclase content results in short diffusion paths and permits significant segregation in a limited period of geological time. The coarse grain size and abundance of quartzofeldspathic lenticles in the low-grade garnetiferous pelite suggests that this readily-migmatized rock already possessed an enhanced tendency to segregate and that the same process operating at higher temperatures and diffusion rates gave rise to the trondhjemitoid migmatites.

7.2.2 *West Highland granitic gneiss*

The West Highland granitic gneiss crops out as a series of discrete bodies within the Glenfinnan and Loch Eil divisions (Figure 7.1). Although Bailey and Maufe (1916, p. 88) originally described the augen facies of the Ardgour body as a deformed pre-metamorphic intrusion, subsequent workers interpreted it as a metasomatized Moine sediment. Harry (1953) considered the trondhjemitoid migmatites to have been produced by Na-metasomatism and the granitic gneiss by subsequent K-metasomatism. Dalziel (1966) accepted that the regional migmatites had formed isochemically, but believed the granitic gneiss to have been produced by K-metasomatism in a major crustal shear zone corresponding to the Loch Quoich line, which was thought then to coincide with the boundary between the Glenfinnan and Loch Eil divisions. The presence of rounded, 'detrital' zircons was considered particularly diagnostic of a metasomatized sediment. Subsequent Geological Survey

Table 7.2 Whole-rock analyses and mass-balance calculations.

Major elements in weight %, trace elements in ppm, H_2O and CO_2 not determined. All analyses carried out on the Liverpool University XRF; further details in Barr (1983*a*). Mesonorms were calculated by assuming biotite to be $K_2(MgFe)_6Si_6Al_2O_{20}(OH)_4$ and muscovite $K_2Al_6Si_6O_{20}(OH)_4$. An% is normative plagioclase composition.

Q1	leucocratic segregation in granitic gneiss, Quoich spillway [NH 071 023].
Q11a	granitic gneiss, Quoich spillway [NH 071 023].
Q13b	granitic gneiss in sinistral shear zone, Quoich spillway [NH 071 023].

Q13r	calculated shear zone restite compositions for melt fractions of 0.15 and 0.3.
Q13m	calculated composition for melt expelled from shear zone.
Q22	antiperthite-bearing migmatitic pelite, Quoich spillway [NH 071 023].
Q22a	coarse leucocratic vein in Q22.
Q22r, Q22m	calculated restite and melt compositions for Q22/Q22a.
LH23	late-D3 pegmatite, Kinlochourn [NG 943 069].
Q31a	late, cross-cutting pegmatite, Quoich spillway [NH 071 023].
S12	semipelitic *schlieren* migmatite, strontian aureole [NM 778 608].

Q13b is considered to have been derived from Q11a through extensive partial melting, with expulsion of some of the melt. From thin section observation, the restite is mainly plagioclase with some quartz, while the interstitial melt consists of albite, K-feldspar and quartz. The extracted melt composition which has zero normative An can be calculated from the relationship:

$$s.\ (\text{shear zone}) + (1\text{-}s).\ (\text{expelled melt}) = (\text{granitic gneiss}).$$

This corresponds to $s = 0.348$. Smaller values of s (more melt) yield unacceptably anorthitic plagioclase (e.g. 0.25 gives An_5) while larger values leave insufficient CaO to combine with P_2O_5 as apatite. The melt is rather similar to the initial granitic gneiss (not surprising, given that rock's near-cotectic composition) but somewhat richer in Q and Or. Restite compositions can be calculated if (a) the interstitial melt fraction m is known and (b) this melt has the same composition as that which was expelled. The mass-balance equation is: $m.$ (melt) + $(1\text{-}m)$. (restite) = (shear zone). Results are tabulated for $m = 0.15$ and $m = 0.3$. The restite consists mainly of plagioclase and quartz with minor K-feldspar and muscovite. Note the strong fractionation of Ba and Rb into the melt but the retention of Sr in the restite. Biotite, magnetite and ilmenite appear to be concentrated in the melt, but this is probably an artefact of the model. Either the original gneiss from which Q13b was derived was lower in mafic minerals than Q11a, or biotite was suspended in the departing melt. Trace element anomalies (negative values for Zr, Th, Y etc.) probably have the same explanation, since heavy minerals are included in biotite. A further complication arises from the fact that the shear zone probably corresponded more closely to a fractional melting than an equilibrium melting environment, so the interstitial melt will have a different composition from the expelled melt, which itself will represent an average of all the increments of expelled melt.

An antiperthite-bearing pelite sample was split into Q22 (a restite-rich palaeosome containing small quartzofeldspathic lenticles) and Q22a (a leucocratic vein containing biotite-rich *schlieren*). Both were considered to represent restite/melt mixtures. A mass-balance model was derived by assuming (on petrographic grounds) all apatite to be contained in the restite and all K-feldspar to be extracted into the melt. Letting W^h_{ap} represent weight fraction of apatite in the host rock, W^m_{ap} weight fraction of apatite in the melt, W^r_{Kf} weight fraction of K-feldspar in the restite, W^v_{Kf} weight fraction of K-feldspar in the vein, etc., then:

$$W^h_{ap} = W^r_{ap} \cdot W^h_r \quad (1a) \qquad W^v_{ap} = W^r_{ap} \cdot W^v_r \quad (1b)$$

$$W^h_{Kf} = W^m_{Kf} \cdot W^h_m \quad (2a) \qquad W^v_{Kf} = W^m_{Kf} \cdot W^v_m \quad (2b)$$

Dividing (1*a*) by (1*b*) and (2*a*) by (2*b*):

$$4.375 W^v_r = W^h_r \quad (3) \qquad 0.3546 W^v_m = W^h_m \quad (4)$$

Noting that $W^v_r = 1 - W^v_m$ and $W^h_r = 1 - W^h_m$, adding (3) and (4) and back-substituting:

$$W^v_m = 0.8395 \qquad W^v_r = 0.1605 \qquad W^h_m = 0.2977 \qquad W^h_r = 0.7023$$

For a given component x,

$$W^v_x = W^m_x \cdot W^v_m + W^r_x \cdot W^v_r \quad (5) \qquad W^h_x = W^m_x \cdot W^h_m + W^r_x \cdot W^h_r \quad (6)$$

By substituting values for W^v_m etc., equations (5) and (6) can be solved for W^m_x and W^r_x. The melt consists essentially of Q, Or, Ab and An; additional components scatter around zero, presumably reflecting sampling heterogeneities and/or analytical errors. Its composition lies close to Winkler's (1979) quaternary cotectic plane for $P_{H_2O} = 5$–7 kbar. Ba and Rb are retained in the micaceous restite while Sr is extracted into the melt (contrast this with the plagioclase-rich restite Q13r). Y, Zr and the REE have negative values in the restite and Th is very low, suggesting that they are retained in inert heavy minerals whose heterogeneous distribution was not adequately sampled.

Table 7.2 (contd.)

	Q1	Q11a	Q13b	Q13r (0.15)	Q13r (0.3)	Q13m	Q22r	Q22	Q22a	Q22m	LH23	Q31a	S12
SiO_2	76.22	75.17	71.92	71.04	69.79	76.90	59.01	63.31	71.16	73.44	72.85	78.87	73.22
TiO_2	0.15	0.28	0.17	0.13	0.09	0.35	1.53	1.08	0.26	0.02	0.14	0.10	0.67
Al_2O_3	13.50	11.92	15.29	16.20	17.51	10.12	12.97	13.63	14.84	15.19	15.07	12.17	12.19
Fe_2O_3	0.85	1.28	0.77	0.59	0.34	1.78	3.13	2.23	0.59	0.11	0.91	0.70	0.83
FeO	0.55	1.38	0.74	0.59	0.37	1.61	9.49	6.81	1.94	0.50	0.35	0.16	2.91
MnO	0.02	0.04	0.02	0.013	0.004	0.06	0.24	0.17	0.04	0.002	0.03	0.01	0.07
MgO	0.29	0.25	0.35	0.38	0.41	0.20	1.80	1.31	0.42	0 16	0.40	0.18	1.24
CaO	1.09	0.88	1.82	2.07	2.43	0.39	2.50	2.57	2.70	2.74	1.62	0.14	1.92
Na_2O	5.38	2.78	4.83	5.38	6.17	1.69	1.72	2.53	4.00	4.43	2.97	2.81	2.24
K_2O	1.28	5.01	3.14	2.63	1.91	6.01	3.51	3.30	2.91	2.79	4.98	4.72	3.38
P_2O_5	0.14	0.24	0.14	0.11	0.07	0.29	0.43	0.30	0.07	0.00	0.05	0.02	0.09
Total	99.47	99.28	99.18	99.15	99.12	99.33	96.76	97.53	98.93	99.35	99.35	99.88	98.77
Q	38.10	39.21	28.31	25.53	21.63	44.12	31.38	31.30	31.17	31.11	33.18	44.21	43.39
Or	2.40	25.67	14.76	11.69	8.07	32.28	0.00	4.72	13.31	15.84	23.13	21.38	10.24
Ab	45.52	22.59	40.86	45.62	52.40	14.16	14.57	21.40	33.84	37.49	25.13	23.77	18.95
An	4.49	2.88	8.11	9.45	11.40	0.00	9.61	10.79	12.94	13.57	7.71	0.56	8.94
An%	9	11	16	16	17	0	38	32	26	25	23	2	31
Musc	6.33	3.98	3.80	3.98	4.38	3.19	10.95	7.58	1.45	−0.36	7.69	8.72	6.14
Biot (Mg)	0.72	0.93	1.21	1.31	1.42	0.71	6.21	4.52	1.45	0.54	1.38	0.62	4.28
Biot(Fe)	0.13	1.08	0.62	0.51	0.34	1.42	16.49	11.89	3.52	1.05	0.00	0.00	4.76
Mag	1.23	2.07	1.12	0.86	0.46	2.43	4.53	3.23	0.86	0.16	0.82	0.26	1.20
Haem	0.00	0.00	0.00	0.00	0.00	0.00	0.00	0.00	0.00	0.00	0.34	0.52	0.00
Ilm	0.28	0.55	0.32	0.24	0.17	0.67	2.91	2.05	0.49	0.03	0.27	0.19	1.27
Apat	0.32	0.58	0.32	0.27	0.17	0.69	1.00	0.70	0.16	0.00	0.12	0.05	0.21
Zirc	0.02	0.046	0.008	−0.002	−0.016	0.059	0.33	0.23	0.046	−0.01	0.008	0.005	0.07

Ba	90	869	487	383	236	1072	891	755	508	435	211	954	778
Ce	33	64	18	5	– 12	89	331	231	49	– 5	14	12	37
Co	4	3	3	3	3	3	11	9	5	4	4	4	9
Cr	6	5	4	4	3	5	12	11	10	10	6	6	38
La	8	30	5	– 2	– 12	44	149	103	19	– 6	4	4	6
Nd	21	45	13	2	– 14	77	232	163	37	0	8	8	10
Ni	5	4	5	5	6	3	9	8	7	7	7	7	14
Pb	18	29	30	30	31	28	18	21	27	29	46	27	19
Rb	57	140	102	92	77	160	142	123	88	78	104	130	95
Sc	2	6	3	2	1	8	42	30	8	2	3	3	10
Sr	219	95	235	273	327	21	204	261	364	394	339	185	266
Th	5	11	2	2	– 4	16	46	33	9	2	8	4	12
Ti	849	1681	1016	836	579	2036	8574	6082	1549	211	959	672	3781
V	14	14	15	15	16	13	84	61	20	8	11	8	58
Y	25	76	22	7	– 14	105	304	210	39	– 11	8	6	18
Zn	29	56	27	19	8	71	284	206	64	22	30	13	55
Zr	100	225	38	– 12	– 85	324	1677	1165	234	– 41	42	25	368
K/Rb	137	299	255	237	206	312	205	226	278	297	399	303	297
K/Ba	118	48	54	57	67	47	33	36	48	53	195	41	36
Ba/Rb	1.6	6.2	4.8	4.2	3.1	6.7	6.3	6.1	5.8	5.6	2.0	7.3	8.2
Rb/Sr	0.26	1.5	0.43	0.34	0.24	7.6	0.70	0.47	0.24	0.20	0.31	0.70	0.38
Ca/Y	212	82	591			27	59	88	495		1450	167	761
Ca/Sr	36	66	55	54	53	133	87	70	53	50	34	5.4	52
Sr/Ba	2.43	0.11	0.48	0.71	1.40	0.02	0.23	0.35	0.89	0.91	1.61	0.19	0.34

mapping (Johnstone *et al.*, 1969) has shown that the granitic gneiss is not restricted to the vicinity of the Loch Quoich line, which in any case is a late, Caledonian structure (Roberts and Harris, 1983), and inherited, 'detrital' zircons have proved to be common in other, undoubtedly intrusive granites (Pidgeon and Aftalion, 1978). Mercy (1963) first noted the essentially igneous chemical composition of the Ardgour gneiss, while Gould (1966) showed that it is uniformly granitic (*sensu* Streckeisen, 1976), but postulated injection of an unusually Al-undersaturated magma in order to retain a significant metasedimentary component. Recent work by the author and others (to be published elsewhere) has clarified the nature and origin of the granitic gneiss, and their conclusions are summarized below.

The granitic gneiss bodies are internally homogeneous and have sharp although locally sheeted contacts with the metasediments. Compositionally, they range from granite to (rarely) alkali granite or granodiorite (Gould, 1966; Table 7.2) and the Quoich gneiss shows a fractionated rare earth pattern with a negative Eu anomaly (Figure 7.6). The typical granitic gneiss is medium-grained (1–2 mm), locally porphyritic and consists mainly of quartz, oligoclase and K-feldspar. Biotite forms 5–15% of the rock, and minor muscovite, garnet, hornblende, ilmenite, pyrite, zircon, epidote, allanite, rutile and monazite have been recorded. The dominant tectonic fabric corresponds to S2 in the Glenfinnan Division and is defined by thin, impersistent biotite laminae. In areas of low D2 strain, an earlier (S1) gneissic fabric is preserved; this is a deformation fabric in which there is no new segregation of leucocratic material. Locally (e.g. Loch Arkaig—D.I. Smith, verb. comm.), larger augen preserve patches of undeformed, igneous-textured granite. MP1 leucocratic segregations with mafic selvages cut the S1 fabric but are folded by F2, and at high strains are transposed into the S2 fabric. Their feldspar compositions accord with those of the host gneiss so these are probably subsolidus segregations, contemporaneous with those in the regional migmatites. These fabrics and the gneiss outcrop itself are folded with bedding around the D3 structures which define the Loch Quoich line.

The granitic gneiss bodies are best interpreted as a suite of granite sills (locally slightly discordant) which have recorded all the tectonic and metamorphic events which affected the surrounding Moines. A significant thermal aureole is not preserved, possibly because most of the sheets are thin (several hundred metres) and were emplaced during the early stages of metamorphism. That they are not slices of pre-Moine basement is demonstrated by the presence as xenoliths of distinctive Moine lithologies and by the fact that the gneiss and the metasediments are cut by a common suite of pre-metamorphic basic intrusions. The granite appears to have been essentially S-type (*sensu* Chappell and White, 1974): it has a restricted range of truly granitic bulk compositions with low Na/K ratios, shows very limited differentiation, is corundum-normative and lies in the quartz field of a Q + Or + Ab + An melting diagram. It contains metasedimentary rather than metabasic xeno-

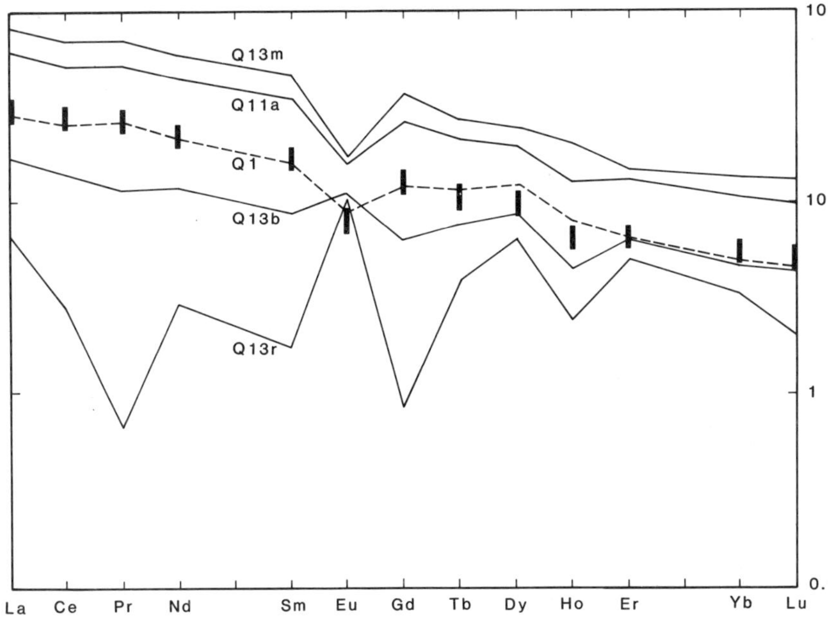

Figure 7.6 Rare earth distributions for rocks in the Quoich spillway [NH 071 023], normalized against the Leedey chondrite (Masuda *et al.*, 1973).

Q11a Quoich granitic gneiss.

Q1 granitoid patch in conjugate fold core. The parallelism of Q1 with Q11a suggests that the REE are entirely retained in some inert phase. The vertical bars define predicted REE concentrations if they were retained in apatite (higher values) or zircon (lower values).

Q13b granitic gneiss in sinistral shear zone. Major element and petrographic data suggest that Q13b consists of the restite after a large degree of partial melting of granitic gneiss, together with some trapped interstitial melt. Its positive Eu anomaly has developed from a rock which originally had a negative anomaly and suggests equilibrium between an extracted melt and a plagioclase-rich restite. Q13m is the calculated composition of the extracted melt—it is very similar to the source rock Q11a, but with higher total REE and a slightly enhanced Eu anomaly. A restite calculated for 30% interstitial melt would have negative values for the low and middle REE. That for 15% melt is plotted and has a flat pattern with a positive Eu anomaly; its extreme 'spikiness' probably reflects analytical errors and sampling heterogeneities which have been amplified by the mass-balance calculation.

liths, has two micas (although muscovite is subordinate and hornblende is locally present) and carries accessory garnet, monazite, ilmenite and sulphides. The Ardgour gneiss has been dated at *c.* 1000 Ma by Rb/Sr (Brook *et al.*, 1976) and U/Pb methods (Aftalion and van Breemen, 1980). It has a fairly high initial $^{87}Sr/^{86}Sr$ ratio (0.709) and its 'detrital' zircons retain a *c.* 1700 Ma isotopic component. It was probably produced at least in part by 'Grenvillian' crustal melting, and the similarity in zircon systematics between the Ardgour gneiss and a Moine metasediment (Aftalion and van Breemen, 1980) suggests that Moine-like rocks were represented in the source terrain. The *c.* 1000 Ma dates obtained by Aftalion and van Breemen (*op. cit*) from an MP1 vein in the gneiss and from a metasediment demonstrate that the 'Grenvillian' event involved metamorphism as well as granite intrusion.

7.3 Late migmatites

7.3.1 *Loch Quoich*

A variety of migmatitic phenomena are represented in K-feldspar-bearing rocks near the eastern end of Loch Quoich (Figure 7.1); they are best displayed in the spillway of the Quoich dam [NH 071 023]. There, the vertical NNE–SSW-trending S2 fabric of the granitic gneiss is deformed in a set of vertical shear zones (Figure 7.7*a*), of which sinistral members trending at $145 \pm 5°$ are the more common. Dextral members trend at 060°, implying compression sub-normal to banding. Some of these shear zones are enriched in leucocratic components; others are biotite-rich and pegmatites extend from them to intrude the granitic gneiss normal to its foliation. At least one such pegmatite agmatizes a concordant amphibolite (Figure 7.7*b*) and others fill boudin necks in the amphibolites. Similar segregations are also developed in the cores of conjugate folds whose geometry implies layer-parallel shortening. One large example occurs near the southeastern margin of the granitic gneiss in the spillway, and within it the S2 gneissic foliation is eliminated, but a 'ghost stratigraphy' is defined by intensely altered (feldspathized) amphibolite sheets. These phenomena are most strikingly developed in the eastern (upper) part of the granitic gneiss. In both shear zones and segregations, the normal equigranular quartz + microcline + oligoclase texture of the granitic gneiss is severely modified: relict, restitic quartz and reverse-zoned plagioclase grains are corroded, surrounded and invaded by a fine-grained quartz + alkali feldspar matrix or enclosed in alkali feldspar megacrysts (Figure 7.8*a*).

Barr (1983*a*) proposed a model whereby both the pegmatite-rich and the restite-rich shear zones could be produced by partial melting. Given the near-cotectic composition of the granitic gneiss, the degree of melting should be controlled by the availability of H_2O, which would be drawn into a deforming shear zone by dilatancy pumping (cf. van der Molen and Paterson, 1979), along with a proportion of any widespread interstitial melt. At low melt fractions, the shear zone would become enriched in the lowest-melting components and also somewhat weakened, which would promote continued deformation. As the melt fraction in the shear zone (locally-derived + introduced) increased beyond *c.* 30%, the strength of the crystal mush would fall by several orders of magnitude (to a few bars—van der Molen and Paterson, *op. cit.*) and the resolved normal stress on the shear zone would tend to expel any additional melt by filter pressing. An equilibrium melt fraction of around 30% would be maintained, but as melting continued, the volume of the shear zone would decrease and its composition would become dominated by restite. Mass-balance modelling of one such restite-rich shear zone (Table 7.2)

(a)

(b)

Figure 7.7 (*a*) Sinistral shear zone deforming the S2 fabric in the Quoich granitic gneiss [NH 071 023]. Note weak dextral shear zone (at *c*. 30° to banding) in top right corner. (*b*) Metabasic dyke agmatized by post-D2 leucosome in the Quoich granitic gneiss [NH 071 023]. Note sinistral shear zone to left of hammer.

(a)

(b)

Figure 7.8 (*a*) Granitic gneiss in shear zone. Twinned, reverse-zoned plagioclase surrounded and invaded by alkali feldspar (low relief) and quartz (often strained). Field of view 2.5 mm [NH 071 023]. (*b*) Antiperthite-bearing leucosome in migmatitic pelite west of the Quoich granitic gneiss. Field of view 5 mm [NH 071 023].

indicates *c*. 75% partial melting with expulsion of 87% of the melt produced (assuming 30% interstitial melt), or *c*. 70% partial melting with expulsion of 93% of the melt produced (assuming 15% interstitial melt, based on thin-section observations). These are probably limiting values, since some of the crystallized melt may not have been recognized petrographically. The calculated melt consists mainly of quartz and orthoclase with lesser albite, while the restite is dominated by plagioclase. Trace element and rare earth (REE) distributions are broadly consistent with this interpretation, Ba, Rb and most of the REE being extracted into the melt, but Sr and Eu being preferentially retained in the plagioclase-rich restite (Table 7.2, Figure 7.6). In contrast, the large fold-core segregation is very albite-rich (Table 7.2), its trace-element composition (apart from the severe depletion in Ba) is broadly explicable in terms of equilibrium between major minerals in it and in normal granitic gneiss and its REE appear to have remained in the restite in some inert phase (Figure 7.6). Either it is a subsolidus segregation, or melt and restite have re-equilibrated, perhaps due to the release of large volumes of fluid on crystallization (which may explain the severe retrogression in the vicinity of this body).

The shear zones are folded round the F3 Beinn Beag synform at [NH 093 025], and can be linked by rotation about a NNE–SSW axis with a suite of normal, WNW–ESE-trending shear zones in the flat-flying Ardgour gneiss. The shear zones and conjugate structures are present (although less common) in the adjacent metasediments, where they are again associated with leucocratic segregations. In addition, certain K-feldspar-bearing pelites close to (NW and SE of) the Quoich gneiss contain coarse, concordant segregations which are less deformed than those of the normal trondhjemitoid migmatites. These leucosomes have thin biotite selvages, consist of quartz, perthite and antiperthite, and have significantly more sodic plagioclase than their host rocks. An analysed leucosome has a near-cotectic composition, but major and trace element characteristics which are quite distinct from those of the granitic gneiss (Table 7.2; cf. Figures 7.6 and 7.9), i.e. the veins are unlikely to be pegmatitic apophyses of the granite which were deformed into parallelism with the foliation in the metasediments. The leucosomes contain biotite-rich *schlieren* and xenocrysts of twinned andesine, while the host rock contains lenticles of fine-grained quartz + perthite + antiperthite (Figure 7.8*b*). The calcic plagioclase (An_{25}), the inclusion of up to 50% of the minor phase and the presence of swap textures where a K-feldspar grades through a perthite into an antiperthite and then a plagioclase, each individual species being in optical continuity, all suggest that the perthites and antiperthites formed by simultaneous crystallization rather than exsolution. The near-cotectic composition of the leucocratic veins makes such an origin plausible. Barr (1983*a*) derived a mass-balance model for one such migmatite (Table 7.2), whereby the host rock consists of 70% restite + 30% melt and the leucosome 16% restite + 84% melt. Coarse veins make up 17% of the outcrop, so that the system as

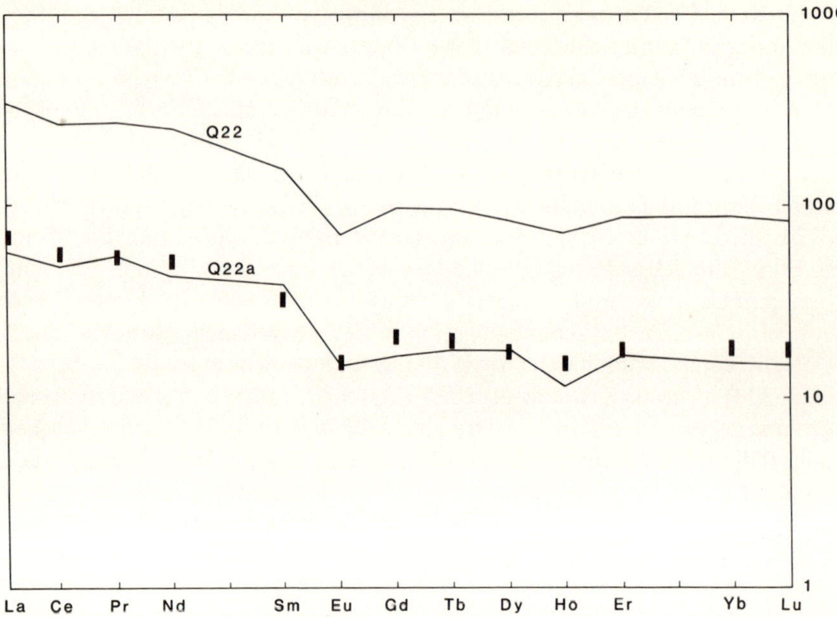

Figure 7.9 Rare earth distributions for antiperthite-bearing migmatitic pelite Q22 and leucocratic vein Q22a [NH 071 023], normalized against the Leedey chondrite (Masuda *et al.*, 1973). See caption to Table 7.2 for mass-balance model. Vein Q22a is sub-parallel to host pelite Q22, suggesting that the REE, including Eu, were retained in inert restite phases. The vertical bars define predicted REE contents in Q22a if all the REE were retained in apatite (higher values) or zircon (lower values).

a whole has undergone *c.* 40% partial melting. The retention of melt in the host rock suggests that its mechanical behaviour can be modelled by van der Molen and Paterson's (1979) experiments—i.e. that its strength decreased dramatically when melting exceeded 30%, and any further melt was expelled to form discrete veins. The REE have been 100% retained in the restite and behave coherently with P_2O_5 and Zr (Figure 7.9). In the immobility of its REE, this migmatitic pelite contrasts with the partially melted granitic gneiss in the shear zone. Deformation may have helped to break down REE-bearing minerals in the granitic gneiss; alternatively, most of the REE in the pelite may be contained in inert, detrital heavy minerals or in discrete REE-rich phases which have formed as a result of its high total REE + Th content (0.1%).

The age of these phenomena is unclear. They post-date the probably-Precambrian D2 and pre-date the certainly-Caledonian D3: a link with D2 would seem the more likely, particularly for the shear zones, which could have been produced by a strain which simply tightened the F2 folds in the remainder of the Glenfinnan Division. Their restriction to a small area around the Loch Quoich dam deserves some comment. The shear zones are presumably more common in the granitic gneiss because it is relatively homogenous and rudely foliated, and less able than the metasediments to deform by folding or interlayer slip; it may also have re-orientated the local stress system. The localization of partial melting may simply reflect the unusual (for the Moines) abundance of K-feldspar-bearing lithologies. The antiperthite-bearing leucosomes and the granitic gneiss itself could melt at relatively low temperatures (below 650 °C at 5–7 kbar, based on Winkler's (1979) data), at the top of the range determined for the kyanite and sillimanite zones of the Moines and below the muscovite-out isograd. Significant melting would require access of H_2O, which would be drawn by dilatancy pumping into zones undergoing active deformation (the shear zones, and perhaps a zone of contact strain at the gneiss/metasediment boundary). It is extremely unlikely that the 70 m thick Quoich gneiss retained sufficient heat of intrusion to perturb the local isotherms through what must have been a lengthy episode of MP1 (subsolidus?) segregation and D2 deformation. The phenomena described here are clearly products of regional metamorphism and do not represent a 'cryptic aureole' to the pre-D1 granitic gneiss.

7.3.2 *Glen Tarbert*

Northwards from Glen Tarbert (Figure 7.1), a little-studied suite of migmatites crops out in the sillimanite + K-feldspar zone near the base of the Loch Eil Division, which in this area comprises a mixed sequence of coarse quartzites, psammites and pelitic gneisses and is transitional to the Glenfinnan Division (Stoker, 1980, 1983). The best exposures are in the Abhainn Coire an Iubhair [NM 922 262]. Early trondhjemitoid migmatites are present, with leucosomes which are folded by structures resembling D2 of the Glenfinnan Division; they lie within the Caledonian steep belt and so have near-vertical, N-S attitudes. The folded leucosomes are disrupted and cut by later pegmatites and irregular aplitic bodies. Semipelites form *schlieren* migmatites and show evidence of flow, agmatizing intercalated quartzites and amphibolites (Figure 7.10). Discrete dykes of *schlieren* semipelite reach 1 m in thickness and cross-cut bedding and banding in the trondhjemitoid migmatites. Similar agmatites near the Strontian 'granite' (section 7.3.6) were attributed by Tyler and Ashworth (1982) to partial melting in the aureole of the intrusion, but this origin is unlikely for the Glen Tarbert rocks as they lie 5 km east of the 'granite'. They probably result from a large degree of partial (dehydration?) melting, whereby the semipelites were mobilized as a crystal mush during the regional metamorphic event which produced sillimanite + K-feldspar. The already-segregated trondhjemitoid leucosomes retained their integrity because partial melting required the addition of water and K-feldspar components from muscovite. Pressure must have been above the 'invariant' point defined by the intersection of the 'muscovite' dehydration and 'granite' melting curves (cf. Tracy, 1978) and so significantly higher than in the Strontian aureole (section 7.3.6). Tyler and Ashworth (1982) regarded the sillimanite + K-feldspar zone

Figure 7.10 20 cm thick hornblende schist intruded by *schlieren* semipelite, Glen Tarbert migmatites [NM 922 262].

as Caledonian, but Stoker (1980) claims that isograds in this area are subvertical and parallel to stratigraphy and so probably Precambrian, and Dalziel and Brown (1965) believed sillimanite to be texturally early. In that case, the Glen Tarbert migmatites may be broadly contemporaneous with the restricted partial melting at Loch Quoich.

7.3.3 *Morarian pegmatites*

The Morarian pegmatites are historically important as the first Precambrian intrusions to be recognized in the Moines, and most studies have been primarily concerned with their age rather than their origin. Although their distribution has not been mapped systematically, they occur widely in the Morar and Glenfinnan divisions, particularly within the zone of trondhjemitoid migmatization. The pegmatites are generally concordant, *c.* 1 m thick, and are deformed by Caledonian structures. They are coarse-grained (several centimetres), consist of quartz + plagioclase + muscovite + garnet + K-feldspar + biotite ± apatite ± beryl ± tourmaline, and are generally considered to be products of regional metamorphism, although their large size and lack of a mafic selvage implies that they did not form strictly *in situ*. The Cairn

Gorm, Knoydart, Sgurr Breac and Loch Eilt pegmatites (Figure 7.10) have all yielded radiometric dates of 715–815 Ma, along with younger dates attributed to Caledonian resetting (Giletti *et al.*, 1961; Long and Lambert, 1963; Fitch *et al.*, 1969; van Breeman *et al.*, 1974, 1978). Techniques used included Rb/Sr muscovite, $^{39}Ar/^{40}Ar$ muscovite, U/Pb zircon and U/Pb monazite. Controversy has centred on whether the *c.* 750 Ma dates are real or are partially-reset older dates, and whether they define a 'Morarian' tectonometamorphic event or merely a minor episode of igneous intrusion (Powell *et al.*, 1983).

An example at Ardnish, in garnet-grade Morar Division metasediments, is clearly dilational and cross-cuts the D2 (Precambrian) structures but is deformed by 'D3' structures loosely correlated with the Sgurr Beag slide (Powell *et al.*, 1983). Powell *et al.* (*op. cit.*) argue that this pegmatite, and by inference the others, is *not* associated with the 'Grenvillian' regional metamorphism of the Morar Division but belongs to an independent, *c.* 800 Ma suite, perhaps of non-orogenic origin. In contrast, Piasecki and van Breemen (1983) have based their model of major Morarian sliding in the Glenfinnan Division and Central Highlands on Rb/Sr muscovite dates obtained from small pegmatitic veins and pods lying within high-strain zones. In their Central Highland examples, they claim that migmatization is restricted to the high-strain zones and so argue that quartzofeldspathic segregation was triggered by sliding and that both are Morarian. In the Glenfinnan Division, their pegmatites are difficult to distinguish from the regional trondhjemitoid leucosomes which are certainly not restricted to slide zones, and it is possible that Piasecki and van Breemen (*op. cit.*) have recorded a partially-reset 'Grenvillian' date. At Kinlochourn, for example, their sample locality is only 80 m east of an upfold of the Sgurr Beag slide (Barr, 1983*a*), and it is likely that the strong fabric present is Caledonian. This conflicting evidence has led Soper and Anderton (1984) to suggest that the Morarian is an extensional event related to lithospheric stretching prior to the opening of Iapetus. If so, its effects should have been most pronounced in the southeast, in the Central Highland and Glenfinnan divisions, and may not have extended as far as the Morar Division (*c.* 100 km further west prior to Caledonian shortening on the Sgurr Beag slide).

7.3.4 *Syn-slide and post-slide migmatization*

Where metamorphic temperatures were high enough, localized migmatization has taken place in zones of unusually high Caledonian strain; most strikingly, in the footwall shear zone of the Sgurr Beag slide. Typical leucosomes are planar veins lying parallel to bedding or to the slide fabric. They range from intensely to weakly deformed, may be folded by syn-slide folds or may be emplaced parallel to their axial planes and yet carry a slide fabric: they are presumably syntectonic and were emplaced during intense progressive deformation. Veins are trondhjemitoid or granitoid, depending on the host

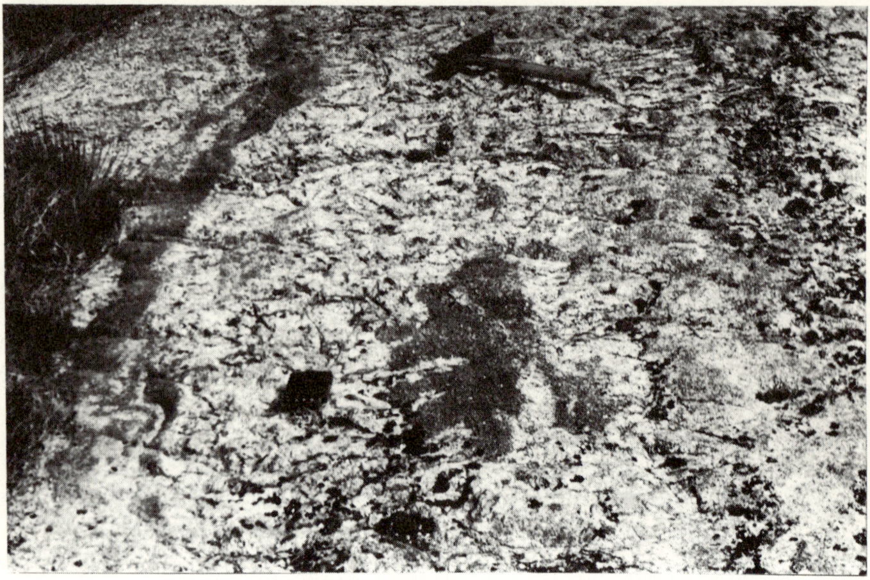

(a)

(b

Figure 7.11 Late pegmatites and associated restite, Lochan Coire Shubh complex. (*a*) Pegmatite (under hammer) containing darker restitic xenolith (to right of compass) and reaction zone against host psammite (to left of compass) [NG 960 043]. The hammer shaft is parallel to bedding in the psammites. (*b*) Pegmatite (white) with unusually wide restite zones (left). Psammite (under hammer) strikes left to right. To right of hammer, leucosome has been expelled to leave only psammite and restite [NG 958 046].

lithology, and often contain epidote. At lower grade, quartz veins show identical structural relations, and in the garnet zone they begin to carry feldspar. Subsolidus segregation is inferred, triggered by intense grain-size reduction (providing a large surface area for reaction) and increased fluid access during sliding. If sliding was fast enough, a thermal contribution is possible from frictional heating or from emplacement of hotter rocks over colder (Brewer, 1981; cf. the inverted metamorphic zones of Powell *et al.*, 1981).

Post-slide migmatization in the eastern Morar and western Glenfinnan divisions comprises syn- to post-D3 (steep belt) pegmatites. Where dilational, these presumably crystallized from mobile fluids or intruding melts. Others (particularly the latest ones) are non-dilational with locally-controlled compositions, have mafic selvages or diffuse boundaries and appear to be subsolidus segregations. In the area northeast of Salen (Figure 7.1), these phenomena show a mappable zonation, whereby *in-situ* segregations give way westwards firstly into a 2 km wide zone in which leucosomes are restricted to rare pegmatitic veins and dykelets, and then into rocks where comparable structural positions are occupied by quartz veins with minor feldspar. An 'isograd' can be defined where quartzofeldspathic (as opposed to quartzose) segregations become abundant; this coincides with a noticeable coarsening of psammites and the complete annealing of all microtextures related to slides and to D3 high-strain zones. A corresponding contrast in the style of obviously-intrusive pegmatites, from planar and sharp-sided to irregular with crystals growing across the boundary, presumably reflects higher wall-rock temperatures in the east. There is no comparable eastern limit to pegmatite intrusion, although from reconnaissance traverses it appears that *in-situ* segregations decline in importance. The 'isograd' cuts across the folded Sgurr Beag slide, so that it lies in the Glenfinnan Division south of Loch Shiel and in the Morar Division north of the loch. On a regional scale, it tends to lie a few kilometres west of the main outcrop of the slide and may be related to thermal recovery following crustal thickening by sliding and D3 (steep belt) folding.

7.3.5 *Late pegmatites*

Late, undoubtedly Caledonian pegmatites are widespread throughout the eastern Morar Division and the Glenfinnan/Loch Eil divisions, and have yielded radiometric dates in the range 430–445 Ma (van Breemen *et al.*, 1974, 1978; Brewer *et al.*, 1979). They are usually represented by sharply cross-cutting, dilational dykes ranging from a few centimetres to a few metres in thickness, and carry two feldspars, one or two micas and (rarely) garnet. Particular suites are geographically restricted, and in some cases can be linked to a specific pegmatite complex.

The Lochan Coire Shubh complex occupies the core of a major F3 (steep belt) antiform 2 km SSE of Kinlochourn (Figure 7.1), and its constituent pegmatites carry a weak NNE–SSW-trending fabric. Coarse, graphic-textured pegmatites form up to 70% of the outcrop, the remainder comprising K-feldspar-bearing micaceous psammite. The earliest pegmatites are very coarse-grained (several centimetres), mica-poor and form concordant lenses and sheets in the core of the major antiform [NG 958 044]. These have sharp contacts, contain disorientated xenoliths and presumably were intruded into voids opening up in the crest of the fold. Similar pods occupy the cores of minor antiforms and carry a vertical, 030°-trending fabric continuous with S3. The next intrusions are common in the area around [NG 958 046], where they form *schollen* migmatites in SW-dipping psammites; these also carry the NNE–SSW-trending foliation. They are cut by a major swarm of broad, NE–SW-trending dykes with NW–SE-trending apophyses. These are often composite, and show marked reaction zones against the host psammites (Figure 7.11*a*). These zones are enriched in mica and garnet and strongly depleted in feldspar, and probably represent restites. All the material in the dyke swarm could not have been derived from the local (relatively feldspar-poor) psammites, but the abundance as cognate xenoliths of similar restite patches, even in dykes which lack reaction zones, suggests derivation from similar rocks at depth. In places, the proportions of pegmatite/restite/psammite are consistent with local derivation, and even with expulsion of pegmatite from the immediate area (Figure 7.11*b*). Many of these veins appear to be intrusive, resulting in the minor misorientation of psammite blocks, and display flow-like patterns of muscovite books. Both the locally-derived and the introduced material was probably molten. Elsewhere (e.g. Figure 7.12), quartzites are agmatized in blind veins and pods of quartz-rich aplite which carry the NNE–SSW-trending fabric; these were probably produced by *in-situ* partial melting. All the above bodies are cut by a second major swarm of coarse, muscovite-rich NW–SE-trending pegmatites. They form individual dykes zoned towards their centres, preserve the regional orientation of banding in septa of country rock, but include rare cognate xenoliths. The dyke swarm defines an open fold whose axial plane coincides with that of the major F3 antiform, and must have been emplaced before final tightening of the structure.

The bulk composition of a distal member of the dyke swarm (emplaced parallel to an F3 axial plane) is given in Table 7.2. It has a virtually cotectic composition and could melt at *c*. 650 °C at 5 kbar, according to Winkler's (1979) data, but has surprisingly calcic plagioclase. Several Glenfinnan Division micaceous psammites have similar bulk compositions (Barr, 1983*a*), and one way to produce such a pegmatite would be to partially melt a psammite until all the feldspars had been consumed and only quartz, micas and garnet remained. This coincides precisely with the composition of the restite at Lochan Coire Shubh. The trace element chemistry of the pegmatite is consistent with such a model. The low Ba, moderate Rb and high Sr contents, the high K/Rb, K/Ba and Sr/Ba ratios and the low Ba/Rb, Rb/Sr and Ca/Sr ratios all suggest equilibration with a restite which is mica-rich but lacks

Figure 7.12 Quartz-rich aplite, apparently derived *in situ*, brecciates host psammite [NG 961 044].

feldspars. The high Ca/Y ratio and low total REE suggest that heavy minerals were important constituents of the restite.

This episode of migmatization took place during the latest stages of D3 deformation and probably records a rise and fall of isotherms, with the presently-exposed rocks spending a shorter period above the solidus than those at deeper levels. The ultimate cause of partial melting was probably tectonic thickening due to D3 deformation, since it is unlikely that the intrusive pegmatites were significantly superheated. Fluids from the dehydration of previously low-grade Morar Division rocks, present at no great depth beneath the Sgurr Beag slide, may have been channelled upwards into the core of the antiform and helped promote partial melting.

Late pegmatites elsewhere in the Glenfinnan Division are similar to those at Lochan Coire Shubh. Further east, most pegmatites are completely post-D3 (although locally slightly deformed) and occur as widespread dyke swarms and sheets lacking an obvious source (e.g. Talbot's (1984) 'pink leucogranites and aplites'). They are distinguished by their pink coloration, the prevalence of microcline and albite and the common presence of two micas. Chemically, their compositions are comparable with those of very acid igneous rocks (Table 7.2). They may represent an early stage of the main Caledonian magmatic event which included major granodiorite vein complexes (Fettes and Macdonald, 1978) and plutons as old as 435 ± 10 Ma (Strontian—Pidgeon and Aftalion, 1978).

7.3.6 *Aureole migmatites*

The Strontian granodiorite (Figure 8.1) is a typical Caledonian I-type 'granite' and was intruded close to the Glenfinnan/Loch Eil division boundary at 435 ± 10 Ma (Munro, 1965, 1973; Pidgeon and Aftalion, 1978; Pankhurst, 1979). In the well-exposed western aureole, folds of bedding and of trondhjemitoid leucosomes become prominent several hundred metres west of the 'granite', tighten eastwards, develop an axial planar fabric and are eventually represented by reclined isoclines. Semipelites in the inner 300 m carry a strong, vertical planar fabric which is parallel to the granite margin and wraps feldspar porphyroblasts and boudinaged quartzites and basic dykes (possibly appinites). This corresponds to MacGregor and Kennedy's (1932) 'breccia' (Figure 7.13). There is a sharp contact between the metasediments and a strongly foliated tonalite which contains boudinaged basic dykes and augened xenoliths and plagioclase phenocrysts. Further into the 'granite', this foliation is weaker and the dykes are defined by trails of blocky xenoliths veined by granitoid material, confirming that the 'granite' was partially molten during deformation and supporting the view that the strain results from forceful intrusion rather than regional tectonic deformation.

The thermal aureole is superimposed on the regional sillimanite zone and appears to distort the regional sillimanite + K-feldspar zone, extending it further west. An inner cordierite + K-feldspar (+ garnet) zone extends 300 m west of the 'granite' and up to 2 km to the east. Ashworth and Chinner (1978) carried out cordierite–garnet geothermometry and geobarometry on rocks from the Strontian aureole, and inferred that all the cordierite + K-feldspar zone migmatites had undergone partial melting at 690–700 °C and 4.2–4.7 kbar. The aureole migmatites were considered to represent a culmination of the regional migmatization process, but this is unlikely since the regional, trondhjemitoid migmatites are Precambrian. In a more detailed study, Tyler and Ashworth (1982) concluded that the trondhjemitoid leucosomes and some granitoid leucosomes were reworked regional migmatites, but that certain microperthite–granitoid leucosomes in the sillimanite + K-feldspar and cordierite + K-feldspar zones had been produced by partial melting. They estimated the muscovite-out isograd at 645 °C, $X_{H_2O} = 0.85$ and the cordierite + K-feldspar isograd at 690 °C, $X_{H_2O} = 0.6$, both at 4.1 ± 0.4 kbar. Because of the low pressure (below the 'invariant' point defined by the intersection of the 'muscovite' dehydration and 'granite' melting curves), dehydration melting did not take place in the muscovite + sillimanite + K-feldspar zone (cf. Tracy and Robinson, 1983). Since the Glen Tarbert migmatites (section 7.3.2) apparently formed above this 'invariant' point, it may prove possible to distinguish the regional and superimposed thermal sillimanite + K-feldspar zones on textural grounds.

Tyler and Ashworth (1982) modelled the cordierite + K-feldspar isograd by the reaction:

biotite + sillimanite + quartz = cordierite + garnet + K-feldspar + H_2O.

K-feldspar and H_2O would be immediately consumed by partial melting, which should therefore be widespread in the inner aureole. The lack of obvious leucosomes was attributed by Tyler and Ashworth to the effects of deformation, which distributed the melt as an intergranular film. The crystal mush thus produced would have had a very low viscosity (van der Molen and Paterson, 1979) and is probably represented by the semipelite matrix of the 'breccia'. This material is K-feldspar-rich and could undergo a large degree of partial melting at a relatively low temperature (Table 7.2).

Similar migmatites occur in the cordierite + K-feldspar zone at Glen Scaddle (Figure 7.1), where Ashworth and Chinner (1978) inferred a slightly higher pressure than at Strontian, although andalusite was recorded. It should be noted that the Glen Scaddle intrusion is not late Caledonian, as suggested by Ashworth and Chinner (*op. cit.*): it post-dates Stoker's (1983) Precambrian 'D1' (possibly equivalent to D1 and D2 of Holdsworth and Roberts, 1984), but is deformed by his major, tight 'D2' and 'D3' structures. In its structural setting, it is more closely comparable to the 456 Ma Glen Dessary syenite (van Breemen *et al.*, 1979; Roberts *et al.*, in press) than to the Strontian granodiorite.

7.4 Sutherland migmatites

7.4.1 *Introduction*

The Sutherland migmatites were among the first to be studied in the Moines (Read, 1931; Cheng, 1944), but apart from their western margin at Ben Klibreck have received little recent attention. Soper and Brown (1971) claimed that all structures in the Sutherland Moines were Caledonian, and that the easterly-increasing calc-silicate and pelite metamorphic zones were also Caledonian and climaxed in migmatization. Subsequent work by Barr (1983*a*) has shown this to be an over-simplification. A major zone of intense ductile strain, probably the northern continuation of the Sgurr Beag slide, separates migmatites whose host rock has Morar Division characteristics from migmatites whose host rock has Glenfinnan Division characteristics. This 'Sgurr Beag slide' forms the eastern front to intense Caledonian reworking, and the 'steep belt' of the southern Moines is not developed. Within the 'Morar Division', lower slides are defined by trails of Lewisian inliers lying within zones of platy psammite; other Lewisian inliers occupy isoclinal Caledonian fold cores. Moorhouse and Moorhouse's (1983) Naver slide lies above the Naver Lewisian (cf. Soper and Barber, 1982) and marks the base of their Bettyhill Assemblage. These migmatitic psammites and semipelites intervene between typical Morar and Glenfinnan division lithologies and may represent an eastern facies of the Morar Division which in the southern Moines is tectonically buried beneath the Sgurr Beag slide.

7.4.2 *Early migmatites*

The 'Glenfinnan Division' consists of garnetiferous pelites and micaceous psammites, both displaying early trondhjemitoid migmatization of lit-par-lit type. Contrary to the views of Soper and Brown (1971), both leucosomes and garnets are augened by the strong fabric in the hanging-wall shear zone of the 'Sgurr Beag slide'. Parts of the Bettyhill Assemblage (particularly the feldspathic psammites) are also highly migmatized. The remainder of the 'Morar Division' lacks early migmatization, except for the poorly-exposed Cnoc Sgriodain pelite, 1 km west of the 'Sgurr Beag slide'. This is the only readily-migmatized 'garnetiferous pelite' in the Sutherland 'Morar Division', and so the absence of early migmatites elsewhere in the kyanite zone is probably lithogeochemically controlled. The Lewisian inliers (hornblendic and acid gneisses) record an episode of pre-Moine migmatization. The intense LS fabric in the footwall shear zone of the 'Sgurr Beag slide' passes downwards into the regional Caledonian S2 fabric which Soper and Brown (1971) have

Figure 7.13 Metabasic dyke boudinaged in *schlieren* semipelite, innermost aureole of Strontian 'granite' [NM 778 608].

traced as far as the mylonites in the Moine thrust belt. This fabric is axial planar to tight, reclined folds, is locally a crenulation cleavage and intensifies into slide zones and at the margins of Lewisian inliers. Garnets in 'Morar Division' semipelites are augened by S2 and preserve inclusion trails which define an earlier S1 fabric. They are probably comparable to the MP1 garnets preserved in the aureole of the *c.* 560 Ma Carn Chuinneag granite (Long and Lambert, 1963; Wilson and Shepherd, 1979) and hence Precambrian. Further west, kyanite, garnet and staurolite in the Meadie pelite appear to be augened by Caledonian structures, and R.E. Holdsworth (verb. comm.) has recognized pre-Caledonian garnets in the Moine mylonites. The gross distribution of migmatization thus results from the foreshortening of an E-W Precambrian metamorphic zonation by Caledonian slides (ductile thrusts) and tectonic emplacement of Precambrian migmatites on to lower-grade 'Morar Division' rocks *after the migmatites had essentially cooled down.*

7.4.3 *Late migmatites*

The simple pattern described above is overprinted by Caledonian metamorphism and in the east by migmatization. From experience in other parts of the Moines (e.g. Tanner, 1976; Barr, 1983*a*) calc-silicates tend to be reset during Caledonian metamorphism, except for the highest-grade assemblages (anorthite + pyroxene) which had been severely depleted in alkalis (Tanner, 1976). Soper and Brown's (1971) calc-silicate zones are probably Caledonian and rise from biotite-equivalent in the west to sillimanite-equivalent in the Bettyhill Assemblage and 'Glenfinnan Division'; rare post-'Grenvillian' markers which record only Caledonian metamorphism (e.g. the Carn Chuinneag aureole and some metabasic bodies) confirm this conclusion. The eastern Morar Division includes several sets of granites (*sensu lato*) and pegmatites which can be categorized as late migmatites on the basis of their field relationships, petrography and structural age:

(i) Coarse, white quartz + albite/oligoclase $\pm$ K-feldspar veins and lenses occur in the 'Sgurr Beag slide' zone. They have mafic selvages and are slightly boudinaged but much less deformed than the trondhjemitoid leucosomes. They probably represent local segregations triggered by sliding.
(ii) Concordant, pink 'granite' sills range from 10 cm to 30 m in thickness and are clearly intrusive, with disorientated xenoliths and pegmatitic apophyses which penetrate the country rocks. They cut F2 folds of bedding but carry an LS fabric parallel to S2 and display boudinage or ptygmatic folding depending on their orientation; they can be regarded as syn- to late-D2 in age. They have granitic or granodioritic compositions, comprise euhedral plagioclase (An_{20}), lesser microcline and quartz and accessory hornblende, biotite, sphene, magnetite and zircon, and die out *c.* 1 km west of the 'Sgurr Beag slide'. The thinner bodies are more

microcline-rich, lack hornblende but occasionally carry minor garnet; quartz is elongated into ribbons and feldspar twins are deformed.

(iii) Slightly discordant 'granite' and pegmatite veins range from a few centimetres to a few metres in thickness and generally have N-dipping contacts. The larger bodies have distinctive 'granitic' cores and pegmatitic margins. Both types consist of quartz, microcline and tabular plagioclase (An_{10}) in equal proportions. They carry a weak S2 fabric in which recrystallized quartz wraps large perthite and antiperthite porphyroclasts. They become common *c.* 2 km west of the 'Sgurr Beag slide' and are progressively more deformed towards the east. In zones of high D2 strain, they are probably represented by a suite of ptygmatically-folded veins whose S-profiles suggest original emplacement as N-dipping bodies.

(iv) Very coarse (several centimetres) graphic-textured quartz + albite + microcline pegmatites showing little internal deformation cut across the platy fabric in the slide zone and the overlying 'Glenfinnan Division'.

Sets (ii)–(iv) are clearly igneous and intrusive, and appear to define a differentiation sequence, which from their degree of deformation is probably also a time sequence. Although some of the sheets analysed by Brown (1967) were peraluminous, the more mafic members have I-type characteristics suggestive of a mantle or lower crustal source. These 'granites' are comparable petrographically to the Vagastie suite of minor intrusions which crops out further west and to the Loch Coire 'granite' 7 km to the south. The 30 m thick 'granite' sill and the Vagastie Bridge 'granite' contain similar early tonalitic to dioritic phases (R.E. Holdsworth, verb. comm.). It is of interest to note that the pre- or post-D2 age of the Vagastie suite has been in dispute (Soper and Brown, 1971; Fergusson, 1978; Moorhouse and Moorhouse, 1978): if these granites were intruded during a lengthy episode of progressive D2 deformation, one would expect to find them both cutting D2 structures and deformed by them. Even so, the 405 ± 11 Ma date obtained for the Vagastie Bridge 'granite' by Pidgeon and Aftalion (1978) using U/Pb zircon and monazite methods is surprisingly young.

7.4.4 *Discussion*

Brown (1967, 1971) and Soper and Brown (1971) assumed a close genetic relationship between the lit-par-lit migmatites and the subsequent sheets and veins. Clearly though, if one is Precambrian and the other Caledonian there can be no such relationship. As Brown pointed out, the trondhjemitoid leucosomes are unlikely products of partial melting and probably formed by subsolidus processes. With the recognition of a major slide boundary separating the 'Morar' and 'Glenfinnan' divisions, there is no need to invoke metasomatism to explain their chemical differences, in that the sedimentary precursors of the migmatites were probably deposited closer to the southern

Glenfinnan Division (where sodic pelites are common) than to the nearby 'Morar Division'. The intrusive sheets and veins probably differentiated from a granodiorite or tonalite magma and include (cumulate?) phenocrysts, so there is no requirement that they have *minimum* melt compositions. Even so, many of Brown's (1967) samples plot close to the cotectic surfaces in the quaternary Q + Ab + An + Or system for $P_{H_2O} = 5$–7 kbar (Winkler, 1979), although their projections on Q + Ab + Or and An + Ab + Or scatter away from the ternary minima. Since the early migmatites were tectonically emplaced on to the 'Morar Division', there is no need to postulate an inverted thermal gradient and it is likely that the genuine Caledonian isograds are right-way-up and dip westwards (although there may have been minor perturbations during sliding which later recovered). Unfortunately, there is insufficient relief west of Ben Klibreck to confirm this.

The remainder of the Sutherland migmatites are inaccessible and poorly exposed, and have been subjected to little recent study. However, extrapolating from Ben Klibreck, one would expect to find early trondhjemitoid Moine migmatites and Lewisian basement displaying pre-Moine migmatization, both reworked in later, Caledonian migmatites. From Read's (1931) descriptions, much of the obvious post-tectonic migmatization appears to involve magmatic leucosomes and much of the outcrop is recorded as granite. Some of this is Caledonian (e.g. Rogart–Soper, 1963), but the *c.* 300 km^2 Strath Halladale complex (Figure 7.1), which post-dates early phases of deformation and migmatization (McCourt, 1980; Storey and Lintern, 1981), is itself at least in part Precambrian (649 ± 30 Ma—M. Brook, quoted in Pankhurst, 1982). The undeformed nature of this complex suggests that the rocks above the 'Sgurr Beag slide' were emplaced as a single rigid sheet and that useful Caledonian structural markers will not be available; on the other hand, the prevalence of magmatic leucosomes should facilitate radiometric dating.

7.5 Conclusions

The Moines display a lengthy history of Precambrian and Caledonian deformation and metamorphism: this has given rise to a complex suite of migmatites which can only be properly interpreted in the light of their structural relationships. Confusion has arisen in the past where a single mechanism has been sought for migmatites of several generations. A crude separation into early (strongly deformed) and late (weakly deformed) migmatites works well in practice, provided that a knowledge of the regional structure permits zones of anomalously low or high strain to be recognized.

The earliest widespread migmatization in the Moines was associated with the peak of metamorphism in the kyanite and sillimanite zones. This gave rise to subsolidus trondhjemitoid leucosomes and is probably of 'Grenvillian' (*c.* 1000 Ma) age. A suite of pre-metamorphic intrusions was also migmatized and now forms the West Highland granitic gneiss. These early migmatites suffered intense deformation before renewed local migmatization. At Loch Quoich, this involved partial melting of K-feldspar-bearing rocks in structurally-defined zones and was probably Precambrian, but may have been early Caledonian. In the sillimanite + K-feldspar zone near Glen Tarbert, more widespread partial melting of semipelites gave rise to *schlieren* migmatites which were mobilized as a crystal mush. Inefficient segregation of melt from restite is common to all these late migmatites, the leucosomes carrying entrained restite material while the melanosomes or mesosomes retain an interstitial melt film. Morarian pegmatites (715–815 Ma) post-date the trondhjemitoid migmatites but their relationships to subsequent events are not clear. Renewed subsolidus segregation was associated with Caledonian ductile thrusting (sliding) and subsequent steep refolding. This was probably controlled by deformation-enhanced grain-size reduction and fluid access, and by perturbations of thermal gradients consequent on crustal thickening. Several suites of pegmatite dykes were emplaced as Caledonian deformation waned, some related to central complexes where there was localized partial melting of Moine country rocks. The Sutherland migmatites show evidence of a similar history, but their late leucosomes involve a greater proportion of magmatic material.

The early, subsolidus, trondhjemitoid leucosomes are extremely stable under renewed high-grade metamorphism and migmatization, unless they are first broken down by very intense deformation. For the most part, deformation transposes leucocratic veins into a new fabric, but causes only slight recrystallization and no renewed segregation. This is the case east of Kinlochourn, where Tanner (1976) recognized early and late migmatites but believed the bulk of the leucocratic material to be late—in fact, only the large, cross-cutting pegmatites are late, most of the leucosomes being Precambrian but transposed into a Caledonian fabric. Trondhjemitoid leucosomes retain their integrity even when re-metamorphosed at higher grade in the sillimanite + K-feldspar zone and to some extent in the cordierite + K-feldspar zone. This is probably because free water was scarce in these already-metamorphosed rocks, and dehydration melting was localized in the muscovite-bearing mesosomes rather than the dry, quartz + plagioclase leucosomes. A similar history of repeated migmatization can be anticipated in other areas of reworked basement (e.g. Grenvillian inliers in the Appalachians), so care should be taken to identify episodes which occurred at broadly the same time in a given area and can plausibly be related.

Acknowledgements

I should like to thank Mike Atherton, Tony Harris and Mike Brotherton, along with my colleagues past and present at Liverpool University, at the British Geological Survey and in the Moine community, for their help, encouragement and lively discussion. I am also grateful to Brian Steveson for permission to use his whole-rock analyses. This work was carried out during the tenure of a Natural Environment Research Council research studentship, and is published with the approval of the Director, British Geological Survey (NERC).

References

Aftalion, M. and van Breemen, O. (1980) U-Pb zircon, monazite and Rb/Sr whole-rock systematics of granitic gneiss and psammitic to semipelitic host gneiss from Glenfinnan, northwestern Scotland. *Contrib. Miner. Petrol.* **72**, 87–98.

Ashworth, J.R. and Chinner, G.A. (1978) Coexisting garnet and cordierite in migmatites from the Scottish Caledonides. *Contrib. Miner. Petrol.* **65**, 379–394.

Bailey, E.B. and Maufe, H.B. (1916) The geology of Ben Nevis, Glen Coe and the surrounding area. *Mem. Geol. Surv. G.B.*

Barr, D. (1983*a*) Genesis and Structural Relations of Moine Migmatites. Unpubl. Ph.D. Thesis, Univ. Liverpool.

Barr, D. (1983*b*) Geochemical constraints on migmatisation in the Moine rocks of Scotland. In *Migmatites, Melting and Metamorphism*, eds. Atherton, M.P., and Gribble, C.D., Shiva, Nantwich, 297–298.

Brewer, J.A. (1981) Thermal effects of thrust faulting. *Earth Planet. Sci. Lett.* **56**, 233–244.

Brewer, M.S., Brook, M. and Powell, D. (1979) Dating of the tectonometamorphic history of the southwestern Moine, Scotland. In *The Caledonides of the British Isles-Reviewed*, eds. Harris, A.L., Holland, C.H. and Leake, B.E., *Spec. Publ. Geol. Soc. Lond.* **8**, 129–137.

Brook, M., Powell, D. and Brewer, M.S. (1976) Grenville age for rocks in the Moine of northwestern Scotland. *Nature* **260**, 515–517.

Brown, P.E. (1967) Major element composition of the Loch Coire migmatite complex, Sutherland, Scotland. *Contrib. Miner. Petrol.* **14**, 1–26.

Brown, P.E. (1971) The origin of granitic sheets and veins in the Loch Coire migmatites, Scotland. *Miner. Mag.* **38**, 446–450.

Butler, B.C.M. (1965) A chemical study of some rocks of the Moine Series of Scotland. *Q. J. Geol. Soc. Lond.* **121**, 162–203.

Chappell, B.W. and White, A.J.R. (1974) Two contrasting granite types. *Pacific Geol.* **8**, 173–174.

Cheng, Y. (1944) The migmatite area around Bettyhill, Sutherland. *Q. J. Geol. Soc. Lond.* **99**, 107–154.

Dalziel, I.W.D. (1966) A structural study of the granitic gneiss of western Ardgour, Argyll and Inverness-shire. *Scott. J. Geol.* **2**, 125–152.

Dalziel, I.W.D. and Brown, R.L. (1965) The structural dating of sillimanite-grade metamorphism of the Moines in Ardgour (Argyll) and Moidart (Inverness-shire). *Scott. J. Geol.* **1**, 304–311.

Fergusson, I.W. (1978) Structural age of the Vagastie Bridge granite, Sutherland. *Scott. J. Geol.* **14**, 89–92.

Fettes, D.J. (1979) A metamorphic map of the British and Irish Caledonides. In *The Caledonides of the British Isles-Reviewed*, eds. Harris, A.L., Holland, C.H. and Leake, B.E., *Spec. Publ. Geol. Soc. Lond.* **8**, 307–321.

Fettes, D.J. and Macdonald, R. (1978) Glen Garry vein complex. *Scott. J. Geol.* **14**, 335–358.

Fitch, F.J., Miller, J.A. and Mitchell, J.G. (1969) A new approach to radio-isotopic dating in orogenic belts. In *Time and Place in Orogeny*, eds. Kent, P.E., Satterthwaite, G.E. and Spencer, A.M., *Spec. Publ. Geol. Soc. Lond.* **3**, 157–195.

Giletti, B.J., Moorbath, S. and Lambert, R. St. J. (1961) A geochronological study of the metamorphic complexes of the Scottish Highlands. *Q.J. Geol. Soc. Lond.* **117**, 233–272.

Gould, D. (1966) Geochemical and Mineralogical Studies of the Granitic Gneiss of Western Ardgour, Argyll. Unpubl. Ph.D. Thesis, Univ. Edinburgh.

Harris, A.L., Baldwin, C.T., Bradbury, H.J., Johnson, H.D. and Smith, R.A. (1978) Ensialic basin sedimentation: the Dalradian Supergroup. In *Crustal Evolution in Northwestern Britain and Adjacent Regions*, eds. Bowes, D.R. and Leake, B.E., *Geol. J. Spec. Issue* **10**, 115–138.

Harry, W.T. (1953) The composite granitic gneiss of Western Ardgour, Argyll, *Q.J. Geol. Soc. Lond.* **109**, 285–308.

Helgeson, H.C., Delany, J.M., Nesbitt, H.W. and Bird, D.K. (1978) Summary and critique of the thermodynamic properties of rock-forming minerals. *Am. J. Sci.* **278A**.

Holdsworth, R.E. and Roberts, A.M. (1984) A study of early curvilinear fold structures and strain in the Moines of the Glen Garry region, Inverness-shire. *J. Geol. Soc. Lond.* **141**, 327–338.

Johnson, M.R.W. (1983) Torridonian-Moine. In *Geology of Scotland*, ed. Craig, G.Y., Scottish Academic Press, Edinburgh, 49–75.

Johnson, M.R.W., Sanderson, D. and Soper, N.J. (1979) Deformation in the Caledonides of England, Ireland and Scotland. In *The Caledonides of the British Isles-Reviewed*, eds. Harris, A.L., Holland, C.H. and Leake, B.E., *Spec. Publ. Geol. Soc. Lond.* **8**, 165–186.

Johnstone, G.S. (1975) The Moine Succession. In *A Correlation of Precambrian Rocks in the British Isles*, eds. Harris, A.L., Shackleton, R.M., Watson, J.V., Downie, C., Harland, W.B. and Moorbath, S., *Spec. Rep. Geol. Soc. Lond.* **6**, 30–42.

Johnstone, G.S., Smith, D.I. and Harris, A.L. (1969) Moinian assemblage of Scotland. *Mem. Am. Ass. Petrol. Geol.* **12**, 159–180.

Kennedy, W.Q. (1949) Zones of progressive metamorphism in the Moine Schists of the Western Highlands of Scotland. *Geol. Mag.* **86**, 43–56.

Long, L.E. and Lambert, R. St. J. (1963) Rb–Sr isotopic ages from the Moine Series. In *The British Caledonides*, eds. Johnson, M.R.W. and Stewart, F.H., Oliver and Boyd, Edinburgh, 217–247.

McCourt, W.J. (1980) The geology of the Strath Halladale–Altnabreac district. *Inst. Geol. Sci. Environmental Protection Unit Rep. No.* **80-1** (open file).

MacGregor, A.G. and Kennedy, W.Q. (1932) The Morvern-Strontian 'granite'. *Summ. Progr. Geol. Surv. G.B. for 1931*, 105–119.

McLellan, E.L. (1983) Contrasting textures in anatectic and metamorphic migmatites: an example from the Scottish Caledonides. *J. Metamorphic Geol.* **1**, 241–262.

Masuda, A., Nakamura, N. and Tanaka, T. (1973) Fine structures of mutually normalised rare-earth patterns of chondrites. *Geochim. Cosmochim. Acta* **37**, 239–248.

Mercy, E.L.P. (1963) The geochemistry of some Caledonian granitic and metasedimentary rocks. In *The British Caledonides*, eds. Johnson, M.R.W. and Stewart, F.H., Oliver and Boyd, Edinburgh, 189–215.

Moorbath, S. and Taylor, P.N. (1974) Lewisian age for the Scardroy mass. *Nature* **250**, 41–43.

Moorhouse, S.J. and Moorhouse, V.E. (1978) Discussion of Fergusson (1978). *Scott. J. Geol.* **15**, 69–70.

Moorhouse, V.E. and Moorhouse, S.J. (1983) The geology and geochemistry of the Strathy complex of north-east Sutherland, Scotland. *Miner. Mag.* **47**, 123–137.

Munro, M. (1965) Some structural features of the Caledonian granite complex at Strontian, Argyllshire. *Scott. J. Geol.* **1**, 152–175.

Munro, M. (1973) Structures in the south-eastern portion of the Strontian granite complex, Argyllshire. *Scott. J. Geol.* **9**, 99–108.

Newton, R.C. and Haselton, H.T. (1981) Thermodynamics of the garnet–plagioclase–Al_2SiO_5–quartz geobarometer. In *Thermodynamics of Minerals and Melts*, eds. Newton, R.C., Navrotsky, A. and Wood, B.J., Springer-Verlag, New York etc., 131–147.

Pankhurst, R.J. (1979) Isotope and trace element evidence for the origin and evolution of Caledonian granites in the Scottish Highlands. In *Origin of Granite Batholiths: Geochemical Evidence*, eds. Atherton, M.P. and Tarney, J., Shiva, Nantwich, 18–33.

Pankhurst, R.J. (1982) Geochronological tables for British igneous rocks. In *Igneous Rocks of the British Isles*, ed. Sutherland, D.S., Wiley, Chichester, 575–581.

Peach, B.N., Horne, J., Woodward, H.B., Clough, C.T., Harker, A. and Webb, C.B. (1910) The geology of Glenelg, Lochalsh and the south-east part of Skye. *Mem. Geol. Surv. G.B.*

Piasecki, M.A.J. and van Breemen, O. (1983) Field and isotopic evidence for a *c.* 750 Ma tectonothermal event in Moine rocks in the Central Highland region of the Scottish Caledonides. *Trans. R. Soc. Edin. Earth Sci.* **73**, 119–134.

Pidgeon, R.T. and Aftalion, M. (1978) Cogenetic and inherited zircon U-Pb systems in granites: Palaeozoic granites of Scotland and England. In *Crustal Evolution in Northwestern Britain and Adjacent Regions*, eds. Bowes, D.R. and Leake, B.E., *Geol. J. Spec. Issue* **10**, 183–220.

Powell, D. (1974) Stratigraphy and structure of the western Moine and the problem of Moine orogenesis. *J. Geol. Soc. Lond.* **130**, 575–593.

Powell, D., Baird, A.W. Charnley, N.R. and Jordan, P.J. (1981) The metamorphic environment of the Sgurr Beag Slide; a major crustal displacement zone in Proterozoic, Moine rocks of Scotland. *J. Geol. Soc. Lond.* **138**, 661–673.

Powell, D. Brook, M. and Baird, A.W. (1983) Structural dating of a Precambrian pegmatite in

Moine rocks of northern Scotland and its bearing on the status of the 'Morarian' orogeny. *J. Geol. Soc. Lond.* **140**, 813–823.

Ramsay, J.G. (1958) Moine–Lewisian relations at Glenelg, Inverness-shire. *Q.J. Geol. Soc. Lond.* **113**, 487–523.

Rathbone, P.A., Coward, M.P. and Harris, A.L. (1983) Cover and basement: a contrast in style and fabrics. *Mem. Geol. Soc. Am.* **158**, 213–223.

Read, H.H. (1931) The geology of central Sutherland. *Mem. Geol. Surv. G.B.*

Roberts, A.M. and Harris, A.L. (1983) The Loch Quoich Line—a limit of early Palaeozoic crustal reworking in the Moine of the Northern Highlands of Scotland. *J. Geol. Soc. Lond.* **140**, 883–892.

Roberts, A.M., Smith, D.I. and Harris, A.L. (in press) The structural setting and tectonic significance of the Glen Dessary syenite, Inverness-shire. *J. Geol. Soc. Lond.*

Siegel, S. (1956) *Nonparametric Statistics for the Behavioral Sciences.* McGraw-Hill, New York.

Soper, N.J. (1963) The structure of the Rogart igneous complex, Sutherland, Scotland. *Q.J. Geol. Soc. Lond.* **119**, 445–478.

Soper, N.J. and Anderton, R. (1984) Did the Dalradian slides originate as extensional faults? *Nature* **307**, 357–360.

Soper, N.J. and Barber, A.J. (1982) A model for the deep structure of the Moine thrust zone. *J. Geol. Soc. Lond.* **139**, 127–138.

Soper, N.J. and Brown, P.E. (1971) Relationship between metamorphism and migmatisation in the northern part of the Moine nappe. *Scott. J. Geol.* **7**, 305–325.

Steveson, B.G. (1968) A Statistical Analysis of Chemical Relationships in the Moine Rocks of Lochailort. Unpubl. Ph.D. Thesis, Univ. Liverpool.

Steveson, B.G. (1971) Chemical variability in some Moine rocks of Lochailort, Inverness-shire. *Scott. J. Geol.* **7**, 51–60.

Stoker, M.S. (1980) The Geology of Eastern Ardgour, Argyllshire. Unpubl. Ph.D. Thesis, Univ. Liverpool.

Stoker, M.S. (1983) The stratigraphy and structure of the Moine rocks of Eastern Ardgour. *Scott. J. Geol.* **19**, 369–385.

Storey, B.C. and Lintern, B.C. (1981). Geochemistry of the rocks of the Strath Halladale–Altnabreac district. *Inst. Geol. Sci. Environmental Protection Unit Rep. No.* **81–12** (open file).

Streckeisen, A.L. (1976) To each plutonic rock its proper name. *Earth Sci. Rev.* **12**, 1–33.

Talbot, C.J. (1984) Microdiorite sheet intrusions as incompetent time- and strain-markers in the Moine assemblage NW of the Great Glen fault, Scotland. *Trans. R. Soc. Edinb. Earth Sci.* **74**, 137–152.

Tanner, P.W.G. (1976) Progressive regional metamorphism of thin calcareous bands from the Moinian rocks of NW Scotland. *J. Petrol.* **17**, 100–134.

Tanner, P.W.G., Johnstone, G.S., Smith, D.I. and Harris, A.L. (1970) Moinian stratigraphy and the problem of the central Ross-shire inliers. *Bull. Geol. Soc. Am.* **81**, 299–306.

Thompson, A.B. (1976) Mineral reactions in pelitic rocks: II. Calculation of some P–T–X(Fe–Mg) phase relations. *Am. J. Sci.* **276**, 425–454

Thompson, A.B. and Tracy, R.J. (1979) Model systems for anatexis of pelitic rocks: II. Facies series melting and reactions in the system $CaO–KAlO_2–NaAlO_2–Al_2O_3–SiO_2–H_2O$. *Contrib. Miner. Petrol.* **70**, 429–438.

Tracy, R.J. (1978) High grade metamorphic reactions and partial melting in pelitic schist, west-central Massachusetts. *Am. J. Sci.* **278**, 150–178.

Tracy, R.J. and Robinson, P. (1983) Acadian migmatite types in pelitic rocks of central Massachusetts. In *Migmatites, Melting and Metamorphism*, eds. Atherton, M.P. and Gribble, C.D., Shiva, Nantwich, 163–173.

Treagus, J.E. and King, G. (1978) A complete Lower Dalradian succession in the Schiehallion district, Central Perthshire. *Scott. J. Geol.* **14**, 157–166.

Tyler, I.M. and Ashworth, J.R. (1982) Sillimanite–potash feldspar assemblages in graphitic pelites, Strontian area, Scotland. *Contrib. Miner. Petrol.* **81**, 18–29.

van Breemen, O., Pidgeon, R.T. and Johnson, M.R.W. (1974) Precambrian and Palaeozoic pegmatites of the Moines of northern Scotland. *J. Geol. Soc. Lond.* **130**, 493–507.

van Breemen, O., Halliday, A.N., Johnson, M.R.W. and Bowes, D.R. (1978) Crustal additions in late Precambrian times. In *Crustal Evolution in Northwestern Britain and Adjacent Regions*, eds. Bowes, D.R. and Leake, B.E., *Geol. J. Spec. Issue* **10**, 81–106.

van Breemen, O., Aftalion, M., Pankhurst, R.J. and Richardson, S.W. (1979) Age of the Glen Dessary syenite, Inverness-shire: diachronous Palaeozoic metamorphism across the Great Glen. *Scott. J. Geol.* **15**, 49–62.

van der Molen, I. and Paterson, M.S. (1979) Experimental deformation of partially-melted granite. *Contrib. Miner. Petrol.* **70**, 299–318.

Wilson, D, and Shepherd, J. (1979) The Carn Chuinneag granite and its aureole. In *The Caledonides of the British Isles–Reviewed*, eds. Harris, A.L., Holland, C.H. and Leake, B.E., *Spec. Publ. Geol. Soc. Lond.* **8**, 669–675.

Winchester, J.A. (1974) The zonal pattern of regional metamorphism in the Scottish Caledonides. *J. Geol. Soc. Lond.* **130**, 509–524.

Winkler, H.G.F. (1979) *Petrogenesis of Metamorphic Rocks.* (5th edn.), Springer-Verlag, New York, etc.

Yardley, B.W.D. (1978) Genesis of the Skagit Gneiss migmatites, Washington, and the distinction between mechanisms of migmatisation. *Bull. Geol. Soc. Am.* **89**, 941–951.

8 Fluid inclusions in migmatites

J. TOURET* and SAKIKO N. OLSEN**

8.1 Introduction

The fluid inclusions provide petrologists with the only direct method for studying a *possible* metamorphic fluid from a high *P–T* environment. Basic principles of a fluid inclusion study are as old as modern petrography (Sorby, 1858), but only in recent years have advances in the technology (heating-freezing stages), and in the understanding of fluid systems at high *P–T* conditions, permitted the studies of fluid inclusions in many rock types. For specific techniques and basic principles of data interpretation, readers are referred to the recent publications on the subjects (e.g. Weisbrod *et al.*, 1976; Touret, 1977; Hollister and Crawford, 1981; Roedder, 1984).

The study of fluid inclusions in migmatites is of special interest for two reasons. First, this is because some fluid phase must be present during migmatization. Formation of most migmatites seems to involve partial melting in some form (e.g. Touret, 1969; Olsen, 1977, 1982, 1984; Johannes and Gupta, 1982). Because of the large solubilities of volatile components, especially H_2O, in a partial melt at the *P–T* conditions reasonable for the middle and lower Crust, it is most likely that volatiles in a fluid phase, in addition to those available in the hydrous minerals in the rock, are necessary for partial melting to occur in significant degree. Not only had some fluid phase to be present during migmatization, but some migmatites may form by metasomatic effects of infiltrating fluids (Olsen, 1984). A knowledge of the composition and nature of the fluid phase is an essential step towards a full understanding of the migmatization process.

The second important reason for the study of fluid inclusions in migmatites is related to the origin of CO_2 in the lower Crust. Migmatites occur in the geological environment transitional between amphibolite and granulite facies. Touret (1977) has found that there is a major change in composition of the fluid at the boundary between the two facies. In the

*Free University, Amsterdam, The Netherlands.
**Department of Earth and Planetary Sciences, The Johns Hopkins University, Baltimore, Maryland 21218, USA.

amphibolite-facies rocks, H_2O, accompanied by CH_4 and N_2 in some cases, dominates, while the dominant fluid in the granulite-facies rocks is pure CO_2. The origin of pure CO_2 fluids is a much debated subject (see e.g. Hollister and Burruss, 1976; Konnerup-Madsen, 1977, for summaries). CO_2 fluid is thought by some to be flooding the lower Crust from Mantle source(s) (e.g. Newton *et al.*, 1980) and by others as locally derived from either graphite or carbonates in the rocks (e.g. Glassley, 1983). Whatever the source is, the widespread occurrence of CO_2, in probable conditions of miscibility in the CO_2–H_2O system, calls for a double mechanism: influx of CO_2 *and* removal of H_2O. As H_2O is more soluble in silicate melts than any other gases, notably CO_2 (by one order of magnitude: Kadik and Lukanin, 1973), the differential solubility of fluids in magmas is an attractive mechanism to explain the separation between H_2O and CO_2 at depth. Indeed, as soon as rocks become migmatized, pure CO_2 is likely to appear (Touret and Dietvorst, 1983). CO_2 in migmatites must therefore hold an important clue to the origin of CO_2 in the lower Crust.

Although the result of the fluid inclusion study of a migmatite is potentially very important, such a study is not a simple one. Migmatites are complex in their chemistry, structure and texture, and in many cases are polymetamorphic in nature. The fluid inclusions in migmatites reflect the complexity of the rocks and are extremely complex themselves in composition and density of the fluid, time of trapping, and relative abundance, all of which can be variable on thin-section or even grain-to-grain scale. In addition, the inclusions are typically small, less than 1 μm to about 30 μm, but the majority being less than 5 μm in diameter. Nevertheless, well-documented recent studies have demonstrated that the complexity is more apparent than real, or at least can be greatly reduced by a careful inclusion study combined with a thorough petrological one. The studies have found that the number of gas species is fairly small: the inclusions contain one or more of the four main gas species, H_2, CO_2, CH_4 and N_2. CO_2 and H_2O seem to be the only fluids present in some migmatitic terrains; for example, Naxos, Greece (Kreulen, 1980), Connemara, Ireland (Yardley *et al.*, 1983), and Colorado, USA (Olsen, 1984), while all four gas species are important in others, for example, Rogaland, Norway (Swanenberg, 1980), Massif Central, France (Kreulen and Schuiling, 1982), and Bamble, Norway (Touret and Dietvorst, 1983). Commonly, several generations of fluid inclusions are present but can be recognized with some experience. Probably the most important and at the same time the most difficult aspect of the fluid inclusion study of a metamorphic rock is determination of the timing of entrapment of the fluid. Without some supporting evidence it cannot be concluded that the fluid inclusions have trapped a synmetamorphic fluid (e.g. Hollister and Burruss, 1976). Because a migmatite is formed when a mobilizate or leucosome is separated from the palaeosome, difference between the leucosome and palaeosome either in the properties of the fluid or abundance of the inclusions would strongly indicate that the fluid was somehow related to the migmatization process. Such differences have been used as major criteria in the studies discussed below, in concluding that some of the fluid inclusions were related to the migmatization process.

Because there is more than one generation of fluid present in all cases, the results obtained by the bulk analyses of the fluid extracted by crushing a sample are very difficult, if not impossible, to interpret. One area where such a method can be useful, however, is in checking for the presence of gas species not detected by other methods. The microthermometric method using a heating–freezing stage is much more suited for studying metamorphic fluid inclusions in general and those in migmatites in particular, and has been used exclusively in obtaining the results discussed below. In the future, analyses by a laser-excited Raman spectroscope (e.g. Rosasco *et al.*, 1975) should become an important supplement to the microthermometric method.

Two studies of fluid inclusions in migmatites are discussed in detail below, one of a migmatite from Bamble Province in Southern Norway, another, although of preliminary nature, of the Front Range migmatites in Colorado, USA. In both cases, migmatization involved partial melting (Touret, 1969; Olsen, 1984) and the fluids present in the inclusions appear to be related to the migmatization process.

8.2 Fluid inclusions in migmatites of Bamble, Norway

8.2.1 Observations

A number of fluid inclusion studies, mostly of the granulite-facies rocks, have already been done (Touret, 1981) in the Proterozoic Bamble Province of South Norway. In addition to the granulite-facies rocks, the area contains well-preserved metasedimentary rocks (quartzites, metapelites, marbles) and metavolcanic rocks (amphibolites) of amphibolite facies. These rocks contain mobilizates or leucosomes as lenses, 100 mm to 1 m in length, occupying 10 to 20% of the whole rock in volume, which have been interpreted as locally-derived partial melts (Touret, 1969). Data of Touret and Dietvorst (1983) will now be summarized, prior to re-interpretation in section 8.2.2.

Two migmatite samples were studied, with a total of nearly 2000 microthermometric determinations of melting temperature, T_m, and homogenization temperature (disappearance of gas/liquid interface), T_h. The two samples (located in Figure 8.1) represent the two major kinds of protolith in the area (sedimentary and volcanic), at different metamorphic grades (measured by distance from the hypersthene isograd which marks the amphibolite/granulite facies boundary). In both cases, palaeosomes remain unaffected by migmatization only millimetres away from the mobilizate (neosome). Hydrous minerals (biotite, amphibole) are less abundant in

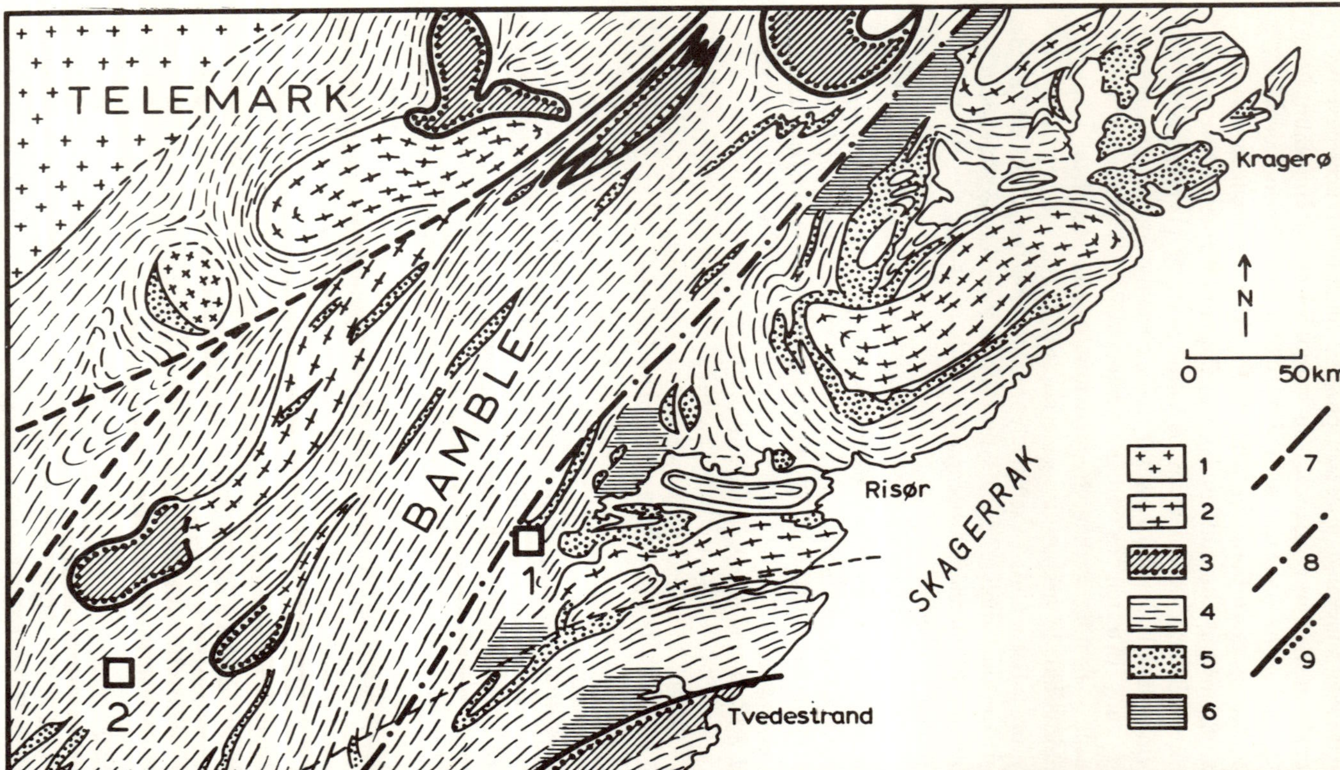

Figure 8.1 Geological setting of the investigated samples from the Bamble Region. Sample location for the Songe amphibolite indicated by square 1; that for the Selås banded migmatite by square 2. Other symbols: 1, Telemark granites; 2, Bamble granitoids and augen-gneisses; 3, granulite-facies area, limited by (9) = hypersthene isograd; 4, undifferentiated migmatites and banded gneisses; 5, quartzites; 6, Al-rich gneisses, mostly metasediments (with cordierite, anthophyllite, etc.); 7, main mylonite zone (the 'Great Breccia'); 8, northern boundary (approximate) of poikilitic garnet mobilizate in amphibolite.

mobilizates than in palaeosomes, and the mobilizates show more alteration, notably of feldspar. The Selås banded migmatite, from locality 2 (Figure 8.1), is a metagreywacke with a granodioritic composition (Table 8.1), and relict graded bedding occurs in some palaeosomes here (Touret, 1966). The Songe amphibolite (locality 1, Figure 8.1), was collected very near the hypersthene isograd, in the transitional domain called Border migmatites by Bugge (1943). The palaeosome is an orthoamphibolite with characteristically igneous chemistry (Touret, 1969) and rarely, relict igneous layering (Touret, 1966). The mobilizate, resulting from partial melting, occurs as pockets <50 cm in diameter. This mobilizate is light-coloured, granoblastic, and both coarser and more equant in texture than the palaeosome. Its boundary cuts across the metamorphic layering in the adjacent palaeosome (Figure 8.2). Grain size decreases from rim to core of the mobilizate. Whereas the palaeosome consists of hornblende and plagioclase (andesine) with minor biotite, quartz and sphene, the migmatitic pockets consists of quartz, plagioclase (andesine) and garnet with minor hornblende and biotite. Garnet occurs as bushy aggregates which can exceed 20 cm in diameter. This type of mobilizate is widespread in Bamble amphibolites in the area within a few kilometres of the regional hypersthene isograd. Prehnite and pumpellyite are found in the mobilizate, associated with altered biotite. This mobilizate of the Songe amphibolite is the only sample for which the fluid inclusion data can be interpreted in detail, but the qualitative conclusions seem to be valid in the case of a number of less thoroughly studied samples, and are believed to be generally applicable.

In this Songe sample, fluid inclusions are more abundant in the mobilizate than in the palaeosome. Since almost all the inclusions are in quartz, the greater abundance and grain-size of quartz in the mobilizate

Table 8.1 Chemical and modal compositions

	Chemical composition				Mode			
Rock type	1	2	4	5		1	2	3
SiO_2	44.37	58.65	70.22	72.60	Quartz	1	31.2	18
Al_2O_3	17.68	17.29	13.23	13.15	K-feldspar	+	—	+
FeO	6.82	11.81	4.80	3.21	Plagioclase	44	31.2	37
Fe_2O_3	7.45	1.30			% An in Pl	(50)	(48)	(55)
MnO	0.13	3.31	0.12	0.05	Biotite	4	2.1	5
MgO	6.68	0.55	1.01	0.67	Garnet	—	32.3	14
CaO	9.07	4.81	2.54	2.29	Amphibole	50	1	24
K_2O	1.12	0.24	4.16	4.27	Prehnite, Pumpellyite	+	0.3	+
Na_2O	2.73	1.56	1.81	1.95				
TiO_2	1.86	1.10	0.71	0.48	Other (*a*)	1	2	2

1, 2, 3: Songe amphibolite; 1, amphibolite; 2, garnet mobilizate; 3, contact zone between (1) and (2) (Dietvorst and Touret, unpubl. data); 4, 5, Selås banded migmatite (Touret, 1969); 4, paragneisses (metagreywacke, average 11 analyses); 5, mobilizate (average 16 analyses); (*a*) opaques, sphene, zircon, chlorite.

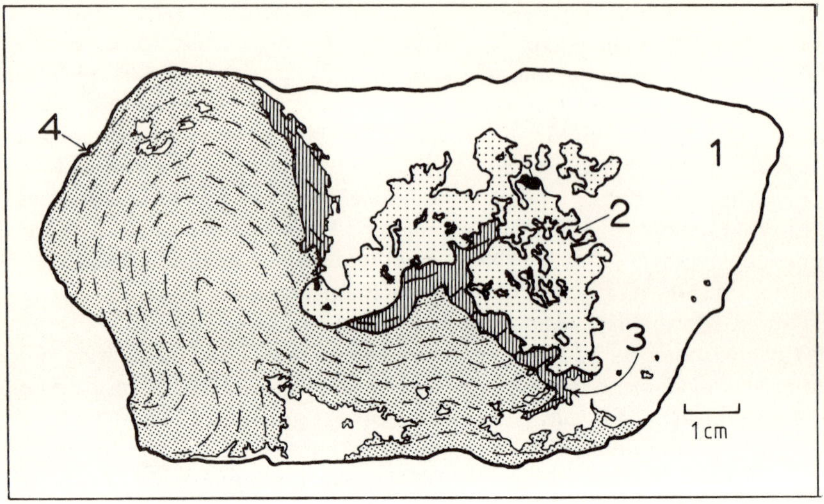

Figure 8.2 Mobilizate/palaeosome relations in the Songe amphibolite (drawing from a polished slab). 1–3, mobilizate: 1, quartz+feldspar; 2, garnet; 3, biotite-rich border zone; 4, amphibolite, here conspicuously folded and transected by the mobilizate; black, opaque grains. Average grain size of the amphibolite: 1–2 mm; of the mobilizate: 1–2 cm (quartz + feldspar) to more than 10 cm (garnet).

may favour their occurrence. However, the quartz grains which do occur in the palaeosome are always clear of inclusions. Thus the abundance of hydrous minerals correlates inversely with the abundance of fluid inclusions in palaesome *v.* mobilizate, suggesting that part (at least) of the fluid in the inclusions may be derived from the breakdown of volatile-bearing minerals.

Inclusions are classified partly by their appearance at room temperature and partly by microthermometry. Two-phase inclusions (with vapour bubble) are easily recognized at room temperature (Figure 8.3). Occasionally there are daughter minerals; halite cubes in these two-phase inclusions (Figure 8.3*b*) and (?) graphite in some single-phase ones (Figure 8.3*c*; see below).

Results of 876 microthermometric measurements are summarized in Figures 8.4 and 8.5. For the inclusions which are two-phase at room temperature, T_m near (or just below) 0°C indicates an aqueous fluid, with variable contents of dissolved salts indicated by the range in T_m (Figure 8.5). Values below the NaCl–H_2O eutectic (−23.3°C) indicate presence of additional species, but are relatively rare (Figure 8.5).

Single-phase inclusions are generally dark at room temperature. Some of these dark inclusions show no phase transitions down to −190°C (the lower temperature limit of the apparatus used), and are 'empty'. But many show well-defined phase transitions (Figure 8.5), and are classified by the range of transition temperature, as carbonic inclusions (above −60°C),

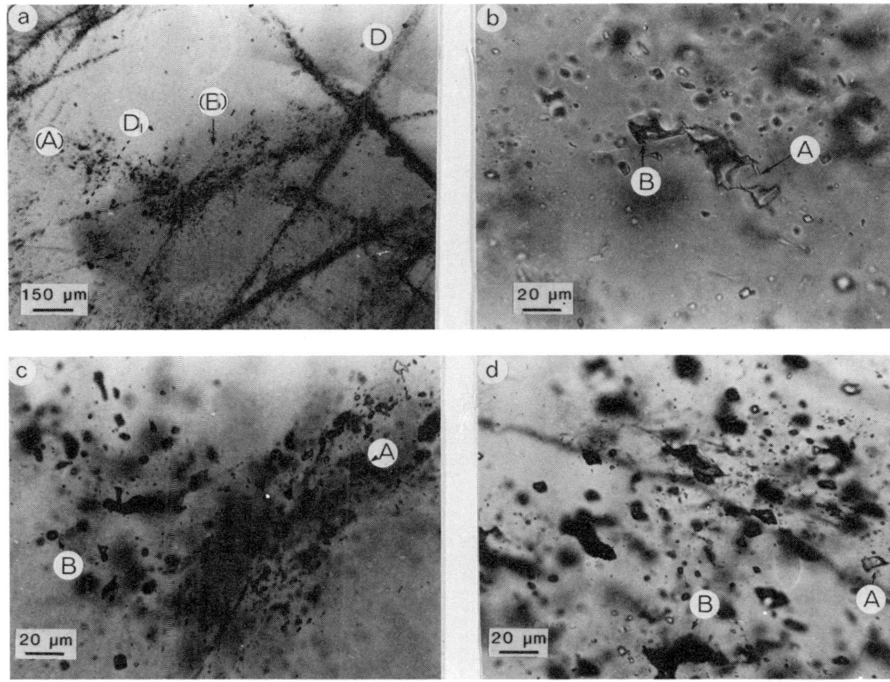

Figure 8.3 Fluid inclusions in mobilizate in Songe amphibolite. (*a*) Intersecting trails of CO_2 inclusions. Note that a relatively old trail (*A*) has been transposed and locally displaced (*B*) along a direction D_1 parallel to the latest trail *D*. (*b*) Aqueous fluid inclusion. isolated, with very irregular contours (early inclusion). It contains a large solid, most probably NaCl(*A*) and a small vapour bubble (*B*). (*c*) CH_4 and graphite-coated inclusions: *A*, graphite, *B*, gas (CH_4) filled inclusion. (*d*) CO_2 and/or N_2 inclusions. They are impossible to distinguish at room temperature (*A*); only microthermometric parameters establish the difference (T_h near $-147°C$ for N_2). Many cavities are empty (*B*).

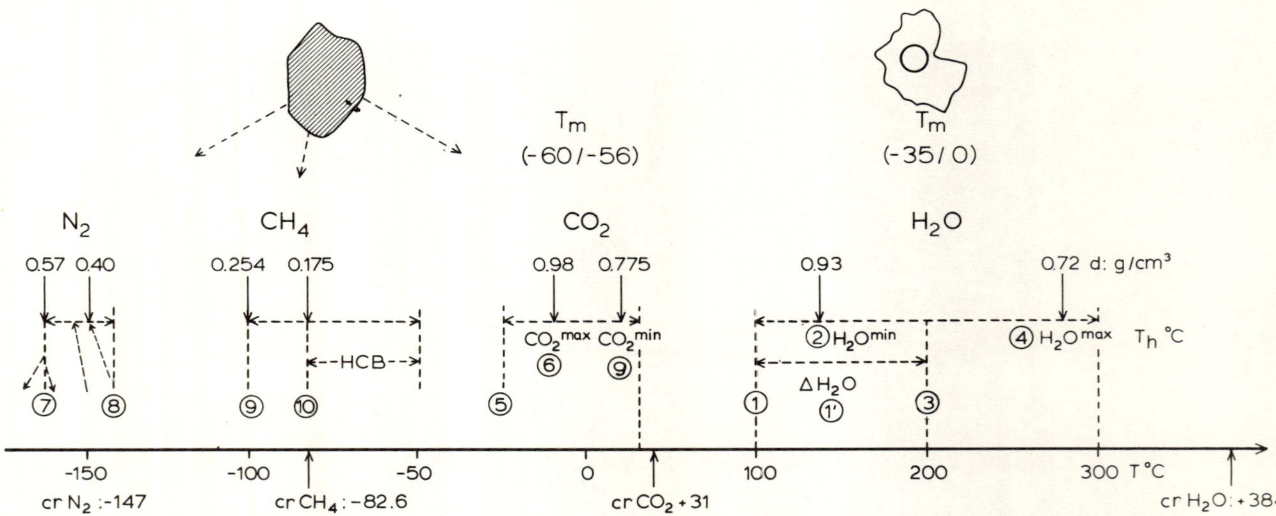

Figure 8.4 Range of melting (T_m) and homogenization (T_h) temperatures. HCB, hydrocarbon; 1–10 (in circles), 'representative isochores' of Figure 8.7; *d*, density in g/cm^3; *cr*, critical temperature; (1′), variation range in ΔH_2O. CO_2 and H_2O max. and min., see text.

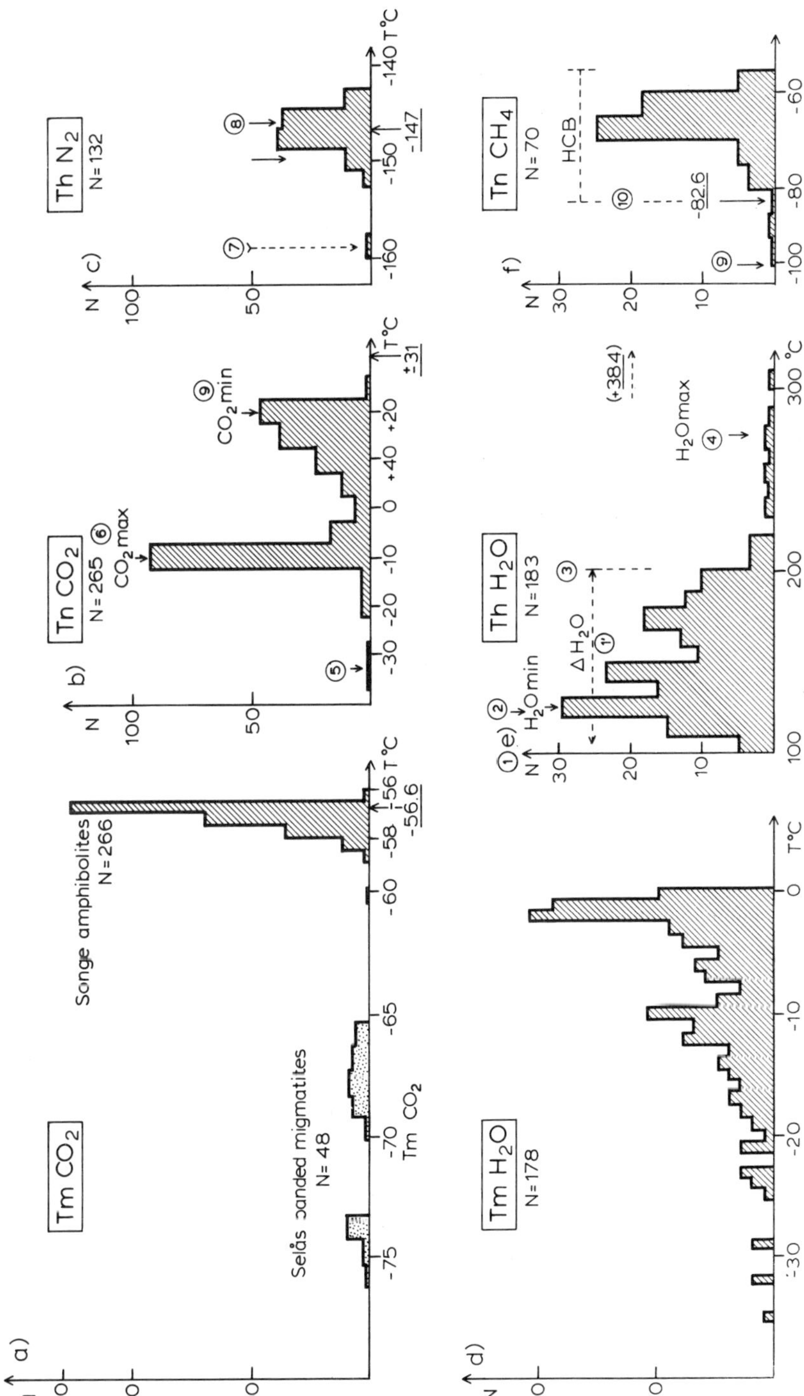

Figure 8.5 Frequency histograms of melting and homogenization temperatures. All homogenization is liquid, unless otherwise specified in the text (N_2, CH_4). 1–10 (in circles) as in Figure 8.4; N = number of measurements. Other symbols as in Figure 8.4.

hydrocarbon inclusions (−50 to −100°C) and nitrogen inclusions (around −150°C). Hydrocarbon inclusions are often coated with a black product, probably graphite.

Carbonic (CO_2-rich) inclusions, of variable density, are indicated by T_h (always liquid homogenization or homogenization to a liquid phase by disappearance of the vapour bubble) between −30 and +25°C, and T_m close to or slightly below −56.6°C (the CO_2 triple point). Such carbonic fluids are known to be very widespread in granulites and mantle-derived rocks (Touret, 1981). The range of T_m for carbonic inclusions in the Songe amphibolite (Figure 8.5) corresponds to pure CO_2 within experimental errors (±1°C). This fact facilitates interpretation, while the low T_m values for the Selås migmatite (Figure 8.5) indicate a mixture probably with CH_4 (graphite is abundant in the palaeosome), making the interpretation of T_h difficult (Hollister and Burruss, 1976; Swanenberg, 1979). Thus, *P–T* interpretation is attempted only for the Songe specimen.

Hydrocarbon and nitrogen inclusions were found only in the Songe amphibolites. Freezing temperatures are too low to be measured, so only T_h data were obtained (Figure 8.5), and unambiguous interpretation is possible only if critical or near-critical homogenization is observed (sudden disappearance of the interphase boundary without relative liquid/vapour volume change). Most of the hydrocarbon inclusions show this effect, but at temperatures below the critical temperature for CH_4 (−82.6°C). The lowest observed T_h (−100°C) represents a distinct liquid homogenization, interpreted as that of a fluid with a density greater than that at the critical point. As in the other types of fluid inclusions, assuming a simple composition (one component), the density of liquid coexisting with vapour at given temperature is uniquely defined and so the density at T_h can be estimated and gives the density of the initially trapped, homogeneous fluid. A reasonable minimum estimate of the density for the $T_h = -100$°C point is obtained by assuming that the hydrocarbon is pure CH_4, and is 0.254 g/cm^3 (Figure 8.4). The nitrogen inclusions mostly have T_h around the critical value of −147°C for pure N_2 (Figure 8.5), but lower values (−150 to −160°C) represent distinct liquid homogenization and are interpreted in terms of high-density N_2, analogously to the similarly behaving hydrocarbon inclusions.

After estimating the density of the fluid in its homogeneous state, and assuming no change in density after trapping, an isochore (line of equal volume) can be plotted in *P–T* space, along which the conditions at the time of trapping should lie. In calculating the isochore, literature *P–V–T* data are used, for the relevant one-component system. These are fairly reliable for CO_2, H_2O and CH_4, but discrepant among different authors for N_2. Calculated isochores for each of the 4 inclusion types (H_2O, CO_2, CH_4 and N_2) are shown in Figure 8.7, but it should be remembered that the evidence for purity of composition is strong only for the H_2O and

CO_2 groups. The isochores were constructed for representative temperatures (1 to 10 in Figures 8.4, 8.5) selected from the T_h histograms, either bracketing a range of values or corresponding to frequency peaks. In the case of CO_2 it is clear that there are two peaks. The H_2O case is less clear, and the low-density H_2O isochore 4 ($d = 0.72\,g/cm^3$, H_2O 'max') is poorly defined. Suffixes 'min' and 'max' refer to metamorphic conditions ('max' being nearer the peak metamorphic conditions).

For further interpretation, it is assumed that the fluid inclusions were trapped at or after the peak of metamorphism, i.e. there are no prograde fluids (Touret, 1981): then the chronological sequence of the different inclusions must be established. In the present observations, it is clear that the 4 inclusion types represent successive pulses of fluid, because the different components are miscible under the relevant *P–T* conditions and would form mixed inclusions if they were coeval. To deduce the relative chronology requires care, but is possible using criteria based on the fact that inclusions in high-grade metamorphic rocks are almost all secondary (Roedder, 1981) and comprise synchronous groups ('clusters' or 'trails') along healed microfractures. Then the chronological criteria are: (1) isolated, single inclusions are regarded as older than those in a larger cluster (Hollister and Burruss, 1976); (2) inclusions along poorly-healed, nearly-open fractures are late; and (3) intra-grain trails are older than those transecting grain boundaries: in addition, the older of two intersecting trails may be revealed by wiping out of some inclusions, replacement of their content, relative displacement, etc. (Touret, 1981). Figure 8.6 shows the results of such studies in the Songe amphibolite, leading to the following chronology:

(1) CO_2 inclusions are the earliest, and, as in all cases studied in the Bamble region (Touret, 1981), high-density inclusions tend to be older than low-density ones.
(2) The ages of N_2 inclusions occupy part of the range represented by CO_2 ones. As in other examples (Touret, 1982), the lack of CO_2 N_2 mixing is problematic.
(3) H_2O inclusions are very late. In contrast to the case of the CO_2 inclusions, high-density H_2O ones are younger than low-density H_2O ones. This might not be the case for supersaturated brines (H_2O–NaCl), which for many of them appear to be relatively early (Figure 8.3*b*). But no general conclusions could be obtained and these will be omitted from the following discussion. H_2O inclusions tend to be spatially associated with prehnite and pumpellyite.
(4) Hydrocarbon inclusions are intermediate in age between CO_2 and H_2O. Some mixed $CH_4 + H_2O$ inclusions are detected.

Thus the complete sequence proposed is: CO_2, $CO_2 + N_2$, CH_4 (hydrocarbons), $CH_4 + H_2O$, H_2O.

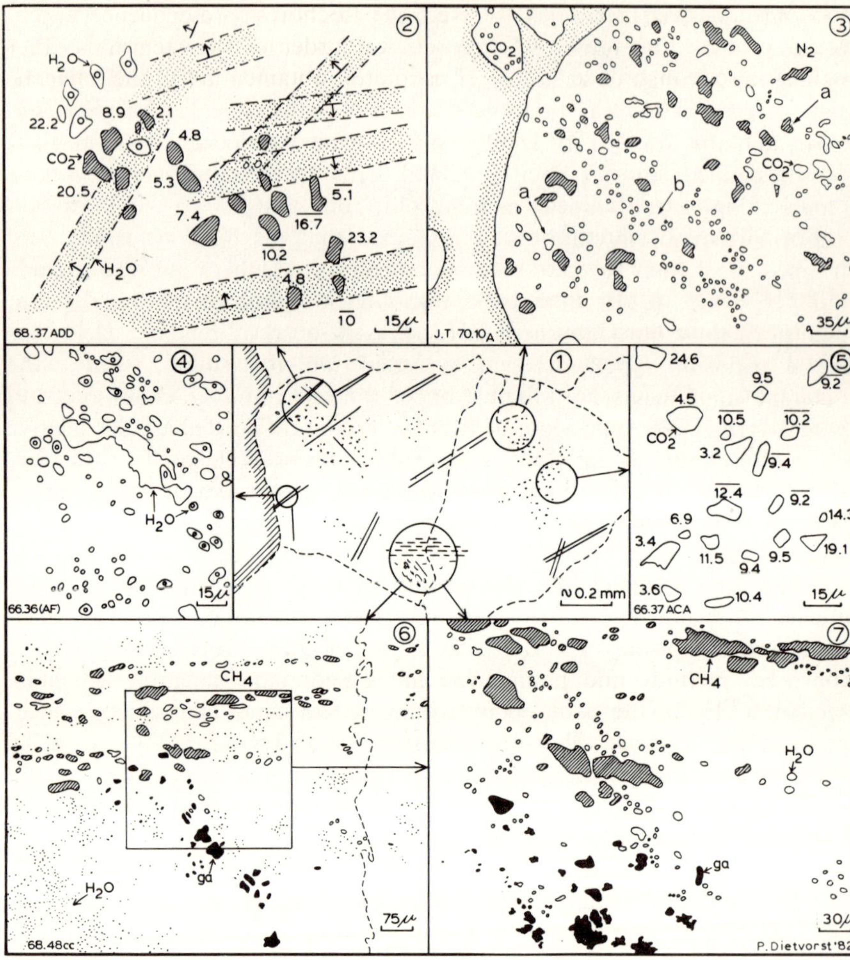

Figure 8.6 Relative chronology of the fluid inclusions (camera lucida drawing) in Songe amphibolite. (1) Sketch indicating the mode of occurrence of the different inclusion trails in quartz grains (actual observations artificially combined). (2) Cluster of CO_2 inclusions (number near the cavity $= T_h$; likewise in (5)) transected by later H_2O trails. Note at the upper left some cavities, probably filled initially with CO_2, later replaced by H_2O. (3) Relatively older N_2 inclusions (shaded) and white CO_2 inclusions, contemporaneous with (*a*) or later than (*b*) the N_2 ones. (4) H_2O inclusions. (5) CO_2 inclusions (with T_h). (6) and (7) CH_4 and hydrocarbon inclusions; *ga*, probably graphite. Sample numbers in the camera lucida drawings refer to double polished plates, each approximately 2×3 cm, all coming from the quartz-feldspar rich parts of the mobilizate. (7) Enlargement of the rectangle in (6) (in a slightly different focusing plane).

8.2.2 Interpretation of fluid inclusion data

The fluid inclusion data for the Bamble Region were interpreted by superimposing a model uplift path on to the set of representative isochores. A valid path must intersect the isochores in a sequence which would produce the observed chronological order of the inclusions. The path is also constrained by *P–T* estimates obtained from the mineral assemblages. The peak metamorphic conditions in the Bamble Region were estimated to be 750°–850°C at 6–8 kbar from the available data (Touret and Dietvorst, 1983, Figure 9). There is one temperature constraint on the retrograde path provided by the presence of the prehnite–pumpellyite pair, stable between 300°C and 400°C (Turner, 1981). Recently, a late kyanite-bearing quartz vein was found a few kilometres north of Songe in the Bamble Region. The retrograde *P–T* path for the region therefore must pass through the kyanite field. The proposed path shown as path 3′ in Figure 8.7 is basically an isobaric cooling path similar to that proposed for the nearby Rogaland rocks (path 1′, Figure 8.7) by Swanenberg (1980) and is very different from the path previously proposed for Bamble (path 2′, Figure 8.7) by Touret and Dietvorst (1983). Paths convex toward the *T*-axis, as path 2′ is, are thought to be much more common than those concave toward the *T*-axis, as path 3′ is (see Hollister *et al.*, 1979).

It has been assumed that the cooling path fits a simple, steady uplift model. The main difficulty with the assumption is that the extent of post-Proterozoic imprint on the region is not known. If, as proposed in Rogaland, prehnite and pumpellyite are Caledonian in age (Priem and Verschure, 1982), the cooling path may be far more complex near the origin of the *P–T* axes than the simple paths in Figure 8.7 indicate. On the other hand, the fact that prehnite and pumpellyite are more abundant in the mobilizate suggests a direct relationship between the formation of these minerals and migmatization.

The previously proposed *P–T* path (path 2′, Figure 8.7) was consistent with the observed features of fluid inclusion data (Touret and Dietvorst, 1983). Despite the apparent large difference of the new path (path 3′) from the old one, the new path is still consistent with most of the data. Briefly restated, the observed sequence of types of fluid inclusions is:

(1) high density, pure CO_2;
(2) CO_2 of decreasing density and independent pulses of N_2;
(3) CH_4 and hydrocarbons, associated with graphite;
(4) CH_4 and H_2O, simultaneous and separate pulses and some mixture;
(5) H_2O.

The new path is consistent entirely with (1), (4), (5), partly with (2) (local episodes of increasing CO_2 density at the beginning of the uplift), but not with (3). It seems more likely that either CH_4 and H_2O should appear simultaneously at first or H_2O (max) should be the first to appear. The discrepancy may not affect the model seriously. If the path was on the slightly higher-temperature side, some CH_4 could precede most of the H_2O inclusions by a short time interval. The data for H_2O (max) are very

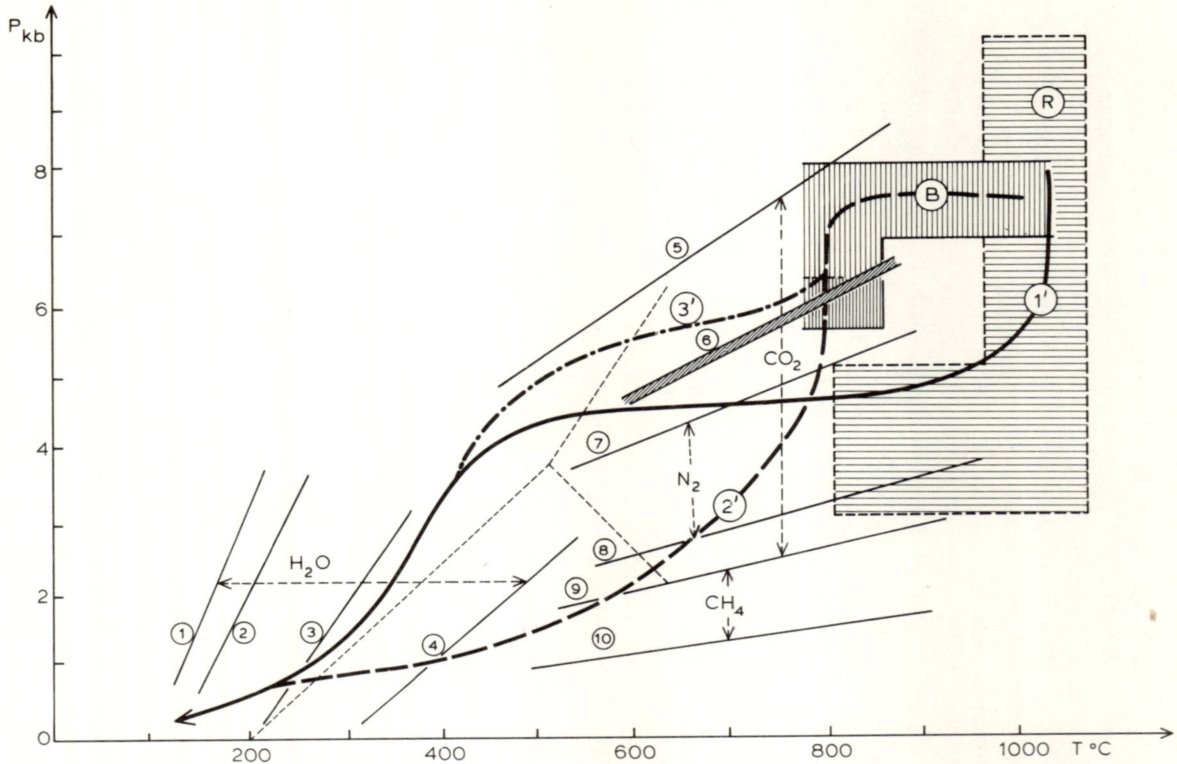

Figure 8.7 *P–T* interpretation of fluid inclusion data. Light lines labelled 1–10, representative isochores (see Figures 8.4 and 8.5). *PVT* data for: H_2O, Burnham *et al.* (1969); CO_2, Touret and Bottinga (1979); CH_4, Jacobs and Kerrick (1981); N_2, Holloway, in Hollister and Crawford (1981) for isochore 7, and Swanenberg (1979) for isochore 8. B and R, *P–T* estimates from mineral assemblages for peak metamorphic conditions in Bamble (Touret, unpublished data; B. Jansen, pers. comm.) and Rogaland (Swanenberg, 1980; B. Jansen, pers. comm.) respectively. Heavy lines, model uplift paths for: 1′, Rogaland (Swanenberg, 1980); 2′, previously proposed for Bamble (Touret and Dietvorst, 1983); 3′, new path for Bamble. Light dotted lines separate stability fields for aluminosilicates (Holdaway, 1971).

scanty and their low densities may have resulted simply from leakage or necking down. Other problems do remain to be resolved. One is the origin of N_2 and the mechanism of the separation of two obviously miscible fluids, CO_2 and N_2. Another is the effect of Caledonian tectonism. In this regard, the sudden appearance of H_2O and the striking coincidence between prehnite/pumpellyite and the H_2O 'max' isochore again poses the problem of the age of these minerals. The proposed *P–T* path indicates a relatively low pressure for formation of pumpellyite, much lower than what is known from present day information (absence of pumpellyite in most geothermal systems, B. Yardley, pers. comm.). This could indicate a Caledonian influence, which would simply displace the trajectory towards higher pressures. Further investigations are needed to solve these problems.

8.3 Fluid inclusions in the Front Range migmatites

8.3.1 Geological setting and sample description

The Colorado Front Range, USA, comprises mainly amphibolite-facies, metasedimentary and metavolcanic rocks of the Precambrian Idaho Springs Formation, and granitic intrusive bodies of batholithic dimensions. The oldest (1700 Ma) intrusive, Boulder Creek granodiorite, has been interpreted to be a syntectonic intrusive at the waning stage of metamorphic episodes that formed migmatites of the Idaho Springs Formation (Peterman *et al.*, 1968). Reasonable estimates for *P–T* conditions of migmatization are 3.5–4 kbar and 650°–700°C (see Olsen, 1982). The other two major plutonic events (at 1400 Ma and 1100 Ma) were not accompanied by regional metamorphism (Peterman *et al.*, 1968). All ages are Rb/Sr whole-rock ages. Some areas of the Front Range were also affected by Tertiary igneous activities and mineralization (Lovering and Goddard, 1950).

The Idaho Springs Formation consists of three main rock types:

(1) biotite–quartz–feldspar gneisses with compositions similar to those of sodic rhyolite to rhyodacite, most of which are probably metamorphosed felsic volcanic rocks;
(2) amphibolite and hornblende gneisses which are mainly metabasalts;
(3) mica schists, some sillimanitic, mostly metapelites.

Three of the four samples used for the fluid inclusion study were taken from a large sample of the biotite–quartz–feldspar migmatite used previously for a petrological study (Sample 10-3-T, Olsen, 1984). The mass-balance calculations on the sample indicated that the leucosomes in this migmatite formed by a metasomatic introduction of K and losses of Na and Mg through a fluid. Introduction of the fluid also caused partial melting of the wall rocks and the present leucosomes must contain both the partial melt and metasomatically introduced material. Another sample of the biotite–quartz–feldspar gneiss was collected from the western end of the same traverse where Sample 10-3-T was collected (Olsen, 1982, Figure 1, Loc. 1).

8.3.2 *Fluid inclusions*

The fluid inclusions found in the matrix quartz of the study samples occur in four major modes:

(1) as solitary inclusions, in small clusters of a few to several inclusions, or in short planar arrays including up to about 30 inclusions;
(2) in d*P*–d*T* fractures (see Simmons and Richter, 1976, and also see below);
(3) in healed fractures that cut across the grain boundary and also the zone boundaries between the leucosome, mafic selvage, and palaeosome: the inclusions in this group have been transposed, or migrated away from the plane of fracture (see Swanenberg, 1980; Touret, 1981);
(4) in poorly healed fractures that cut across grain and zone boundaries as in (3): the inclusions are not transposed.

The order of groups above is deduced to be a chronological one from the oldest (1) to the youngest (4).

A preliminary microthermometric study of four sections of three samples has found the following types of fluid; pure or nearly pure CO_2, aqueous solution, and H_2O–CO_2 mixture. The aqueous inclusions include both dilute and apparently highly saline ones. Highly birefringent colourless solid inclusions are found in many of the CO_2 and brine inclusions. Most CO_2–H_2O inclusions probably formed by mixing of the two fluids: H_2O–CO_2 most commonly occur where an evidence for interference between H_2O and CO_2 trails exists. One such place is the intersection between H_2O and CO_2 trails. Another is between the toe of some fan-shaped d*P*–d*T* fractures and the grain boundary from which the fracture originates: the inclusions at the toe are filled with CO_2, those near the grain boundary with H_2O, and those in between with H_2O–CO_2 fluid (see also Swanenberg, 1980).

These fluids are present in the four types of inclusions listed above, as follows.

Type (1): pure or nearly pure CO_2, most with high densities;
Type (2): low density CO_2, H_2O, and mixed CO_2–H_2O;
Type (3): low density CO_2, H_2O, and rare CO_2–H_2O;
Type (4): H_2O.

Type (1) inclusions are the only group which has been studied in some detail to date. These inclusions, which are inferred to be the earliest, have been grouped together because they have the following in common:

(a) they are all CO_2 inclusions, mostly monophase with a high density; they must contain pure or nearly pure CO_2 because their final melting temperatures (−56.5 to −57.2°C) are very close to the triple point of pure CO_2 (−56.6°C);
(b) most have negative crystal shapes;
(c) a cluster never crosses grain boundaries.

These earliest inclusions are fairly common in the leucosome quartz but are quite rare in the mafic selvages and palaeosome (host rock) quartz (Figure 8.8). Relative abundance of the inclusions in the leucosome deduced from Figure 8.8 would be underestimated because not all T_h of such inclusions in the leucosome were measured while T_h of every measurable inclusion found in the mafic selvage and palaeosome was. The leucosome to mafic selvage + palaeosome ratio in total amount of quartz in the sections studied is estimated to be approximately 60:40.

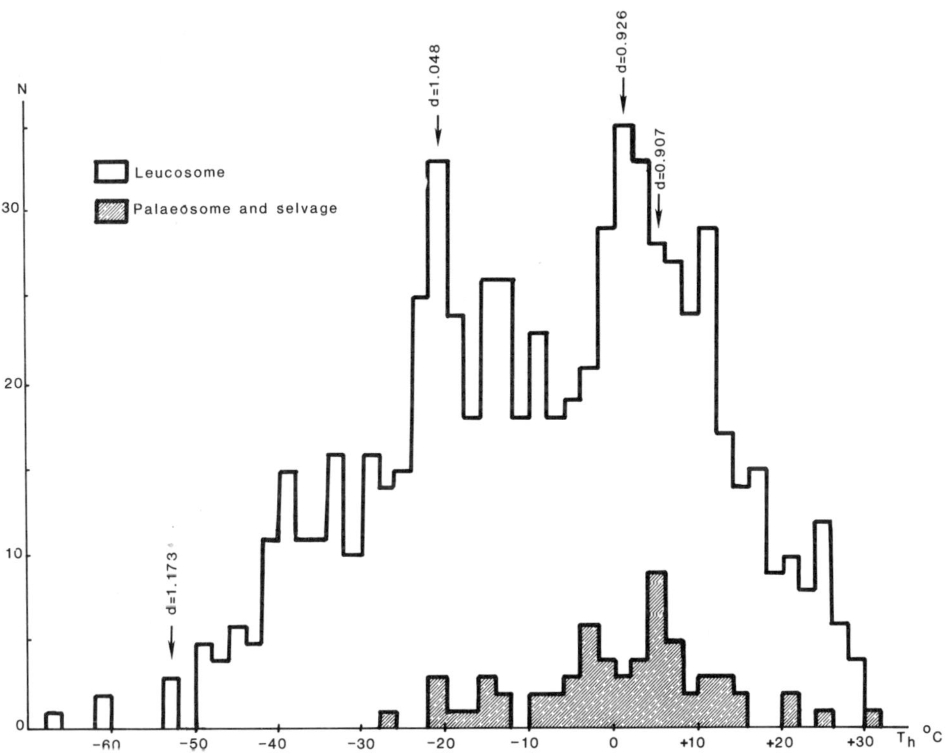

Figure 8.8 Frequency histograms of homogenization temperature for Type 1 CO_2 inclusions (see text) in the Front Range migmatites. N = number of inclusions.

In addition to the difference in abundance of the inclusions, there is a difference in CO_2 density distribution between the leucosome and other zones. T_h for the leucosome inclusions ranges from −67° to +30°C with a maximum at about $+1°C$ and another possible maximum at about $-21°C$ (Figure 8.8). The inclusions in the palaeosome and mafic selvage have T_h ranging from $-27°$ to $+31°C$ with a maximum at about $+5°C$. The difference in temperature of the maximum is probably not significant but the lack of superdense inclusions with $T_h \leqq -30°$ in the palaeosome and selvage probably is. These differences between the migmatitic zones which occur on the scale of a few centimetres suggest that the formation of these early inclusions was related to the migmatization process. It is difficult to explain the difference by a post-migmatization process such as channelling of a fluid along the leucosome.

If the fluid inclusions contain a synmetamorphic fluid, the density of the fluid must be what is expected at the P–T conditions of the metamorphism. The isochores for the superdense inclusions in the leucosome are at much higher pressures than the 4 kbar estimated for the peak condition at 700°C (Figure 8.9). It should be noted, however, that the isochores for the maxima in the T_h histogram for both the leucosome and palaeosome pass through the P–T estimates. In some of the studies of fluid inclusions in high-grade metamorphic rocks it has been found that the isochores for the earliest inclusions pass through the P–T conditions estimated by mineral geothermometers and geobarometers (e.g. Hollister and Burruss, 1976; Touret and Dietvorst, 1983). In many other cases, however, the densities are too

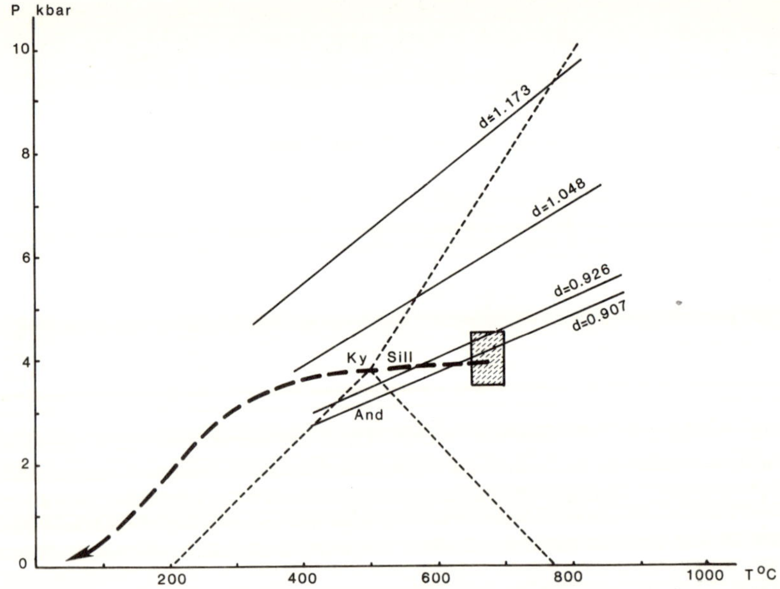

Figure 8.9 Representative isochores for CO_2 inclusions in Figure 8.8 (light, solid lines with densities). *PVT* data from Touret and Bottinga (1979). Hatched box, *P–T* estimates for migmatization in the Front Range. Light dotted lines, aluminosilicate stability fields after Holdaway (1971). Heavy dashed line, a possible isobaric uplift path.

low for the peak metamorphic conditions (e.g. Kreulen and Schuiling, 1982; Crawford *et al.*, 1979). If there is some supporting evidence for the fluid to be synmetamorphic, such as agreement of the observed fluid composition with that calculated from the mineral assemblage (e.g. Crawford *et al.*, 1979), the low density is generally explained as the result of leakage from the inclusion during uplift. In only one case besides that of the Front Range rocks are the densities too high for the estimated *P–T* conditions (Swanenberg, 1980). Swanenberg attributed the high density in some of the inclusions in Rogaland rocks to the change to a higher density due to isobaric cooling and a possible Caledonian influence, an increase in lithostatic pressure at low temperatures as the result of overthrusting. Some mineral data seem to support his postulates. Such evidence has not been found in the Front Range, at least to date. Nor is there any relict high-pressure mineral assemblage that could explain the high densities of the inclusions. There is, however, an evidence for post-entrapment change(s) in density. One healed fracture which is confined within a quartz grain contains CO_2 inclusions that exhibit almost a full range in T_h (-49 to $+24$°C) as exhibited by the entire population of the early inclusions measured (Figure 8.10, compare Figure 8.8). It is tentatively concluded that the superdense inclusions formed as the result of modification of inclusions due to isobaric cooling (see Figure 8.9). The reason for such inclusions to occur only in the leucosome is not clear.

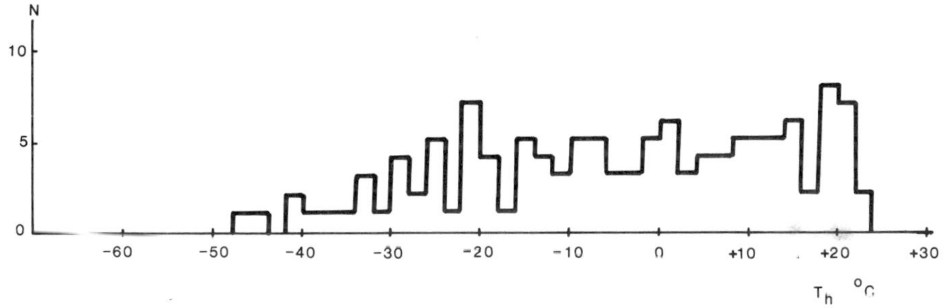

Figure 8.10 Frequency histogram of T_h for Type 1 CO_2 inclusions in an intragranular healed fracture in a quartz grain. Compare Figure 8.8. N = number of inclusions.

To conclude that these early CO_2 inclusions contain a synmigmatization fluid, another discrepancy needs to be explained. At the *P–T* conditions estimated, P_{H_2O} must be fairly high for partial melting to occur in significant amount: the fluid could not have been pure CO_2. One possible explanation is that CO_2 fluid is the residual of H_2O–CO_2 fluid from which H_2O has been lost by some mechanism. H_2O is much more strongly partitioned into the silicate melt than CO_2 is, at the *P–T* conditions of migmatization. Such partitioning has been invoked to explain the presence of CO_2 inclusions in migmatites (e.g. Touret and Dietvorst, 1983). This, however, does not explain the fact that aqueous inclusions of the same generation are not present in the leucosome: a large amount of H_2O must be released at the time of crystallization of the melt and should have been trapped as inclusions. Although H_2O may be incorporated into hydrous minerals in the mafic selvages commonly bordering the leucosome (L. Hollister, pers. comm., 1983), the selvages show little evidence for having been an anhydrous assemblage. In these Front Range migmatites, H_2O may have been consumed in subsolidus hydration reactions which must have succeeded the crystallization of the melt: sericitic and clayey alteration products of feldspar, especially of plagioclase, are extremely common in the leucosome but virtually absent from the palaeosome and selvage. The fluid concentration in a leucosome could have been a cause and/or an effect of the presence of melt in the leucosome: abundant fluid could have caused selective partial melting or a fluid could have been released upon crystallization of the melt.

8.4 Possible melt inclusions in migmatites

Certain small solid inclusions, with or without associated small amounts of brine, have been found in quartz and plagioclase grains of some granitoids and mobilizates in migmatites by Russian workers (e.g. Tomilenko and Chupin, 1983), who interpreted them as relict silicate melt inclusions. These inclusions (Figure 8.11*a*, *b*) a few to 10 μm in diameter, are roughly spherical or negative-crystal shaped and consist mainly of unknown silicate (?) phases. Characteristically, each contains a vapour bubble which occupies 10 to 15% of the inclusion in volume. To date, nothing seems to be known of the chemical composition of any of the phases found in these inclusions. Silicate melt or glass inclusions, in many cases devitrified, are well known from unmetamorphosed igneous rocks (e.g. Weisbrod, 1981; Touret, 1984). Figure 8.11*d* shows melt inclusions, which clearly have negative-crystal forms typical of these inclusions. A silicate melt may be quenched into glass even in a migmatitic environment if there is a sudden decrease in P_{H_2O}. Such melt inclusions, if formed, must be completely devitrified and recrystallized during the slow and thorough re-equilibration process that seems to characterize most migmatitic environments, as indicated by the texture of the rocks. Although such relict melt inclusions, if any, should be much more abundant in the leucosome, they could also be present in the palaeosome: there is no conclusive evidence we know of that completely rules out the presence of a melt in the palaeosome. Some of the small solid inclusions found in quartz and feldspar of migmatites are feldspar (both K-feldspar and plagioclase) or alteration products (mostly carbonates and sericite). The remaining, unidentified solid inclusions (Figure 8.11*c*) may be relict melt inclusions. Such a possibility will have to be investigated by future studies.

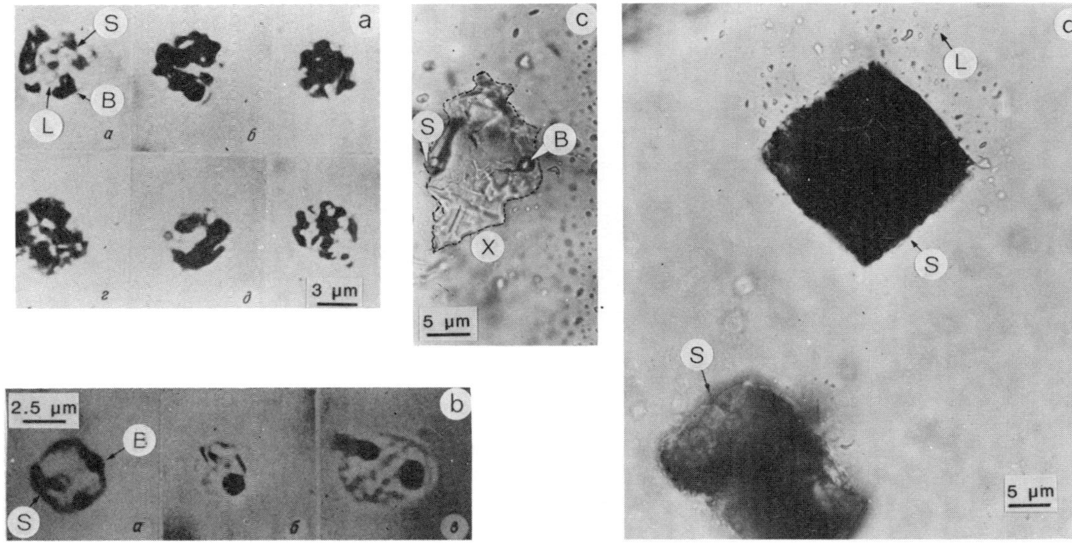

Figure 8.11 'Melt' inclusions in quartz from granitoids.

(*a*) 'Melt' inclusions in quartz from leucosome in granulite-facies migmatites, reproduced from Figure 29 of Tomilenko and Chupin (1983). They contain aggregates of unknown silicates (*S*), a small quantity of interstitial liquid (*L*) and one or several dark vapour bubbles (*B*). Some of the black spots may also be opaques, possibly from immiscible magmatic liquids.

(*b*) 'Melt' inclusions from leucosomes in granitoids, artificially heated to an unspecified temperature, probably about 1000°C, and slowly cooled. The content of the inclusion is more homogeneous, the contours tend to a negative crystal shape (left photograph). Some solids (opaques ?) and one or several gas bubbles (*B*) remain. (Reproduced from Figure 30 of Tomilenko and Chupin, 1983).

(*c*) Possible relict melt inclusion (*X*), associated with brine (*B* = vapour bubble, *S* = salt cube) in a pegmatoid mobilizate near Narestö, Southern Norway.

(*d*) Relict melt inclusion (*S*) in a subvolcanic granite from Morvan, Massif Central, France. Note the aqueous light inclusions (*L*) above the isotropic solid inclusions, possibly the fluid expelled from the melt. Crossed polars.

8.5 Discussion

Recent studies have demonstrated that, despite the complex nature of fluid inclusions in the migmatites, there are some systematic and coherent patterns to the occurrence of these inclusions.

(1) Chemically different fluids are limited to H_2O, CO_2, CH_4, N_2 and their mixtures;
(2) in the cases studied in detail, the oldest seem to be pure or nearly pure CO_2 inclusions; and the youngest, H_2O;
(3) the fluid inclusions, especially the oldest CO_2 inclusions, are much more abundant in the leucosome, indicating a relationship between the formation of these inclusions and migmatization process.

On the other hand, many unresolved problems remain. Two of the most important ones are the source or sources of the fluids, especially CO_2 and N_2, and the reason why these fluids, which should be miscible, are often closely associated yet separated. In the case of migmatites, the occurrence of pure or nearly pure CO_2 in the earliest inclusions poses a special problem. H_2O is needed as a synmigmatitic fluid for partial melting to occur, yet early H_2O inclusions are absent. Although the partitioning of H_2O into the melt phase could enrich the vapour phase in CO_2, it does not explain the separate pulses of CO_2. Neither does the mechanism explain the very late occurrence of H_2O which appears in the rock only at a much lower temperature than is reasonable for the crystallization of the partial melt: for the Bamble rocks, these temperatures are near or below 400°C and near or above 650°C respectively. The mechanism that formed the superdense CO_2 inclusions found in some rocks is also a problem that has not been completely resolved. They seem to be restricted to certain geological environments, deformed and sheared rocks in Rogaland and leucosomes in the Front Range rocks. Possible presence of relict melt inclusions should also be investigated. One aspect of the fluid inclusion problem that has not been investigated to date is internal buffering of f_{O_2} by the oxides, and the effect of changing f_{O_2} on relative abundance of species in the C–H–O system with changing metamorphic conditions. All of these questions will have to be answered by further studies. Any future study which combines microthermometric methods, and perhaps the help of Raman spectroscopic analysis, with careful petrographic studies can only add to our knowledge of both fluid inclusions and migmatization processes.

References

Bugge, J. A. W. (1943) Geological and petrological investigations in the Kongsberg Bamble formation. *Norges Geol. Unders.* **160**.

Burnham, C. W., Holloway, J. R. and Davis, N. F. (1969) Thermodynamic properties of water to 1000°C and 10,000 bars. *Spec. Pap. Geol. Soc. Am.* **132**.

Crawford, M. L., Kraus, D. W. and Hollister, L. S. (1979) Petrologic and fluid inclusion study of calc-silicate rocks, Prince Rupert, British Columbia. *Am. J. Sci.* **279**, 1135–1159.

Glassley, W. E. (1983) The role of CO_2 in the chemical modification of deep continental crust. *Geochim. Cosmochim. Acta* **47**, 597–616.

Holdaway, M. J. (1971) Stability of andalusite and the aluminum silicate phase diagram. *Am. J. Sci.* **271**, 97–131.

Hollister, L. S. and Burruss, R. C. (1976) Phase equilibria in fluid inclusions from the Khtada Lake metamorphic complex. *Geochim. Cosmochim. Acta* **40**, 163–175.

Hollister, L. S. and Crawford, M. L. (eds) (1981) *Short Course in Fluid Inclusions: Applications to Petrology*. Min. Assoc. Canada, Calgary.

Hollister, L. S., Burruss, R. C., Henry, D. L. and Hendel, E.-M. (1979) Physical conditions during uplift of metamorphic terrains, as recorded by fluid inclusions. *Bull. Minéral.* **102**, 555–561.

Jacobs, G. K. and Kerrick, D. M. (1981) Methane: an equation of state with application to the ternary system H_2O–CO_2–CH_4. *Geochim. Cosmochim. Acta* **45**, 607–614.

Johannes, W. and Gupta, L. N. (1982) Origin and evolution of a migmatite. *Contrib. Miner. Petrol.* **79**, 114–123.

Kadik, A. A. and Lukanin, O. A. (1973) The solubility dependent behaviour of water and carbon dioxide in magmatic processes. *Geochem. Internat.* **10**, 115–129.

Konnerup-Madsen, J. (1977) Composition and microthermometry of fluid inclusions in the Kleivan granite, South Norway. *Am. J. Sci.* **277**, 673–696.

Kreulen, R. (1980) CO_2-rich fluids during regional metamorphism on Naxos (Greece): carbon isotopes and fluid inclusions. *Am. J. Sci.* **280**, 745–771.

Kreulen, R. and Schuiling, R. D. (1982) N_2–CH_4–CO_2 fluids during formation of the Dôme de L'Agout, France. *Geochim. Cosmochim. Acta* **46**, 193–203.

Lovering, T. S. and Goddard, E. N. (1950) Geology and ore deposits of the Front Range, Colorado. *U.S. Geol. Surv. Prof. Paper* **223**.

Newton, R. C., Smith, J. V. and Windley, B. F. (1980) Carbonic metamorphism, granulites and crustal growth. *Nature* **288**, 45–50.

Olsen, S. N. (1977) Origin of the Baltimore Gneiss migmatites at Piney Creek, Maryland. *Geol. Soc. Am. Bull.* **88**, 1089–1101.

Olsen, S. N. (1982) Open- and closed-system migmatites in the Front Range, Colorado. *Am. J. Sci.* **282**, 1596–1622.

Olsen, S. N. (1984) Mass-balance and mass-transfer in migmatites from the Colorado Front Range. *Contrib. Miner. Petrol.* **85**, 30–44.

Peterman, Z. E., Hedge, C. E., and Braddock, W. A. (1968) Age of Precambrian events in the northeastern Front Range, Colorado. *J. Geophys. Res.* **73**, 2277–2296.

Priem, H. N. A. and Verschure, R. H. (1982) Review of the isotope geochronology of the high-grade metamorphic Precambrian of S.W. Norway. *Geol. Rdsch.* **71**, 81–84.

Roedder, E. (1981) Origin of fluid inclusions and changes that occur after trapping. In *Short Course in Fluid Inclusions: Application to Petrology*, eds. Hollister, L. S. and Crawford, M. L. Min. Assoc. Canada, Calgary, 101–137.

Roedder, E. (1984) *Fluid Inclusions. Reviews in Mineralogy* **12**, Mineralogical Society of America.

Rosasco, G. J., Roedder, E. and Simmons, J. H. (1975) Laser-excited Raman spectroscopy for nondestructive partial analysis of individual phases in fluid inclusions in minerals. *Science* **190**, 557–560.

Simmons, G. and Richter, D. (1976) Microcracks in rocks. In: *The Physics and Chemistry of Minerals and Rocks*, ed. Strens, R. G. J., Wiley, New York, 105–137.

Sorby, H. C. (1858) On the microscopic structure of crystals, indicating the origin of minerals and rocks. *Q. J. Geol. Soc. Lond.* **14,** 453–500.

Swanenberg, H. E. C. (1979) Phase equilibria in carbonic systems, and their application to freezing studies of fluid inclusions. *Contrib. Miner. Petrol.* **68,** 303–306.

Swanenberg, H. E. C. (1980) Fluid inclusions in high-grade metamorphic rocks from S.W. Norway. *Geologica Ultraiectina, Utrecht* **25**.

Tomilenko, A. A. and Chupin, V. P. (1983) *Thermobarogeochemistry of Metamorphic Complexes*. Nauka Pub., Novosibirsk **524**.

Touret, J. (1966) Sur l'origine supracrustale des gneiss rubanés de Selås (formation de Bamble, Norvège méridionale). *C. R. Acad. Sci. Paris, Série D* **262**, 9–12.

Touret, J. (1969) Le socle Précambrien de la Norvège Méridionale (Region de Vegårshei-Gjerstad). Thèse Etat., 3 vol., Nancy, A. O. CNRS 2902.

Touret, J. (1977) The significance of fluid inclusions in metamorphic rocks. In: *Thermodynamics in Geology*, ed. Fraser, D. G., D. Reidel Pub. Co., Dordrecht, Holland, 203–227.

Touret, J. (1981) Fluid inclusions in high grade metamorphic rocks. In *Short Course in Fluid Inclusions: Applications to Petrology*, eds. Hollister, L. S. and Crawford, M. L., Min. Assoc. Canada, Calgary, 182–208.

Touret, J. (1982) An empirical phase diagram for a part of the N_2–CO_2 system at low temperature. *Chem. Geol.* **37**, 49–58.

Touret, J. (1984) Les inclusions fluides: histoire d'un paradoxe. *Bull. Minéral.* **107**, 125–137.

Touret, J. and Bottinga, Y. (1979) Équation d'état pour le CO_2: application aux inclusions carboniques. *Bull. Minéral.* **102**, 577–583.

Touret, J. and Dietvorst, P. (1983) Fluid inclusions in high-grade anatectic metamorphites. *J. Geol. Soc. Lond.* **140**, 635–649.

Turner, F. J. (1981) *Metamorphic Petrology*, 2nd edn., McGraw-Hill, New York.

Weisbrod, A. (1981) Fluid inclusions in shallow intrusions. In: *Short Course in Fluid Inclusions: Application to Petrology*, eds. Hollister, L. S. and Crawford, M. L., Min. Assoc. Canada, Calgary, 241–271.

Weisbrod, A., Poty, B. and Touret, J. (1976) Les inclusions fluides en géochimie-pétrologie: tendances actuelles. *Bull. Soc. Fr. Min. Crist.* **99**, 140–152.

Yardley, B. W. D., Shepherd, T. J., and Barber, J. P. (1983) Fluid inclusions studies of high-grade rocks from Connemara, Ireland. In: *Migmatites, Melting, and Metamorphism*, eds. Atherton, M. P and Gribble, C. D., Shiva, Nantwich, 110–126.

Index of authors cited

Italics indicate a reference list at the end of a chapter. Only the *first* such listing of a particular paper is indexed. Personal communications, and internal citations (of one chapter in another within this book) are not indexed.

Subject index

Italics indicate a figure or table.